G. Schweitzer (Editor)

Magnetic Bearings

Proceedings of the First International Symposium
ETH Zurich, Switzerland, June 6-8, 1988

Springer-Verlag Berlin Heidelberg New York
London Paris Tokyo

Prof. Dr. G. Schweitzer

Institute of Mechanics
ETH Zurich
CH-8092 Zurich
Switzerland

ISBN 978-3-642-51726-6   ISBN 978-3-642-51724-2 (eBook)
DOI 10.1007/978-3-642-51724-2

Library of Congress Cataloging-in-Publication Data
Magnetic bearings : proceedings of the first international symposium.
ETH Zurich, Switzerland, June 6-8, 1988 / G. Schweitzer, editor.

1. Magnetic bearings--Congresses.
I. Schweitzer, G. (Gerhard),
II. Eidgenössische Technische Hochschule Zurich.
TJ1073.7.M34   1988
621.8'22--dc 19   88-39971

Softcover reprint of the hardcover 1st edition 1989

2161/3020 543210

Preface

Magnetic Bearings are bearings where the suspension forces are generated magnetically without any contact.

The advantages to modern machinery are obvious: no mechanical wear, no lubrication, potential for high rotor speed, accuracy, and high dynamic performance, new constructional solutions to a classical problem in machine dynamics.

The realization of such bearings is in rapid progress. Examples for application areas are

> turbomachinery, centrifuges, vacuum techniques, machine tool spindles, chemical industry, medical devices, robotics, high speed drives, spacecraft equipment, contactless actuators, vibration isolation.

The Symposium is demonstrating the current state of the art in this developing field of mechatronics, showing actual research efforts, reporting on applications in the various areas, and discussing open questions. The main purpose of the Symposium has been to establish a common information basis for people working on magnetic bearings. It will point to promising areas, and it will help to facilitate decisions on research and development projects, and on investments for applications.

The Advisory Committee for the Symposium consisted of

Prof. P.E. Allaire, University of Virginia, Charlottesville, USA
H. Habermann, Société de Mécanique Magnétique, Vernon, France
Prof. T. Higuchi, University of Tokyo, Tokyo, Japan
Prof. R. Holmes, University of Southampton, Southampton, UK
Prof. G. Schweitzer, ETH Zurich, Zurich, Switzerland, (chairman)

The Symposium has been drawing considerable interest from research and from industry. Only 36 papers of a much larger number of submitted contributions could be considered for presentation. The 180 participants came from 15 countries with an especially large delegation from Japan. This Symposium was the first of its kind. There are plans to have the Second Symposium in Japan in July 1990, hosted by the University of Tokyo, and an invitation for the Third Symposium comes from the University of Virginia, Charlottesville, USA.

For the proceedings the original manuscripts submitted by the authors are reproduced. Therefore varying quality of the papers can be observed in this book. Only limited editing has been done as to preserve the individual flavor of each contribution. The contributed papers reflect the scientific, industrial and commercial interests of the authors. The lectures were devoted to the following main topics:

> Applications in Space, Applications in Physics, Applications in Robotics, Special Bearings, Stabilization of Rotor Motion, Vibration Control of Flexible Rotors, Industrial Applications, Identification of Bearing and Rotor Parameters, Unbalance and Disturbance Control, Digital Control, Signal Processors and Applications

Some of the papers are related to more than one of these main topics, but in order to assist the reader we have structured this volume according to the main topics, thus maintaining the structure of the Symposium.

The lectures were extensively discussed, and it is expected that the Symposium will have a stimulating effect on further research in this challenging and promising field of mechatronics. The discussions in the audience and statements of the members of the Advisory Committee indicate that there are necessary and promising directions where future efforts will have to go:

- further industrial applications
- power electronics
- digital control, at least for high level tasks (diagnostics, adaptive control, operational aspects)
- safety design and retainer bearings to improve customer acceptance
- permanent magnets for low energy or low price solutions

The Symposium was sponsored and financially supported by

Eidgenössische Technische Hochschule Zürich
(Swiss Federal Institute of Technology Zurich)
Schweizerische Gesellschaft für Mikrotechnik
(Swiss Society of Microtechniques)
European Research Office of the US Army

This help is gratefully acknowledged.

The Local Organizing Committee consisted of

Mrs. G. Junker, Prof. J. Hugel, Prof. G. Schweitzer (chairman), Dr. A. Traxler

A main contribution to the success of the Symposium is due to the help and excellent work of the staff of the Institute of Mechanics of the ETH.

The editorial work of the Proceedings was supported by the Institute of Mechanics of the ETHZ. Thanks to the Springer-Verlag are due for an agreeable and efficient cooperation.

Zurich, July 1988 G. Schweitzer

Contents

Applications in Space

Applications in Physics

Applications in Robotics

Special Bearings

Stabilization of Rotor Motion

Vibration Control of Flexible Rotors

Industrial Applications

Identification of Bearing and Rotor Parameters

Unbalance and Disturbance Control

Digital Control

Signal Processors and Applications

Applications in Space

Research and Development of Magnetic Bearing Flywheels for Attitude Control of Spacecraft

Atsushi NAKAJIMA

Space Technology Research Group
National Aerospace Laboratory, Tokyo, Japan

Summary

Magnetic bearings for space and industrial applications have undergone considerable development in laboratories and industries. Since 1978, several types of magnetic bearings have been manufactured and tested in National Aerospace Laboratory (NAL), concentrating on the development of the magnetic bearings utilizing permanent magnets and their applications for space. Two degree-of-freedom (DOF) radially controlled magnetic bearings is a candidate of practical use as attitude control actuator of spacecraft. In order to evaluate the characteristics of this type magnetic bearings, on-orbit experiments under zero g condition were performed, i.e. magnetic bearing flywheel developed by NAL was launched by H-I rocket TF#1 into a 1500 km circular orbit on summer of 1986 and several experiments were carried out successfully. This paper summarizes the activities of NAL for research and development of magnetic bearings.

1. Passively stabilized stiffness

Utilization of permanent magnets in the magnetic bearings not only provides the passive stabilization but it also helps the control system to overcome the inherently nonlinear relationships between the electromagnetic quantities (e.g. coil current, gap flux density etc.) and the resultant forces by a technique called "flux density modulation". Another advantage of permanent magnet utilization is that it enables the "Virtually zero power" method capable of minimizing (or nulling) the power consumption in the control coil by balancing the constant external force (e.g. gravitational force) with the unbalance force produced by permanent magnets. For practical uses of this type bearings, it should be pointed out that the procurements of a stable high speed rotation, sufficiently large stiffness and oscillation damping about passively stabilized axes are particularly important.

The passive stiffness is produced by the sum of restoring force acting on the magnetic pole pieces. Fig.1(a) shows the magnetic flux distribution chart calculated by finite element method(FEM). The magnetic fluxes generated by the permanent magnets pass through all magnetic pole pieces and concentrated on the central pieces. A specially magnified detail drawing of a central piece is shown in Fig.1(b). Magnetic force vector of each corresponding magnetic flux is expressed by a straight line with an arrow head and the vector sum is decomposed into attraction force and restoring force. The attraction force is controlled by electro magnets which is not shown in this figure and the restoring force supports the rotor mass passively. The force distribution on the magnetic pole-face is shown in Fig.1(c). The restoring force is highly centered on an extreme edge of magnetic pole pieces. As a result of this calculation, the passive stiffness becomes large with the increasement of the edge of magnetic pole pieces.

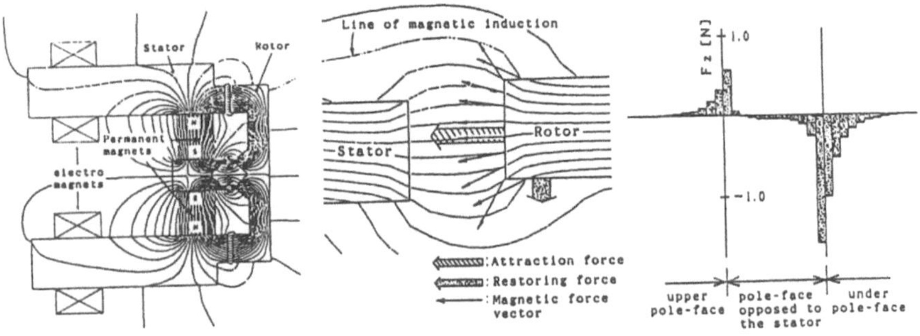

(a)Magnetic flux (b)Magnetic force (c)Restoring force

Fig.1 Magnetic force produced by permanent magnets.

The vibration damping in the passively stabilized axis is produced consequently as eddy current losses caused by the relative movements of stator and rotor in the magnetic fields. Under ordinary circumstances, the value of damping ratio will take an order of 0.05, which is strongly dependent on the magnetic parameters such as materials of magnetic poles, magnetic flux density, levitating position and so forth.

2. Configurations of magnetic bearings

2.1 One DOF active magnetic bearing

One DOF active magnetic bearing has the most simplest construction and the highest reliability because of the minimum

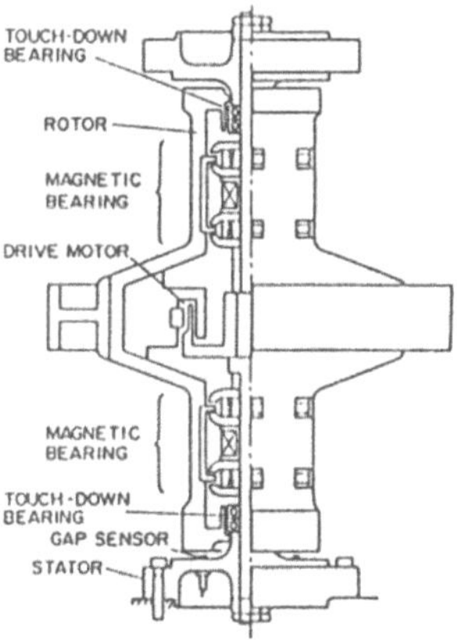

Fig.2 One DOF MBMW.

Table 1 Specifications of one DOF active MBMW (joint development between NAL and MELCO)

items		measured values
Axial stiffness	Ka[N/m]	1.78×10^{6}
Radial stiffness	Kr[N/m]	2.12×10^{5}
Orthogonal stiffness	K_θ[Nm/rad]	4.62×10^{2}
Dimension	[mm]	$\phi 260 \times 400$
Rotor weight	[kg]	6.55
Moment of inertia	[kg·m²]	
Iz(rotation axis)		3.28×10^{-2}
Ix=Iy		3.83×10^{-2}
Angular momentum (at 10,000 rpm)	[Nms]	34

electronic circuit. Fig.2 shows a laboratory model of One DOF active magnetic bearing momentum wheel jointly developed by NAL and Mitsubishi Electric Corporation (MELCO). The purpose of this development is to investigate the rotational characteristics of this type magnetic bearings with two radial and two conical passively stabilized axes. Table 1 shows the specifications of this model. From the solution of the rotor dynamics equations, the natural frequency of the cylindrical mode is 28 Hz. On passing through this rotation speed, the vibration amplitude was 0.47 mm p-p , which could be decreased by balancing the mass unbalance of the rotor more precisely. After the passage of this natural frequency, this cylindrical mode occurred again at the rotation speed over 41.8 Hz. It is well known that the energy dissipation in the rotating body would cause the instability at a high speed rotation. From the results of this instability analysis [1] , the damping ratio of the cylindrical mode becomes as follows :

$$\zeta_{cy} = \zeta_{ocy} + \zeta_{icy}\ (1-w/w_{ncy}) \qquad (1)$$

where ζ_{ocy} and ζ_{icy} are the damping ratios of stator part and rotating part, w is rotaion speed and w_{ncy} is natural frequency of the cylindrical mode, respectively. From the measurements of ζ_{cy} at several rotation speeds, ζ_{cy} decreased linearly in inverse proportion to the rotaion speed and became zero at 41.8 Hz (Fig.3). From Eq.(1) and the experimental data shown in Fig.3, ζ_{ocy} and ζ_{icy} were estimated as 0.0016 and 0.0034, respectively.

In order to avoid this instability, special attention should be paid to design the damping mechanism of passively stabilized axes. In our laboratory model, this problem was solved by introducing the anisotropic radial stiffness [2] and stable rotation more than 10,000 rpm was obtained as a result.

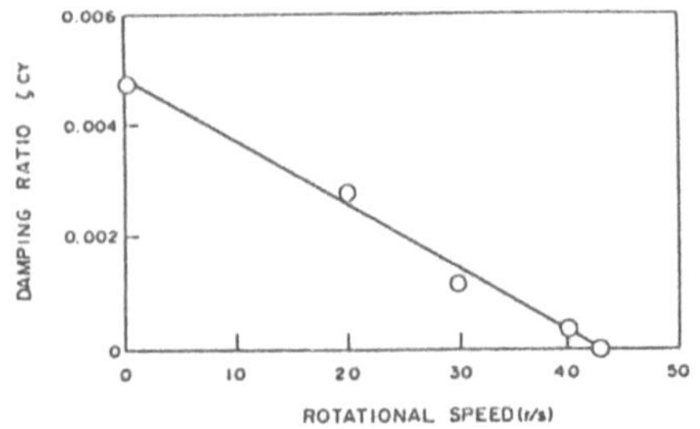

Fig.3
Relation between damping ratio of cylindrical oscillation and rotational speed for isotropic radial stiffness.

2.2 Two DOF active magnetic bearing

Two DOF active magnetic bearing has a relatively simple arrangement and easy construction. The most preferable feature of this type is its virtually flat-shaped form, which enables achieving a high moment of inertial/mass ratio flywheel. From this point of view, it is possible to realize a high capacity angular momentum with minimum weight and manufacturing cost. Since 1980, the author and others have studied and manufactured by way of trial several types of two DOF magnetic bearings [3,4] and found that they have a high potential for space applications such as reaction wheel.

Fig.4 shows a flat shaped reaction wheel jointly developed by NAL and Toshiba Corporation (TOSHIBA). The design goals of this model were to reduce weight, dimensions (especially height) and power consumption, and to increase angular momentum, bearing stiffness and reliability. The magnetic bearings were installed at the outside of the stator, then the mass concentration to the outer part of the wheel could be satisfied. This installation have

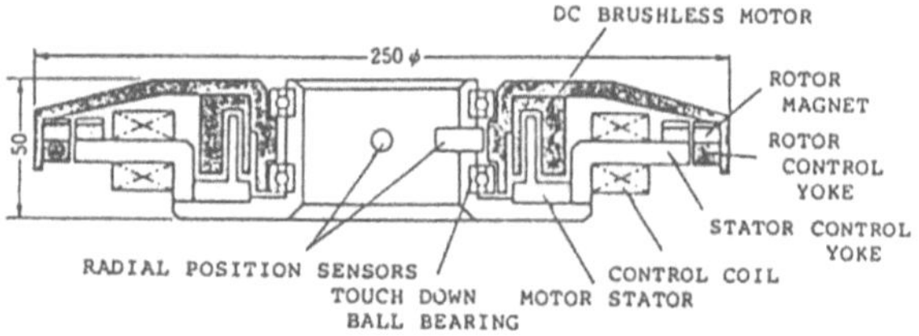

Fig.4 Flat shaped reaction wheel (joint development between NAL and TOSHIBA).

Table 2 Specifications of two DOF active MBRW

items		measured values
Axial stiffness	Ka[N/m]	3.2×10^5
Radial stiffness	Kr[N/m]	7.6×10^5
Orthogonal stiffness	K_θ[Nm/rad]	1.5×10^3
Dimension	[mm]	$\phi 250\times50$
Rotor weight	[kg]	2.4
Angular momentum (at 3,000 rpm)	[Nms]	7.5

brought about high moment/mass ratio and high stiffness of the magnetic bearings. The values of stiffness were measured dynamically by modal analysis using the impulse hammering test, the results of which are shown in Table 2. The damping ratios were also measured and their values are over 0.3 for radial directions and over 0.01 for axial direction and the axes about radial directions.

In order to promote the practical use of magnetic bearings for space, research flight model (REM) of magnetic bearing flywheel was manufactured [5] and launched into orbit to obtain several technical data of magnetic bearings operated under zero g condition. Experimental items are:

(1) manufacturing of a magnetic bearing flywheel for space flight model
(2) functional test of the launch lock mechanism
(3) measurements of the damping characteristics of passively stabilized axis
(4) comparison of the ground tested data with the space experimental results, especially related to the levitation characteristics.

RFM of two DOF active magnetic bearing flywheel was manufactured by MELCO by the contract of NAL. Cross sectional view is shown in Fig.5. It was composed of axially magnetized samarium-cobalt permanent magnets, electromagnetic coils for x- and y-axis control, two pairs of eddy current type position sensors to detect the radial displacements of rotor, a flywheel to

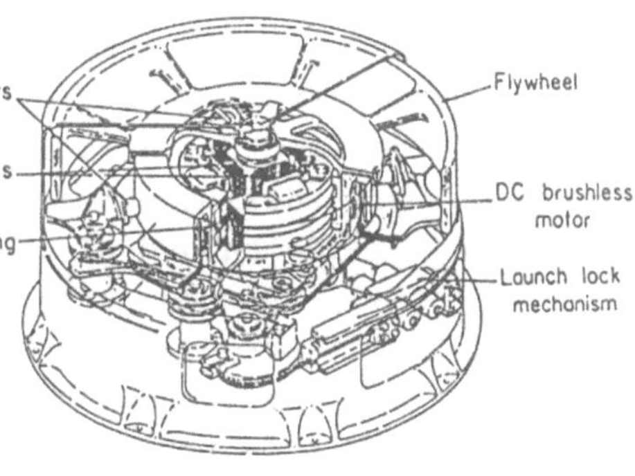

Fig.5 Magnetic bearing flywheel for space experiment (developed by NAL).

Table 3 Spicifications of two DOF active MBFW

items		measured values
Axial stiffness	[N/m]	8.3×10^4
Radial control force	[N/A]	4.8×10
Dimension	[mm]	270(dia.)×127 (machine body) 200×252×151 (electronics)
Weight	[kg]	7.23 (machine body) 5.10 (electronics)
Rotation speed	[rpm]	1,000 (only for this exp.)
Power consumption	[W]	3(suspension), 7(at 1000rpm)

accumulate an angular momentum, emergency ball bearings and launch lock mechanism which fasten the rotating part with elastic metal plates and 2 mm diameter twisted wire. Additional four position sensors are mounted on the bottom base of the stator to detect the axial displacement of rotor by adding the four output signals and to measure the rotor's tilting motions about x- and y -axis by subtracting each pair of sensors. A pair of electromagnetic coils are additionally wound around stator yokes to excite a tilting oscillation for the purpose of damping characteristics measurement. Several environmental tests were performed before system integration. Random vibration level for each direction is about 14.5 g rms and the temperature range is -20 - +50 °C. Dimensions and specifications are shown in Table 3. In order to measure several features of magnetic bearings, this RFM was not designed in conformity to the purpose of the direct application for spacecraft from the view point of dimension and weight because of the additional sensors and coils for experiments.

On the early morning of August 13th, 1986, MAgnetic Bearing flywheel Experimental System (MABES) was launched by H-1 rocket test flight number 1 (2nd stage) into a 1500km nearly circular orbit. MABES was given a pet name "JINDAI" after launch. The total system of JINDAI were already shown at the previous paper [6]. The experiment's duration time was three days and the termination of this experiment was brought about the current consumption of the onboad Ni-Cd batteries. Planned experiments were perfectly performed. After about two hours from launch, 136 MHz VHF telemetry data were transmitted from JINDAI and the first experiment has begun according to the command signal of the ground control center ; the launch lock release was successfully functioned, which was proved by the evaluation of position sensors' output signals. Secondary, magnetic levitation was confirmed after the corresponding command transmission and various telemetry data which indicate the operating conditions of magnetic bearings were recorded on the magnetic tape and converted simultaniously into the physical quantities for quick-look monitoring. The rotation speed of the flywheel was limitted to 1,000 rpm because of the minimization the amount of energy to be used. Fig.6 shows the damping characteristics of the tilting oscillation of the rotor about x-axis perpendicular to the rotation axis caused by the command signal. For application of magnetic bearings with passively stabilized stiffness, it should be focused our attention on the damping characteristics under different circumstances such as zero g condition. Fig.6(a) denotes the output signal of peak hold circuit while (b) is the original analogue value. Under the limitations of the telemetry transmitting capacity, the original analogue signals were sampled every 30 ms and digitized. Example telemetry data are shown in (c). From these data, oscillation period and its damping ratio were calculated. Fig.7 shows the levitation characteristics compared of the ground tested data with the space experimental results. The relative positions of actively controlled x- and y-axis were not different each other as inevitable results. On the other hand, z-axis position has displaced in the axial direction by 0.22 mm which corresponds to the rotor mass. The results of the tilting motion experiment are shown in Fig.7(c). A group of data obtained on the ground are

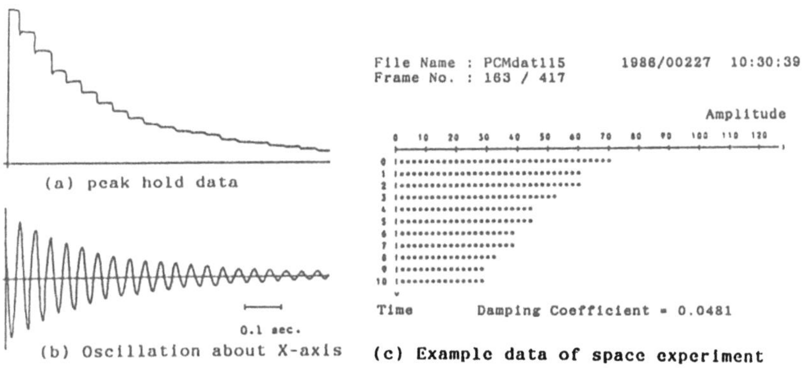

(a) peak hold data

(b) Oscillation about X-axis

(c) Example data of space experiment

Fig.6 Damping characteristics about X-axis oscillation.

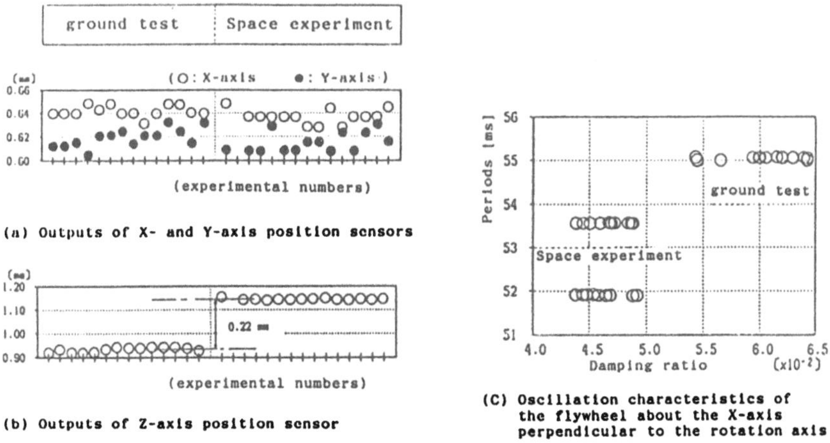

(a) Outputs of X- and Y-axis position sensors

(b) Outputs of Z-axis position sensor

(C) Oscillation characteristics of the flywheel about the X-axis perpendicular to the rotation axis

Fig.7 Levitation characteristics (Comparison of the ground tested data with the space experimental results).

plotted on the upper right of the figure, on the contrary, space experimental data are plotted on the lower left. The oscillation periods under zero g condition become smaller than that of the ground data, which express the increase of stiffness about x-axis. As the increse of the stiffness is closely related to the increase of the axial stiffness, the experimental result can be explained by the fact that the axial stiffness has the maximum value at null position. The seperation of space experimental data (51.9 ms and 53.6 ms) is inexplicable matter at present. The damping ratios under different environment are seperated clearly. It was proved that the oscillation damping ratio of passively stabilized axis became smaller than the measurement's data on the ground. That phenomenon can be explained in physics as follows ; the magnetic fields around the magnetic pole pieces are different under different relative position of stator and rotor. The magnetic flux distribution will effect the damping characteristics which are closely depend on the eddy current losses. The magnitude of the difference was about 20 %. Fig.8 shows the relationship between the axial displacement of rotor and the restoring force. Close agreement between measurement on the ground and calculated values was obtained. Z-axis displacement shown in Fig.7(b) is coincide with the results shown in Fig.8.

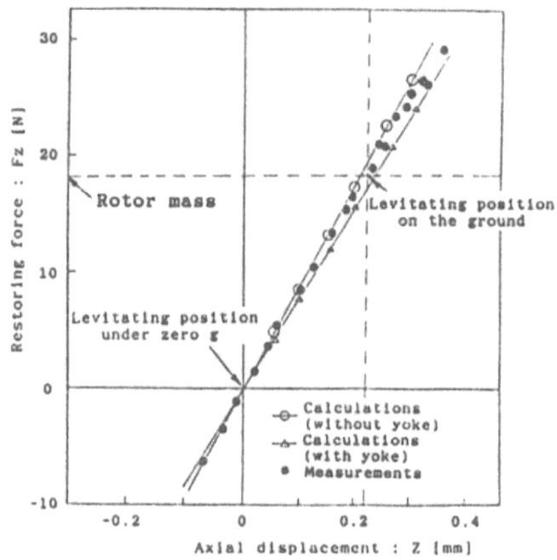

Fig.8 Fz and z relation.

2.3 Three DOF active magnetic bearing

The gimballing capability is possible by adopting the magnetic bearings with more than three DOF active type. Three

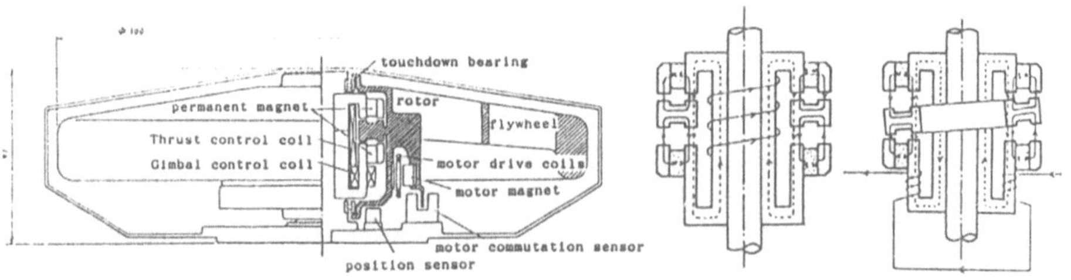

(a)Cross sectional view (b)Axial control (c)Gimbal control

Fig.9 Control principle of three DOF MBMW.

DOF active magnetic gimballed momentum wheel was designed and is under developement by the cooperation with NAL and MELCO [7], which is shown in Fig.9. The stator is composed of four equivalent parts, each of those has a gimbal control coil, permanent magnets and an axial position sensor. The mechanism is much simpler than that of the five DOF active type. For the axial control, a thrust control coil wounded around the four stators is used. The axial position of the rotor is measured by four sensors and the control current modulates the flux densities in the four inner gaps by the same quantity as shown in Fig.9(b). For the gimbal control, two pairs of gimbal control coils are used. The gimbal deflection of the rotor is measured by subtracting a pair of axial position sensors and the control current yields two control forces with opposite directions ; upward force in the left quarter, downward force in the right quarter(Fig.9(c)).

2.4 Four DOF active magnetic bearing

Four DOF active magnetic bearing has also a potential for high speed rotating machine. The basic construction of this type is similar to that of the two DOF active type and has a easy handling for assembly and disintegration. For the purpose of high speed rotation, a laboratory model of this type magnetic bearings was manufactured and some rotational tests were performed. Magnetic characteristics of this model are shown in Table 4.

Table 4 Specifications of four DOF active MB for high speed rotation (developed by NAL)

items		measured values
Axial stiffness	Ka[N/m]	2.46×10^5
Radial control force	[N/A]	$(4\sim5)\times10$(variable)
Dimension	[mm]	app. ϕ 90 × 150
Rotor weight	[kg]	5.7
Rotation speed	[rpm]	32,000

2.5 Five DOF active magnetic bearing

Five DOF active magnetic bearing has also jointly developed by NAL and MELCO [8], which is different from other types in that no permanent magnets are used. In order to improve the mass property of this type magnetic bearings, axial electromagnetic coil was omitted, while two

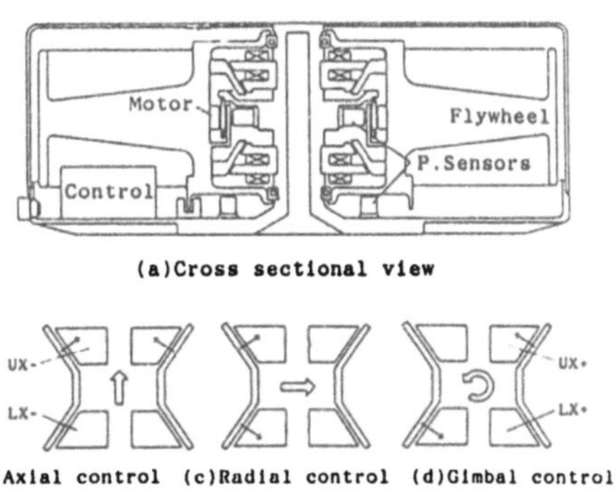

Fig.10 Five DOF MBMW.

pairs of upper and lower electromagnetic coils were installed in a conical shape for five axes active control as shown in Fig.-10. By activiating an equivalent pair of coils, axial, radial and gimbal control could be obtained. The characteristics of this model are shown in Table 5.

Table 5 Specifications of five DOF active MBMW (joint development between NAL and MELCO)

items		measured values
Orthogonal stiffness K_θ	[Nm/rad]	1.9×10^{2}
Dimension	[mm]	$\phi314\times125$
Weight	[kg]	5.2(total) 2.6(rotor)
Angular momentum (at 10,000 rpm)	[Nms]	18
Moment of inertia	[kg·m²]	
Iz(rotation axis)		1.7×10^{-2}
Ix=Iy		1.1×10^{-2}
Gimbal angle	[deg]	±0.5

3. Applications in space

Several types of magnetic bearings are described in this paper. These models are not necessarily manufactured for the purpose of the space use directly, but some of them would reach the stage of becoming practical uses in space. As a result of the space experiment in JINDAI, two DOF active magnetic bearings developed by NAL could be available as a reaction wheel for three-axis stabilized satellite. Five DOF active type jointly developed by NAL and MELCO could also be a candidate as a momentum wheel with gimballing capability.

References

1. Hagiwara. S. et.al.:Rotational tests of magnetically suspended flywheel for spacecraft(2nd report). JSME No.810-4 (1981),pp.67-69 (in Japanese).
2. Tsuchiya, K. et.al.:Damping Characteristics of Passively Stabilized Rotor Suspended by magnetic Bearing. (to be published).
3. Murakami, C. et.al.:A Flat Magnetic Bearing Reaction Wheel. 14th ISTS, Tokyo, pp647-654, 1984.
4. Murakami, C. et.al.:A Compact Type Model of 2-Axis Controlled Magnetic Bearing Reaction Wheel. 15th ISTS, Tokyo, pp.799-804 1986.
5. Nakajima, A. et.al.:Research and Development of Magnetic Bearings. 16th ISTS, Sapporo, 1988.
6. Murakami, C. et.al.:An On-Orbit Experimental Plan of a Magnetic Bearing Flywheel. 15th ISTS, Tokyo, pp.793-798 1986.
7. Yabuuchi, K. et.al.:A Compact Magnetic Bearing for Gimballed Momentum Wheel. Proc. of 17th AMS, JPL, PP.333-342, May, 1983.
8. Inoue, M. et.al.:Five-axis Active Magnetic Bearing Flywheel. Proceedings of the 28th Space Sciences and Technology Conference, Sendai, 2D8 pp.342-343, 1984 (in Japanese).

A 3(5) Degree of Freedom Electrodynamic-Bearing Wheel for 3-Axis Spacecraft Attitude Control Applications

U. Bichler, T. Eckardt
TELDIX GmbH, Germany

Abstract

This paper deals with two related types of electrodynamically suspended rotors, acting either as momentum or reaction wheels for spacecraft attitude control purposes. One is of the three-axis active type, the other involves five actively controlled degrees of freedom.
The wheels are of a pancake design with the control forces acting on the outer rim where the main rotor mass is concentrated thus avoiding low structural resonances. The rim is decoupled from the hub by titanium springs. Both wheels provide a vernier gimballing capability enabling three-axis attitude control of spacecraft with one rotating mass only.
A special decoupling tilt control loop design eliminates all gyroscopic effects, such as precessional and nutational oscillations.
The electrodynamic bearing principle is advantageous because main parts of the mass can be concentrated on the outer rim of the rotor contributing to the desired inertia. Force generation is linear involving only very small lags and practically no cross-coupling exist between the orthogonal axes facilitating a control strategy to prevent disturbing unbalanced forces from affecting the stator of the bearing for the use in so-called 'micro-gravity' environment. The wheels are actually in the development phase and in the paper the principles, the design, the control loops and first results are presented.

Introduction

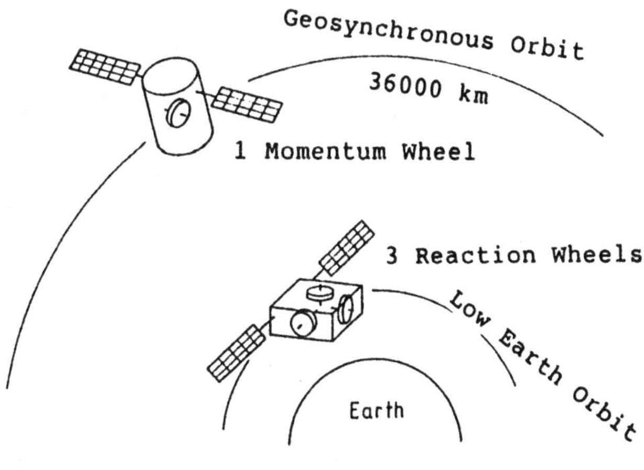

Fig. 1 Satellites with 1 Momentum Wheel or 3 Reaction Wheels

In principle, momentum wheels are gyroscopic actuators designed for compensating periodic disturbing torques which act on geostationary communications satellites. A momentum wheel mounted in a double-gimbal system offers the additional possibility of tilting the satellite with respect to the wheel. Satellites in a low earth orbit are equipped with a set of at least three perpendicular reaction wheels providing attitude control about all three axes.
Almost all momentum and reaction wheels in orbit are equipped with ball bearings but wheels with magnetic bearings offer some special features which are of interest with regard to the increasing attitude performance requirements of modern spacecraft and with environments where only extreme low disturbing accelerations can be tolerated [1].
In magnetic bearings, the achievable speed is not limited by friction and fatigue effects of ball bearings. They provide low friction and no stiction at zero speed, are insensitive to hard vacuum and need no lubricant. If a vernier gimballing capability is implemented an active satellite nutation damping and a three axis attitude control for small roll and yaw angles is enabled with one rotating mass only. Unbalance forces are - as far as no permanent magnets are involved for direct attraction/repulsion - transmitted to the stator only at an amount which can be kept near zero when a special control law is applied.

Principle

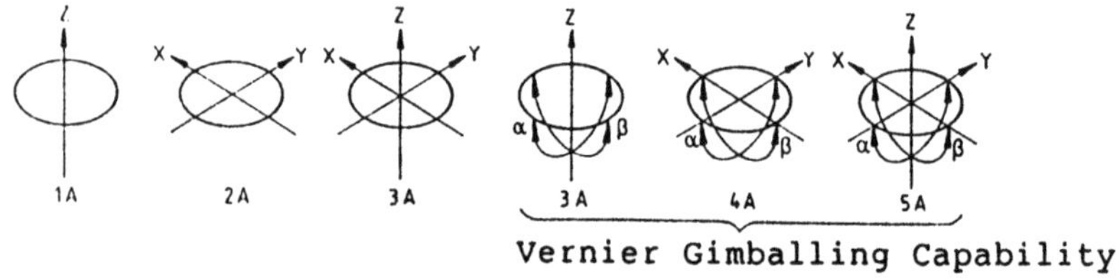

Fig. 2 Different Types of Magnetic Bearings

Several designs of one- and two-active-axis magnetic bearings are known [2,3,4]. They all use a permanent magnet flux which is superimposed for control purposes by a variable coil flux. The disadvantages involved are that the magnets must be very homogeneous to avoid drag torque due to eddy current and hysteresis losses, that the passively stabilized degrees of

freedom have a low stiffness and that practically no inherent damping is present. Especially the passive tilt suspension forms a weak spring and leads, together with energy losses in a passive damper, to satellite nutational motion [5,6] requiring additional nutation dampers. The permanent magnets cannot be 'switched' to a lower flux mode in micro-g environments, and unbalance disturbance isolation is only possible by active means.
Magnetic bearings with three and more active axes have the important advantage of the tilt axes controllability offering the vernier gimballing capability [7]. The top magnetic bearing is the five-axis active one, which does not need any permanent magnets in direct attraction or repulsion. This type is perfect, very stiff and controllable in all axes with excellent damping.

Design

Because of the mentioned features, a modular design of an axial and tilt-controlled bearing with both a passive and an active radial suspension, resulting in 3- and 5-axis active magnetic bearings, was chosen.
Additional overall design requirements have been an easy assembling procedure, a high momentum/mass ratio, obtainable by arranging most of the mass at the outer rim of the rotor, and control force injection and position detection at this location, too, to avoid low resonances between hub and rim.
The magnetic bearing rotor is essentially composed of a rim which is connected to the hub by five titanium springs. Two slots in the rim which are equipped with Neodym-Iron magnets form, together with corresponding air-coils on the stator, the axial/tilt actuators and the drive motor. The hub contains the radial center bearing, either passive or electromagnetically active, and carries the emergency bearings.
The electrodynamic force generation was chosen as best suited for a large diameter and the associated high circumferential speeds, which resulted from the design requirements. The ironless armature coil consists of four 90° segments. The axial forces are provided by applying a similar current to all coils. The tilting torques about the two perpendicular radial axes are yielded by exciting two opposite coils alternatively.

3 Axis Active Magnetic Bearing Wheel

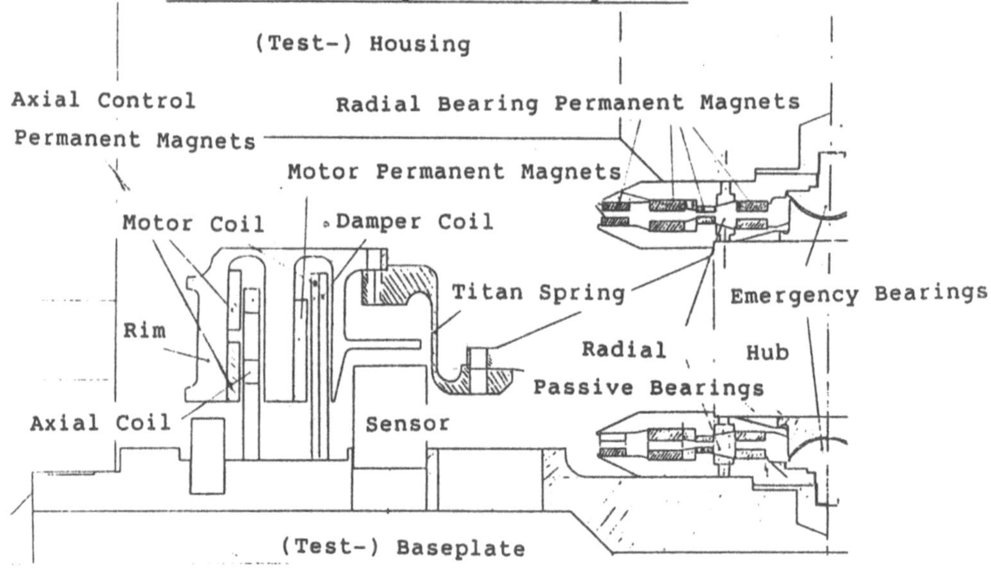

5 Axis Active Magnetic Bearing Wheel

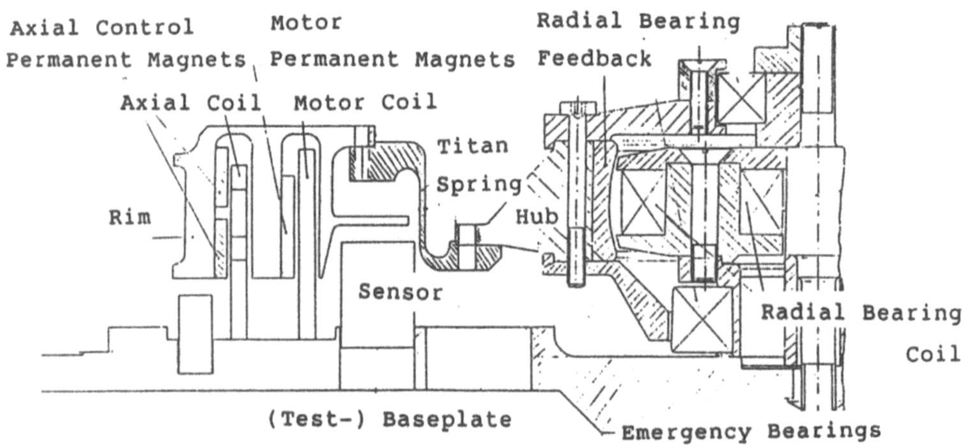

Fig. 3 Schematic Diagrams of the Magnetic Bearings

Redundancy is assured because only three coils are necessary to produce the desired forces and torques.

The radial bearing of the 3-axis wheel consists of two sets of concentric permanent magnet rings. The axial stiffness is, of course, unstable, the tilt stiffness has been designed to zero around the neutral position. For the radial bearing of the 5-axis active system, the electromagnetic force generation principle was chosen because of the small circumference in the hub.

Any deviation from the nominal rotor position is detected in a redundant way by a set of four position sensors, each located besides a stator coil, and - using three control loops - balanced by forces generated in the coils. The sensors are of the eddy current type and involе only very small lags.

The connection of the rim with the hub is achieved by titanium springs, which provide the decoupling of the different strains of rim and spokes due to centrifugal forces and temperatures by means of radial weakness of a single spoke, but a high stiffness in all directions against relative motion of hub and rim.

The task of the emergency bearing is to provide a nondisturbing landing of the (spinning) rotor in the cases of a failure in the suspension electronics or in the power supply and with externally applied forces or torques exceeding the active stiffness capability within the given mechanical limits. The emergency bearing in the 5-axis active system consists of two ball bearings the locating of which is determined by circumferential speed limits and some considerations concerning gyro- and centrifugal effects. In the three-axis active system a spherical surface sliding bearing will be tested.

The drive motor is of the brushless, ironless d.c. type which allows an excellent controllability of torque while providing high efficiency. It is located in the outer rim because in this position the mass distribution between rotor and stator is advantageous. The ironless armature consists of a three- phase stranded winding embedded in an epoxy material to avoid radial destabilizing forces as well as cogging torques caused by armature slots. The commutation is performed using three reflex light barriers which are also utilized to detect the sense of rotation.

In the 3-axis bearing, a special short-circuited winding is

introduced into the motor slot which uses the motor magnets to damp radial oscillations. The rotational drag is negligable if a sufficient uniformity of the induced voltage per magnet can be achieved.
Both types of magnetic bearings have been built and the 5-axis configuration has already been tested statically and up to 7000 rpm. The main data of the 5-axis bearing are:

Diameter	:	0.3	m	Height	:	0.1	m
Rotormass	:	6.65	kg	Total Mass	:	10	kg *
Max. Speed	:	14000	rpm	Nom. Speed	:	10000	rpm
Inertia	:	0.1	kgm^2	Momentum	:	100	Nms
Axial Gain	:	55	N/mm	Tilt Gain	:	3000	Nm/rad
Gimballing Angle	:	± 1.5	°	Slew Rate	:	17	°/s **
Motor Torque	:	± 0.35	Nm				

Angular Momentum Cross Component: 5.3 Nms

* : Preliminary value obtained from laboratory model with test-baseplate, protecting housing and breadboard electronic

** : Short time.

<u>Control</u>

The elements of the axial and tilt control loops of the nonrotating wheel are shown in Fig. 4. The plant consists of a double integration with the mass/ transverse inertia of the rotor as integration constant. The actuator consists of a current-controlled amplifier and the electrodynamic force generation system. The actuator and also the position sensor can be modeled as first order lags. The controller must provide at least one 90° phase shift to achieve stability. In the case of the passive radial bearing, an unstable spring feedback is introduced into the plant. As controller also a simple PD-type can be used, but the stability conditions become more severe in this case. By means of a superimposed zero-power controller, the system becomes selfadjusting into the axial position of minimum (zero) bearing current. If a high static stiffness is required, an integrator term can be added, the controller is then of the PID-Type. Of course, a PID- and a Zero-power controller cannot be used simultaneously in this configuration.

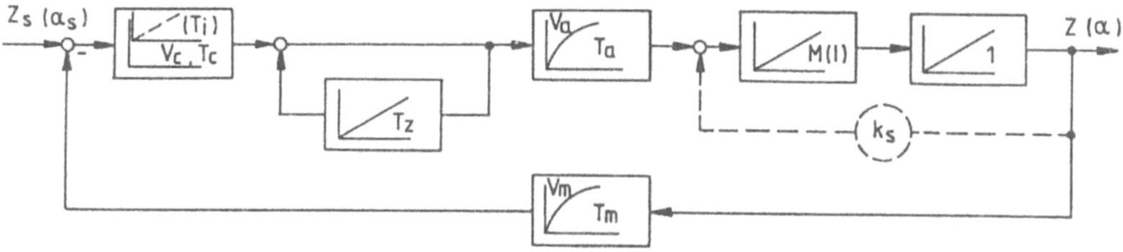

Fig. 4 Block Diagram of the Axial and Tilt Control Loop
Dashed Elements: only with the 3-Axis Active Bearing

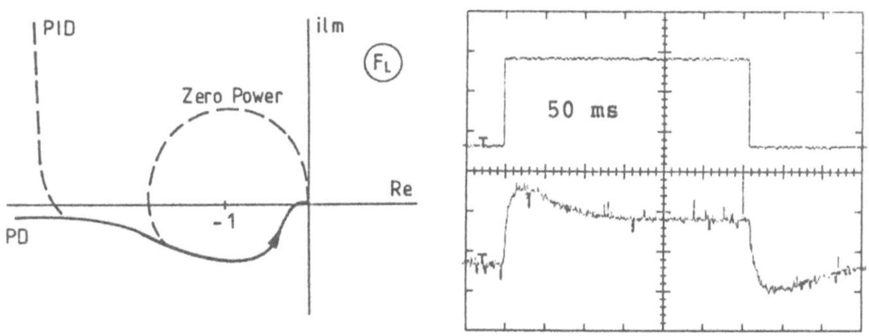

Fig. 5 Nyquist Plots and Step Response of the Axial Loop

The radial control loop of the 5-axis active bearing is principally the same as that of the axial loop of the passive radial bearing. The difference lies in two nonlinear dependencies of the operating point: the destabilizing feedback is proportional to the square of the bearing current while the actuator gain is inverse square proportional to the radial elongation of the rotor. These nonlinearities can be defused with the help of a linearizing network or a bias current. In the radial loop, a PD-controller or a PID-controller can be used. Fig. 6 shows the stability region for the controller parameter which is moving and size-varying with the operating point.

If the controller is provided with a variable parameter set, the active stiffness of the radial loop can be changed while keeping an optimal damping. A stiff controller can be used at normal operation. In zero-gravity conditions where no strong external forces are applied, a very smooth operating parameter set can be chosen which keeps the average rotor position without transmitting high-frequency sensor disturbances, generated by the rotor spinning about its center of gravity, to the stator.

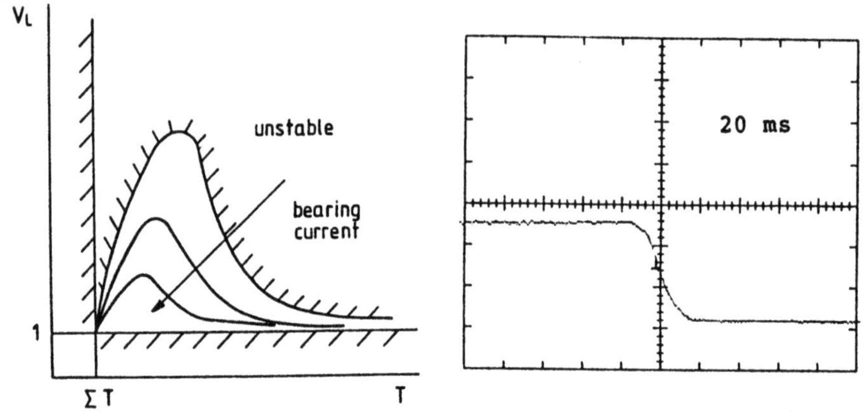

Fig. 6 Stability Area and Step Response PD-controlled Radial Loop

When the wheel is rotating, additional gyro-couplings between the tilt axes of the wheel arise which are proportional to the speed and which enable the precessional and nutational oscillations. The highest nutational frequency of a flat rotor is about twice the speed - a tilt controller, which copes with these frequencies, must be of high bandwidth and therefore causes an increased power demand in the actuator due to noise. One possibility to overcome this problem may be the introduction of a gyro-decoupling network, as depicted in Fig. 7. If the decoupling parameter matches the plant parameter, the resulting plant at all speeds is the same as that of the wheel standing still. In the step responses in Fig. 8 it can be clearly seen that the cross coupling between the tilt axes is removed and the damping is improved.

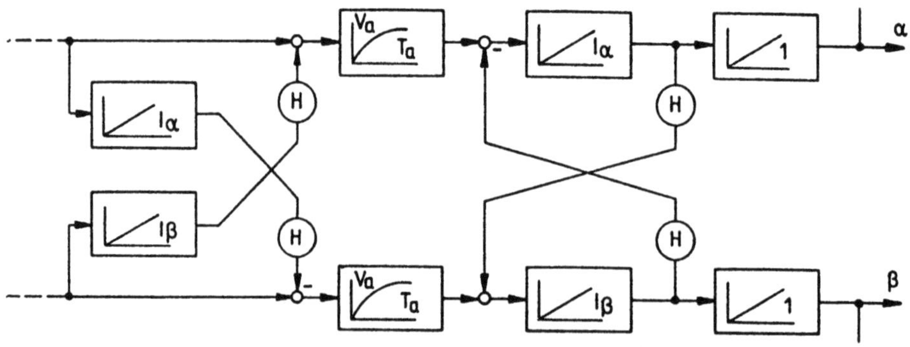

Fig. 7 Block Diagram of the Gyro-coupling and -decoupling

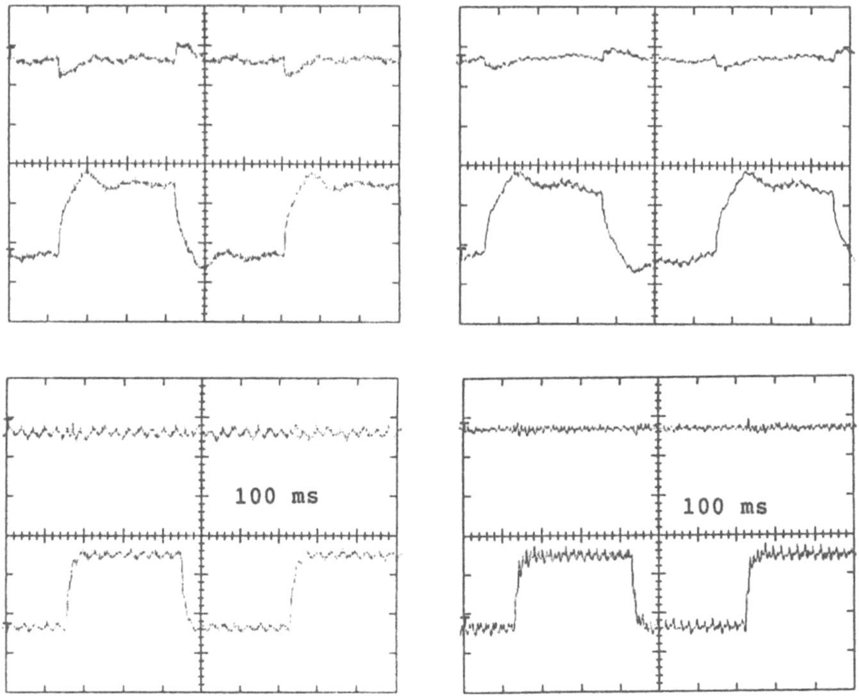

Fig. 8 Step Responses of the Tilt Axis at Different Speed
a: without decoupling, b: with decoupling

The influence of such a decoupling on the overall satellites stability must be considered separately.

<u>Conclusion</u>

Two magnetic bearing wheels of low profile design have been built, a 3-axis active version and a 5-axis active one. Both provide a vernier gimballing capability due to electrodynamically controlled tilt axes. This is of interest with respect to the increasing attitude control requirements of modern spacecraft. With the use of four sensors and four bearing coils, redundancy is built in inherently. The wheels are actually in the test phase and the five degree of freedom wheel has been already stable levitated and rotated up to 7000 rpm. A special tilt control loop design eliminates all gyroscopic couplings and improves the dynamic performance of the active gimballing. Using a control loop with variable active stiffness, the wheel can be switched in a smooth mode preventing unbalance disturbances from affecting the satellite.

Fig. 9 The 5-Axis Magnetic Bearing

References

1. Ball Bearing Versus Magnetic Bearing Reaction and Momentum Wheels as Momentum Actuators.
W. Auer, AIAA 'Global Technology 2000', Baltimore 1980.
2. Satellite Momentum and Reaction Wheels with Magnetic Bearings. J-P. Baudry, P. Poubeau, A. Robinson
3. An Advanced Low Cost 2 Axis Active Magnetic Bearing. Flywheel. M. Gauthier, A. Robinson, J.P. Roland, H. Vaillant. Third European Space Mechanisms & Tribology Symposium Madrid 1987.
4. A Flat Magnetic Bearing Reaction Wheel.
C. Murakami, A. Nakajima, S. Shingu, Y. Shimamoto
14. International Symposium on Space Technology and Science Tokio 1984.
5. Magnetic Bearing Design for High Speed Momentum Wheels.
T. Lange, First European Space Mechanisms & Tribology Neuchatel, Schweiz Oct. 1983.
6. Impact of Magnetic Bearing Rotor Design on Satellite Nutational Stability. G. Heimbold, AIAA Guidance and Control Conference, San Diego 1982.
7. Magnetic Bearing Momentum Wheels with Magnetic Gimballing Capability For 3-Axis Active Attitude Control and Energy Storage.
R.S. Sindlinger, AOCS Conference, Noordwijk, Oct. 1977

Applications in Physics

Radial Shear Force Permanent Magnet Bearing System with Zero-Power Axial Control and Passive Radial Damping

Johan K. Fremerey

Institut für Grenzflächenforschung und Vakuumphysik,
KFA-Jülich GmbH, Postfach 1913, D-5170 Jülich
F.R.G.

Summary

The bearing system described was developed starting from the idea of a simple radial magnetic floatation system with axial pin stops, i.e., mechanical contacts, as devised by L.I. Mendelsohn for application in watt hour meters (General Electric Patent, 1953). The Mendelsohn system provides positive radial and, without the stops, negative axial bearing stiffnesses due to magnetic forces acting between cylindrical magnets with axial magnetization. The magnets are mounted to the rotor and stator assemblies in a face-to-face configuration for mutual attraction. A novel stabilization unit has been developed making the Mendelsohn system a non-contacting device. The unit compensates for the axial force instability and additionally provides radial damping forces. Axial stabilization is accomplished by interaction of electric coils with the end faces of two axially spaced, cylindrical permanent magnets fixed to the rotor shaft, while radial damping is provided by interaction of the rotor magnets with a copper disk projecting into the axial flux between the rotor magnets. A high-speed neutron beam chopper cascade (four units) equipped with the new bearing system is running in continuous operation at ILL-Grenoble since 1985. A turbomolecular pump prototype for industrial application was set up in 1987.

Introduction

The design of a magnetic bearing system is a challenging and enjoyable task in view of the manifold structures that can be realized by various combinations of permanent- and electromagnets, attractive and repulsive forces, radial, axial, or mixed control, coupling and transformation of orthogonal parameters, analog and digital electronics, etc. Each application demands careful analysis of the individual system requirements and optimum adaptation of the bearing parameters toward those requirements. On the other hand, transfer of a magnetic bearing device from the laboratory status into an industrial product sensibly depends on parameters such as manufacturing cost, compatibility with existing systems, specific styling, etc., which do not necessarily meet the ideas of the designer.

The magnetic bearing system described in the present paper has been developed at KFA-Jülich for high-speed rotor application in laboratory as well as in industrial environments. No active components are used for radial force control of the system, and axial control becomes active only in case of dynamic axial load. Due to the simplicity of the bearing and electronics structure, manufacturing cost is substantially lower than that of the available active radial control systems. A large potential for laboratory and industrial applications which do not require the dynamic load capacities of a multiaxis active control system may be exploited by the introduction of passive radial bearing concepts.

The shear force permanent magnet bearing concept

Magnetic forces generated by a tangential displacement of opposing magnetic poles are defined here as magnetic shear forces. In particular, the restoring force arising from a radial displacement of the rotor shaft in Figure 1 is a radial magnetic shear force, and the corresponding bearing structure will be termed a radial shear force permanent magnet (PM) bearing.

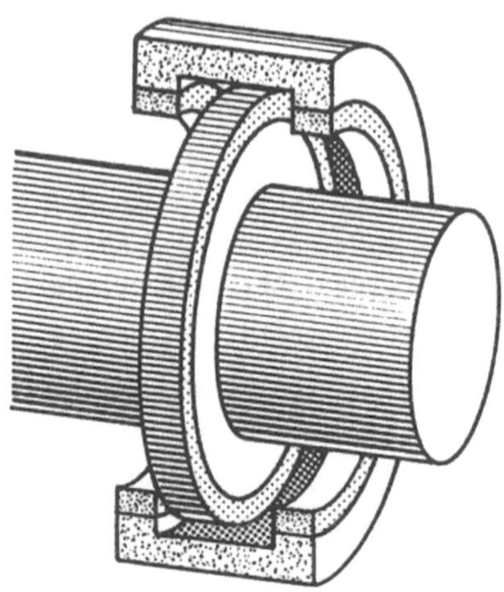

Figure 1. Radial shear force permanent magnet (PM) bearing

This type of bearing has been proposed some 35 years ago by Mendelsohn [1] for application in watt hour meters. The effect of axial unbalance of the rotor magnet between the adjacent stator magnets was eliminated by axial pin stops on either shaft end. The stops were adjusted so as to keep the rotor magnet close to the axial magnetic force equilibrium position.

By using state-of-the-art permanent magnet material such as samarium-cobalt or neodymium-iron-boron alloys, appreciable bearing forces and stiffnesses can be generated by a magnet configuration according to Figure 1. Radial load capacities of more than 100 times the magnet weight can be provided without pulling the system apart, and similar forces are available along the axial directions, still keeping the axial force equilibrium point within the axial play of the rotor magnet.

A radial shear force PM bearing for support of a high-speed rotor system can be utilized advantageously by recognizing and taking into account both the intrinsic capabilities and the weaknesses of the bearing structure. Capabilities have already been mentioned. The axial unbalance stiffness and, in particular, the poor inherent damping of the radial shear force PM bearing are considered the main drawbacks of the structure. A suitable bearing design has been achieved by introducing the asymmetric support and damping concept to be outlined in the following.

The asymmetric support and damping concept

When considering high-speed rotor application a major item of the system design is the damping mechanism for the control of precessional motions. In view of their spatial characteristics these are often termed "conical" motions. In fact, during precession the rotor spin axis moves on a double cone surface with the node point being bound to the center of mass of the rotating assembly. A radial damper will be most efficient when it is located at some distance from the node or center-of-mass point. On the other hand, the efficiency of the support bearing with the primary purpose of balancing the rotor weight will be high when this is located close to the center of mass. These considerations have guided the development of a bearing concept according to Figure 2.

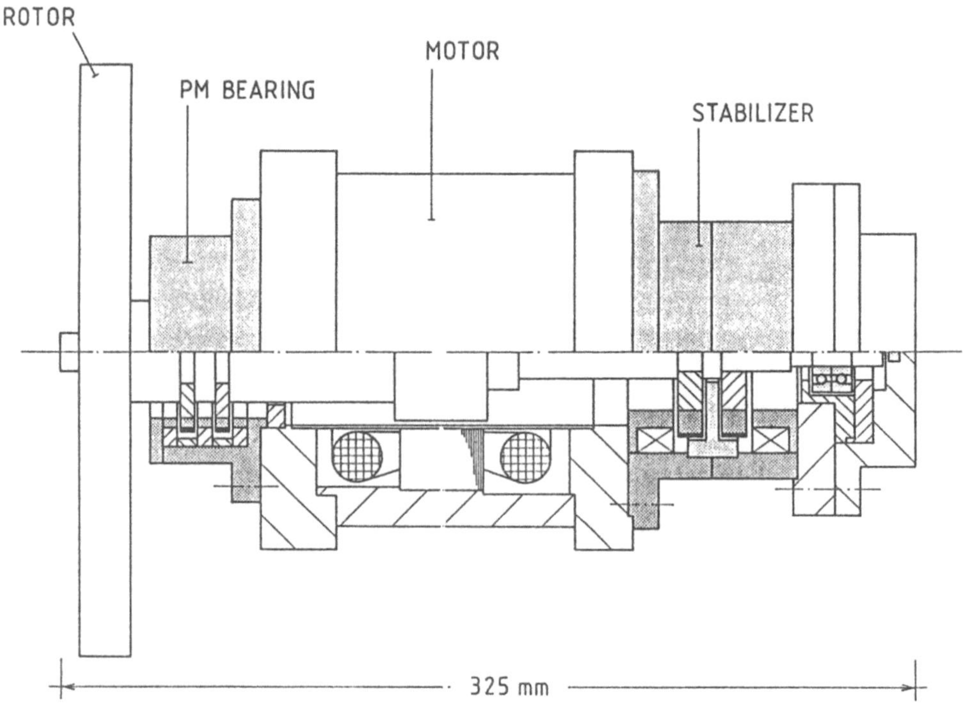

Figure 2. Passive radial bearing system with asymmetric support and damping

The main support is accomplished by a radial shear force PM bearing near the center of mass of the rotor system. Two rotating magnets are provided for higher bearing stiffness instead of one as described before.

A passive radial damping mechanism of the eddy current type is incorporated as an integral part in the stabilizer unit located near to the remote shaft end. The eddy current mechanism gets effective on radial vibration of the rotor shaft due to displacement currents generated within a massive copper disk. The disk is penetrated by the strong axial magnetic field extending between mutually attracting pole faces of rotating permanent magnets. No eddy currents are generated by the shaft rotation when the magnetic field is symmetric with respect to the spin axis. In fact, the overall eddy current losses observed on a 12-kg rotor system (Figure 3) running at 20,000 rpm was only about 3 watts including the losses generated within the support bearings and the motor.

Except for the damping effect the stabilizer maintains the non-contacting condition of the rotor system at the axial force equilibrium point of the support bearing, hence replaces the function of the mechanical pin stops as used in the original Mendelsohn concept. The axial tuning forces are generated by electrical coils acting on the damper magnets. The coils are driven by an electronic amplifier according to the output signal of an axial position sensor looking onto the end face of the rotor shaft. An automatic zero-power stabilization means drives and keeps the rotor system to the axial position where the static axial forces cancel. At that point the coil currents approach zero. The complete stabilization electronics is carried on a 70-cm^2-area printboard.

Besides the radial damping and axial control the stabilizer also serves for passive radial centering of the rotor shaft end. The centering is effected by radial shear-type restoring forces acting between the damper magnets and the ferromagnetic pole faces of the axial control magnets. The asymmetric support and damping concept provides a clear separation of the support and stabilization functions of the bearing system by locating either function in a bearing component of appropriate design. The functional separation is similar to the one applied for aircraft stabilization.

Figure 3. Neutron beam chopper prototype (1984)

Applications

The system displayed in Figure 2 was built in early 1986 as a test facility for high-speed chopper application [2] . Similar systems with support bearings on either side of the disk rotor have been realized at KFA-Jülich since 1981. Figure 3 shows a neutron beam chopper prototype under test before delivery to Institut Laue-Langevin, Grenoble, in 1984. A neutron monochromator consisting of four synchronized choppers was installed at ILL-Grenoble in 1985 for materials research. The system is operated in a 24-h service since that time.

Figure 4 shows an advanced test prototype of a 300-l/s turbomolecular pump which was developed on the basis of a cooperation project with Leybold AG. The pump is operated at a nominal speed of 51,600 rpm of the 1.5-kg turbo rotor. No adjustment is required when changing the pump axis orientation to any arbitrary direction. The bearing system has emerged by adaptation from the one outlined in Figure 2.

Figure 4. 300-l/s turbomolecular pump advanced prototype (photograph by courtesy of Leybold AG)

Acknowledgement

The author gratefully acknowledges continuous support of the magnetic bearing group by Professor George Comsa. As a research scientist in vacuum and surface physics he early recognized the potential benefits of the magnetic bearing technique for application in this field, and consequently incorporated the magnetic bearing activities in his institute.

References

1. US Patent 2,725,266
2. R. Hackenberg and W. Ebert (these proceedings)

High Precision Particle Beam Choppers

R. Hackenberg, W. Ebert

Uranit GmbH, Jülich

Summary

The special requirements of fast rotating particle beam choppers comply best with magnetic bearings. Three types of magnetic bearings for rotor weights between 1 kg and 12 kg are described, which are used in chopper systems. The most important advantages of this bearings are a very low power consumption, a high phase stability and the suitability for high vacuum.

Chopper systems are used in many laboratories and in industry to cut particle beams into packages of known size or energy. This technique is used in the fields of fundamental research, material research and of surface analysis of semiconductors for example.

Usually the particle beams are cut by fast rotating discs, which allow the beams to pass through a couple of windows. The speed of rotation may be in the order of 1 kHz.

Chopper systems are often part of much bigger and very expensive devices. In some cases they may be activated by neutron beams. So it is obvious, that the chopper bearings shall be designed for high life time and high reliability and that they shall show no bearing wear.

In many fields of application it is necessary to have chopper rotors running under vacuum. In these cases it is better to use bearings without lubrication oil.

The above mentioned requirements on chopper bearings disqualify the use of ball bearings. On the other hand active magnetic bearings fit very well the requirements mentioned above.

But in many cases there are additional specifications which

favour special types of magnetic bearings among many others. For example there is often the claim for an extremely high accuracy of phase control, if several choppers are used together in a chopper cascade. This can only be fullfilled, if the required motor power is very small compared with the energy of rotation. In this case it is of advantage to use the zero power magnetic bearings, which have been developed by the Kernforschungsanlage in Jülich and which have been manufactured under license by our company for serveral years.
The very small power loss of this bearing type is possible, because the rotor weight is fully compensated by permanent magnets. Only small electric power is necessary to stabilize the rotor position in the center of the bearing.

Eddy current losses in the rotor are small, because the field of the permanent magnets and of some electromagnets has only small deviations to an axially symmetrical shape. That is why the rotors of our chopper systems have a homogeneous not laminated structure which is suitable for the use in high vacuum.
Another condition for the application of choppers in high vacuum is the great radial gap between rotor and stator magnets, which is necessary to separate rotor and stator by a vacuum tube. The stator packet can easily be removed from the vacuum tube to clean the vacuum surfaces by heat treatment.

The figures 1 to 7 show the design of the two magnetic bearings type I and II, which have been manufactured by us for several aplications. The bearing type III, which has been developed also by the KFA (fig. 8), is now built for the Hahn Meitner Institute.

On fig. 1 you can see a cross section through the chopper type I, which has two magnetic bearings symmetrically arranged.

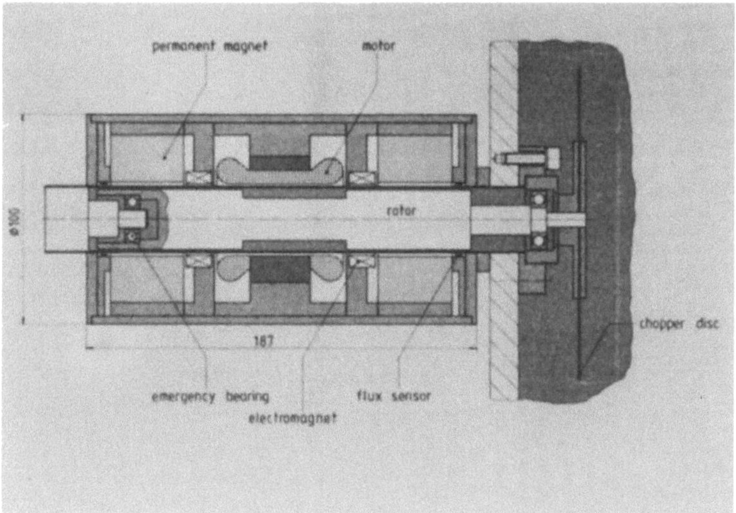

Fig. 1. Cross section of the chopper type I

Fig. 2 shows a total view of this chopper but without the chopper disc. Different kinds of discs may be mounted on the left side to adapt the chopper to the special requirements of molekular beams or neutron beams for example.

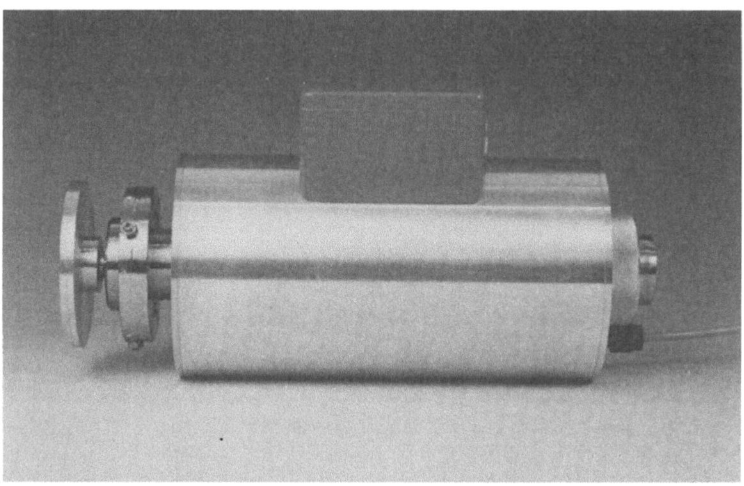

Fig. 2. Total view of the chopper type I

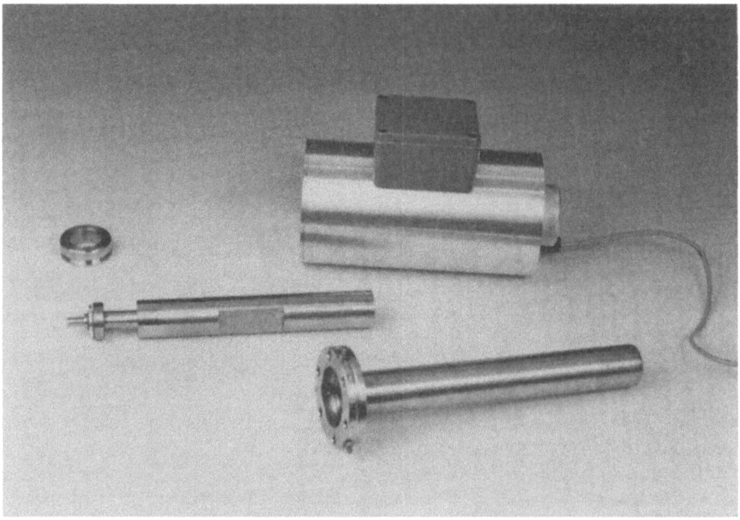

Fig. 3. Vacuum tube, rotor and stator packet of the chopper type I

Fig. 3 demonstrates, that the bearing-motor packet may easily be removed as mentioned above. The rotor may be pulled out of the tube very quickly, if a change of the chopper disc is necessary for example.

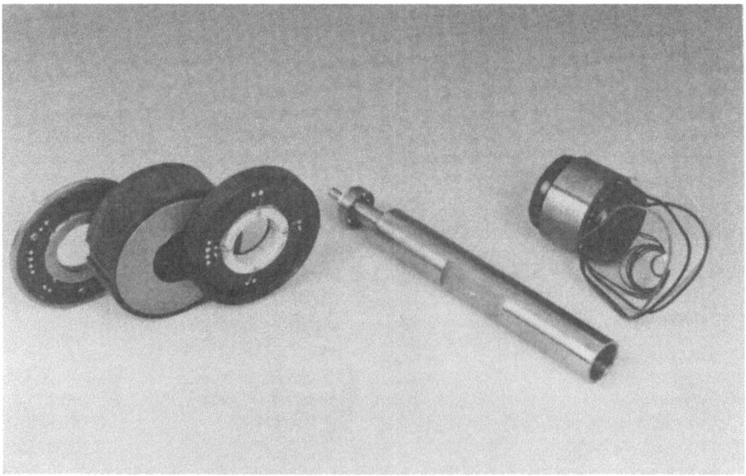

Fig. 4. Bearing and motor parts of the chopper type I

The details of the magnetic bearings are shown in fig. 4. You can see from left to right the magnetic flux sensor, the permanent magnets with a pair of iron sheet discs in front of it, the electromagnet, the rotor and the motor. The iron discs have excentric holes and compensate the rotor wheight by field deformation, if the chopper is arranged horizontally. The centric position of the rotor will be stabilized by the segmentated electromagnet, controlled by the signal of the flux sensors.

The total electronic supply of the choppers type I and II is integrated in one common electronic unit, which contains the DC power supply, the motor power supply, the operation status control unit, two control units of the magnetic bearings and a floating battery power supply, which is sufficient for a complete run down of the chopper.

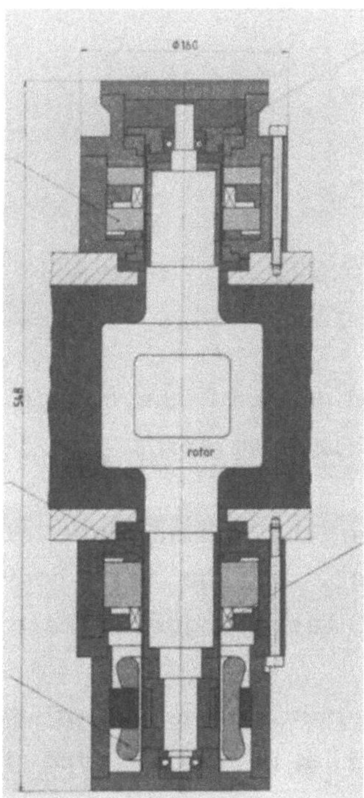

Fig. 5. Cross section of the chopper type II

The following gives some details of the bigger chopper type II (fig. 5 to 7). Fig. 5 shows a cross section of this chopper and fig. 6 a total view.

Fig. 6. Total view of the chopper type II

On fig. 7 you can see the arrangement of the two magnetic bearings and the rotor. As the main body consists of aluminium, some rings of ferro magnetic material are fitted to the rotor.
The details of the magnetic bearing of the chopper type II are very similar to the parts of type I (fig. 4) but of greater size. As type II has been realized only in a vertical arrangement, the pair of excentric iron discs is not necessary.

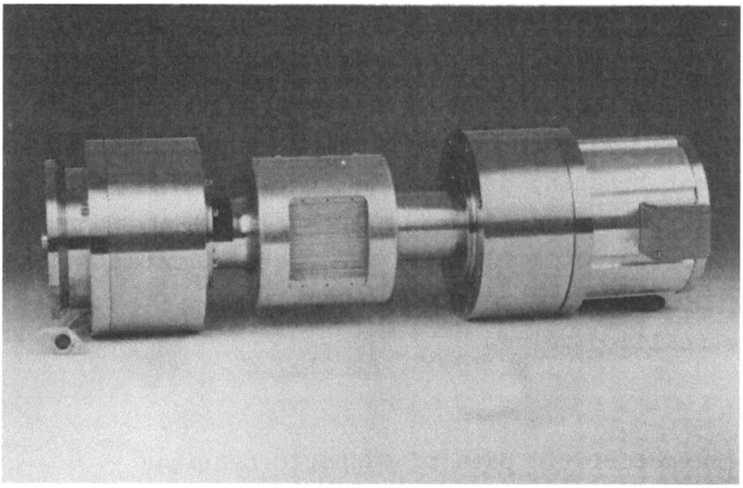

Fig. 7. Rotor/Bearing subassembly of the chopper type II

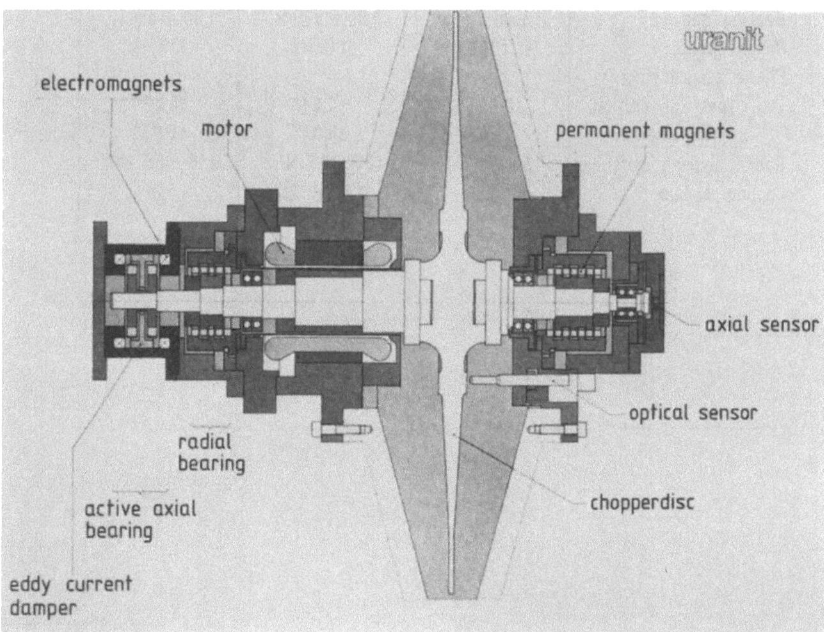

Fig. 8. Cross section of the chopper type III

The third chopper type is shown in fig. 8. This type is now built for the HMI with some modifications. The changes concern the arrangement of the bearings and the chopper disc, which shall be mounted in an overhung position. This is necessary, because the choppers shall be used in pairs with contercurrent rotating discs close together.
The most important data of the three types of chopper bearings are presented in table 1. The speed of rotation and the rotor weight in this table are not ultimate figures but typically realized values.

Tab. 1: Characteristic data of magnetic bearings

	Type I	Type II	Type III
Radial bearings	active	active	passive
Axial bearing	passive	passive	active
Arrangement	vertical/horizontal	vertical	horizontal
Rotor weight	1 kg	8 kg	12 kg
Bearing gap	3,5mm radial	2,0mm radial	1,0mm axial
Rotor speed	≦ 1000 Hz	600 Hz	333 Hz
Phase accuracy:			
per convolution	±0,025°	±0,01°	±0,01°
max.deviation	±10°	±0,01°	±0,01°
Radial bearing stiffness	160 000 N/m	100 000 N/m	210 000 N/m
Cooling medium	air	air	air

Wide-Gap, Electro-Permanentmagnetic Bearing System with Radial Transmission of Radial and Axial Forces

Karl Boden

Kernforschungsanlage Jülich GmbH, Institut für Grenzflächenforschung und Vakuumphysik, Postfach 19 13, D-5170 Jülich, F.R. Germany

Abstract

The unique features and applications of this magnetic bearing essentially result from two facts: (1) The only bearing components attached to the rotor are non-laminated ferromagnetic steel collars or cylinders; (2) all radial as well as axial forces are transmitted via radial gaps. The stator assembly needs no axially overhanging components that might impose restrictions on the rotor design and axial mobility. Due to the use of flux modulation technique wide radial gaps (25 mm are realised) can be provided at virtually zero-power conditions. The large gaps allow for effective encapsulation and shielding of rotors at elevated (or low) temperatures, corrosive environment, voltage potentials, and pressure. A 2-kg X-ray rotary anode was operated under high-vacuum conditions at + 100 kV anode potential, 600°C temperature at the bearing collars and speed 18,000 r.p.m. with 13 mm radial gaps. The complete X-ray tube is easily exchangeable by removing it from the stator assembly along the axial direction. Similar handling conditions were achieved with hollow cylindrical rotors (up to 500 m/s periphal speed at the bearing gap), spin turbines and choppers (up to 90,000 r.p.m.), and with long-stroke supporting shafts for crystal pulling plants.

Introduction

The magnetic bearing system described is part of development activities, which started in KFA Jülich in 1974 on the basis of previous work and experience of the involved staff. At the beginning most effort was directed to vacuum compatible magnetic bearing configurations but in the meantime various types of magnetic bearings for a wide field of applictions were developed, manufactured and tested, ranging from one-axis to five-axis actively controlled degrees of freedom.

This paper presents a special one-axis passiv stabilizing magnetic bearing, its particularity and some applications.

Magnetic bearings

For rotating equipment, five degrees of freedom of the rigid body have to be constraint, three attached to the translational motion and two degrees of freedom attached to the rotational motion about the cross-axes, i.e. perpendicular to the rotational axis. Only the motion in the rotational axis should not be constraint in order to allow free rotation of the suspended body around this axis and keep the drag torque as small as possible. A lot of different magnetic bearing systems are developed, including actively and passively stabilizing bearings, which fulfil the mentioned conditions. According to Earnshaw's theorem at least one degree of freedom must be actively controlled. Of course, a magnetic bearing with only one active stabilization solves all bearing problems for a few applications and seems to be the best configuration at all. But, taking in consideration simplicity of the rotor design, problems with high speed, damping, elevated temperatures, corrosion environment and assembling aspects the presented axially passive and radially active magnetic bearing type meets all this requirements, so that we prefer this special bearing configuration in many cases [1].

Passive axial bearing

The bearing stator has approximately a square cross section with a central hole for the moving body (Figure 1). It comprises at least one axially magnetized annular permanent magnet, which magnetize the compact mild steel part of the moving body via two radial air gaps, the upper gap and the lower one. The mean part of the permanent magnetic flux passes symmetrically one after another a flux concentrating flat steel plate, the upper air gaps, the mild steel collar on the moving part and the lower air gap.
In this radial air gaps the magnetic flux generates attractive forces between the bearing stator and the moving part in axial and radial directions. This forces only depend on the variation of the magnetostatic energy of the system. The system is autonomous - it doesn't comsume energy.
In the axial direction the force to distance characteristic shows a stable working area. In this area a constant force opposite to the magnetic force (i.e. the weight) stretches this "magnetic spring" to a stable equilibrium point. When the magnetic force of the carrier magnet is so dimensioned, that it can safely carry the weight of the moving body, the permanent magnetic

flux is able to provide a stable passive magnetic suspension of the moving body in axial direction without any consumption of external energy.
Since the carrier magnet holds the moving body stable in axial direction, the radially flowing part of the permanent magnetic flux generates an unbalanced force characteristic in the radial directions. The symmetric annular magnet configuration provides a central equilibrium point, where the opposing permanent forces are compensated. At the lightest deflection the moving body is attracted in this direction by the carrier magnet - the equilibrium is unstable.

Active radial stabilization

To stabilize this unstable equilibrium in the radial direction a stabilizing device is required. This device comprises

- contactless sensors for measuring deviations of the rotor position from the equilibrium point or the desired position and means for producing electrical sensor signals in accordance with the deviation
- electronic control devices for amplifying and phase-displacing this sensor signals to produce output signals
- electromagnetic means for producing substantially radially directed magnetic fields which generate restoring and damping forces to the rotor.

The position sensor

The position sensors are preferably small-sized magnetoresistive plates, which are mounted on the inside of the flux concentrating plate in the upper gap, so that they are cut mainly by a radial component of the permanent magnetic flux of the carrier magnet. The change in field strength as a function of the distance produces a change in resistance and forms the sensor signal. The mentioned flux concentration results in a high sensitivity. When operated in differential mode the nonlinear flux-resistivity function is changed to a linear function in the midrange and a null-position is created in the bearing centre.
This sensor system is rather simple and cheap. Beyond it no additional sensor specific rotor peaces are necessary.

Force coil

The electromagnetic coil system (Figure 2) [2] preferably consists of four segment coils, which are wound on a common laminated ring. (Only two segments are figured in the drawing.) They operate in pairs with diametrically opposite coils developing opposing fluxes. The coils of each pair are connected in series to yield the same radial control current. The flux generated by the control current is forced one after another across a first air gap, through the ferromagnetic bearing collar of the moving body, across the opposite air gap and short circuited by the laminated ring, which forms a parallel path for the control fluxes.
The control fluxes dont have to pass the permanent magnet, so that the permanent magnet does not contribute to the control circuit reluctance. On the other hand, the coil system is arranged immediately next to the permanent magnet, so that the permanent magnetic flux passes the lower gap parallel to the control flux. Thus the control flux modulates the normally symmetric permanent magnetic flux differentially and produces controlable radial forces in two perpendicular directions.

Radial gap

The gap flux density depends among other things on the volume of the permanent magnets. The placement of the magnets in the stator permits relatively large area magnets, so that cheap materials with low magnetic energy density can be used and large gaps can be realized without essential restrictions on high bias flux level, which is advantageous for both the sensor sensitivity and the effectiveness of the electromagnetic control coil system.
The radial unbalance stiffness decreases and the homogenity of the permanent magnetic field increases with increasing gap width.
Large gaps allow the arrangement of housing cylinders consisting of non-magnetic material with high electric resistance like stainless-steel, heat-insulators or glasses for effective encapsulation and shielding of the rotor at elavated temperatures, in corrosive environment, in vacuum or at elevated pressure. As there should be an additional clearence between rotor, stator and housing cylinder, both can easily removed to allow withdrawal, exchange or assambling. This clearances also permit axial movement of the rotor

and/or stator during operation, so that both can be adjusted at several desired axial positions, axially transported without touching one another over long axial distances or oscillated by using additional axial field coils or linear motors.

The biasing effect of the control flux by the permanent magnetic flux besides the conversion of the radial instability into a stable one, has the desirable benefits of

- linearizing the force-versus-current relationship
- multiplying the effectiveness of the control ampere-turns
- zero force at zero current in the passive equilibrium position, which consequently represents the reference axis for the position control loop
- minimizing of electric power consumption
- the realization of large radial gaps between rotor and stator.

Control loop

The need of two control loops per bearing complicates the electronic device somewhat. On the other hand due to the use of flux modulation technique the control electronic is freed from linearization networks for sensor and force signals and high-power consumption.

As described before, the control current is fed-back, so that the output command voltage is proportional to the output current or control force. This feedback makes the effect of the coil inductance negligible.

The bearing stiffness is adjustable within the stable area of the control loop. It usually has a frequency characteristic, which allows reduction of unbalancing forces with increasing speed. The damping of the system can also be adjusted by variation of the phase-shifting network.

Electronic balancing of the radially symmetric unbalance forces enables the virtually zero-power method, which is used for minimizing the power consumption and for compensation of the sensor signal drift. Consequently, the electronic control device is characterized by low energy consumption and simple structure, which can be flexibly adapted to the application requirements.

Rotor collar

As already shown, the radial stabilizing control force generated by the electromagnetic coil system is multiplied by the biasing permanent magnetic field strength. The larger the biasing field, the smaller the required ampere-turns of the control system, i.e., that the magnetic polarization of the ferromagnetic rotor collar is the more homogeneous the higher the biasing field can be arranged. Consequently, the hysteresis- and eddy-current losses in the rotor are diminished. The diminution of these losses lowers the generation of heat and permits the use of compact mild steel collars on rotor even in high-speed rotor application. Compared with rotating permanent magnets, steel collars have the essential advantages of high yield strength, high Curie-temperature and corrosion-resistance. The additional fact that no overhanging bearing parts between rotor and stator are needed permits substantial freedom in the rotor design.

Complete bearing system

For the entire suspension of the moving body a second coaxial bearing is needed in order to constraint all radial degrees of freedom (Figure 3). This second bearing can be of identical construction as the one described above [3]. The passive forces in axial direction can be directed to support the first bearing in order to enlarge the axial load capacity in vertical-axis applications. In horizontal-axis applications (Figure 4) the bearings should be arranged to counteract in axial direction, so that one bearing preloades the other one in order to produce a stable equilibrium in the absence of external axial holding forces.

There are a lot of variations and combinations of the magnetic configuration developed [1,2,3,4,5] for example, the arrangement of

- pole-shoes in order to optimize the permanent magnetic flux loop
- additional magnetic force coils in order to influence the axial exerted permanent magnetic force or to stabilize or to vary the axial position
- the bearing stator inside a hollow cylindrical rotor
- multiple gaps in order to enlarge the load capacity and/or the stiffness
- radial magnetization of the permanent magnets
- permanent magnetic material with high energy product.

Rotor drive

For the rotor drive normally a cylindrical rotating field stator is used, which is mounted on the stator. The structure of the rotor armature depends on the motor system. Synchronous and asynchronous drive motors are compatible with the magnetic bearing system. As described above the drag torque is rather low, so that in the absence of external torque very low drive power is required.

Examples of Application

Figure 5	Rotating hollow cylinder (1969)	
	Rotor speed	60,000 rpm
	Velocity in the bearing gap	500 m/s

The cylinder material is high yield strength steel and allows high rotational frequency. The tangential velocity is essentially limited by the yield strength. Due to the absence of axially overhanging rotor and stator parts the rotor is easily dismountable by moving it along the axial direction. This leading feature is used at most of the following applications.

Figure 6	Spinning turbine (1973)	
	Working speed	48,000 rpm
	Fast speed	90,000 rpm
Figure 4	Molecular chopper (1974) [9]	
	Rotor speed	24,000 rpm
	Vacuum environment	
Figure 3	Neutron chopper (1975) [9]	
	Rotor speed	36,000 rpm
Figure 7	Fast light chopper (1983)	
	Rotor speed	84,000 rpm
Figure 8	X-Ray-tube with rotating anode (1979) [6]	
	Rotating frequency	18,000 rpm
	Radial gap	13 mm
	Rotor temperature at bearing parts	$< 600^{\circ}C$
	Anode potential	> 100 kV

This arrangement has the function of a current switch. The rotor can be moved to two different axial positions by means of an additional axial field coil arrangement. [4].

Figure 9 High speed UHV-Chopper (1986)
Installed at Philips Eindhoven
Corrosive environment
Rotor speed 24,000 rpm

Figure 10 Neutron beam chopper (1986)
Installed at Rutherford Appleton Lab., UK
Rotor speed 36,000 rpm

Figure 11 Apparatus for pulling crystals from the melt (1986) [7,8]. Both, crystal and crucible are carried and transported by magnetically suspended shafts. Since these parts move in clean vacuum or pressure environment inside the housing the stators are arranged outside the housing, and can be axially moved by means of a linear drive. Even with comparatively thick-walled vessels, the gap width (24 mm) is sufficient for accommodation of additional tempering elements, i.e. elements for controlling the temperatures of the housing wall up to 700°C, and for cooling the stator unit. Consequently this machine permits for the first time to make use of the "hot-wall"-method for the growth of monocrystals from so called III-V compounds, especially GaAs and InP at an industrial scale. In a test setup several crystals of semiconductor material with a weight up to several kilogramms have been pulled.

Conclusion

A practical non-contacting magnetic bearing system using flux modulation techniques has been developed. It has been worked very effectively, yet the bearing parts are very simple and inexpensive. For example, flat annular speaker magnets, made of cheap Bariumferrite-material, simple coils and sensor systems and mild steel collars make up the entire parts of stator and rotor.

The bearing configuration permits

- wide radial gaps
- very simple rotor structure, far-reaching freedom in rotor design
- no axially overhanging bearing components of rotor and stator unit
- high rotor speed
- low and high rotor temperature
- clean and corrosive atmosphere
- vacuum and pressure environment

- axial mobility and position control of rotor and stator during rotation
- easy installation
- low power consumption

The bearing concept thus covers a lot of complex requirements and sets up new fields of design and application.

References

1. German patent 1750602, 1933031, 1967085, 1967086
2. " " 2213465
3. " " 2312001
4. " " 2420814
5. " " 2420825
6. " " 2716079
7. " " 3432467
8. " " 3530417
9. R. Hackenberg, W. Ebert, "High precision particle beam chopper" paper presented at First Int. Symp. on Magnetic Bearings, June 6-8, 1988

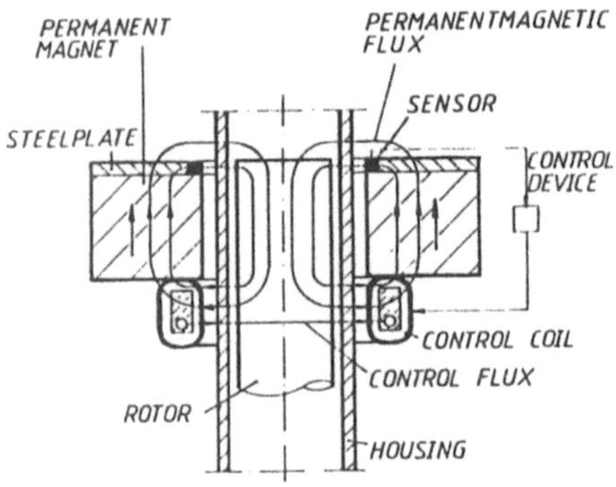
PERMANENT
MAGNET
PERMANENTMAGNETIC
FLUX
SENSOR
STEELPLATE
CONTROL
DEVICE
CONTROL COIL
CONTROL FLUX
ROTOR
HOUSING

1

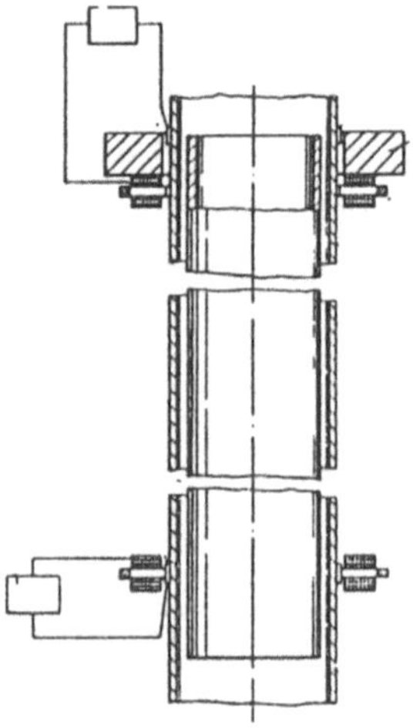

5

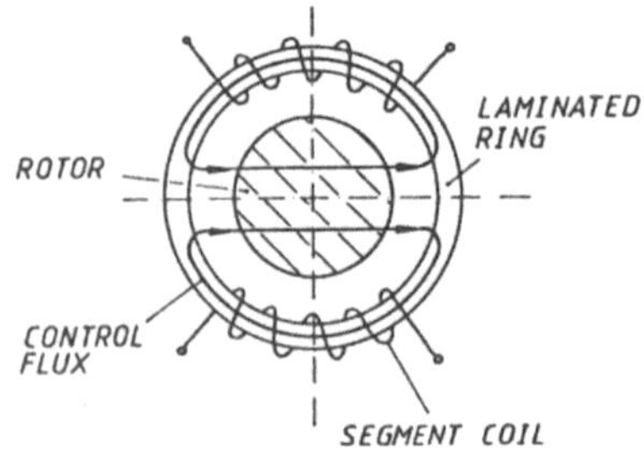
LAMINATED
RING
ROTOR
CONTROL
FLUX
SEGMENT COIL

2

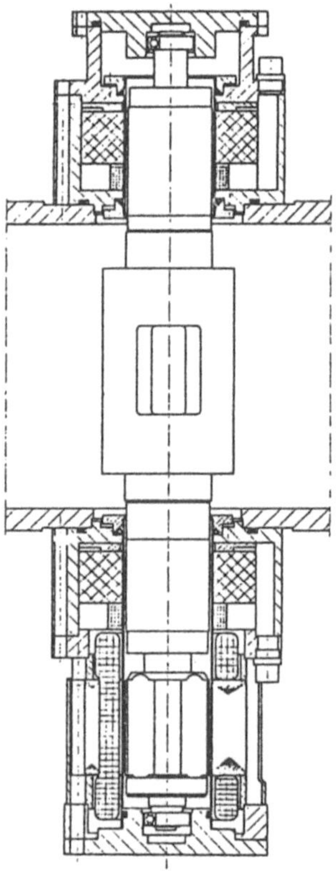

3

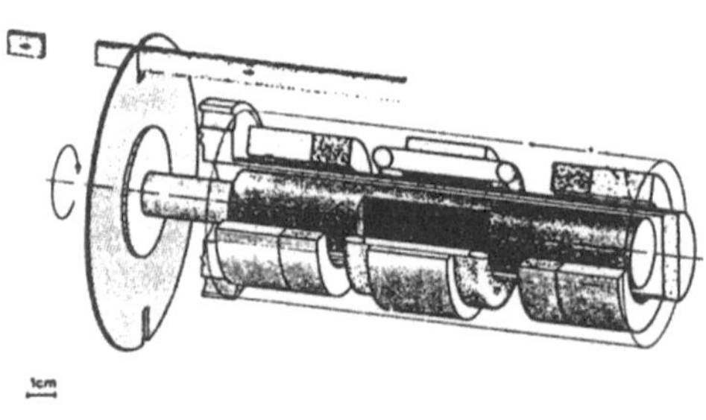
1cm

4

6

7

8

9

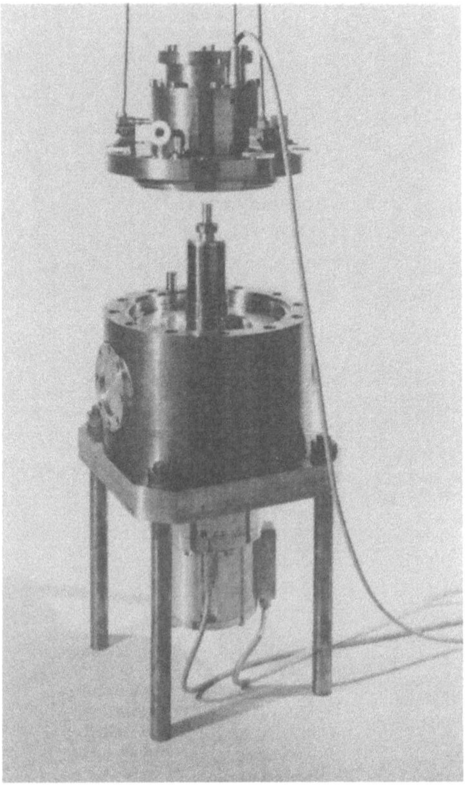

10

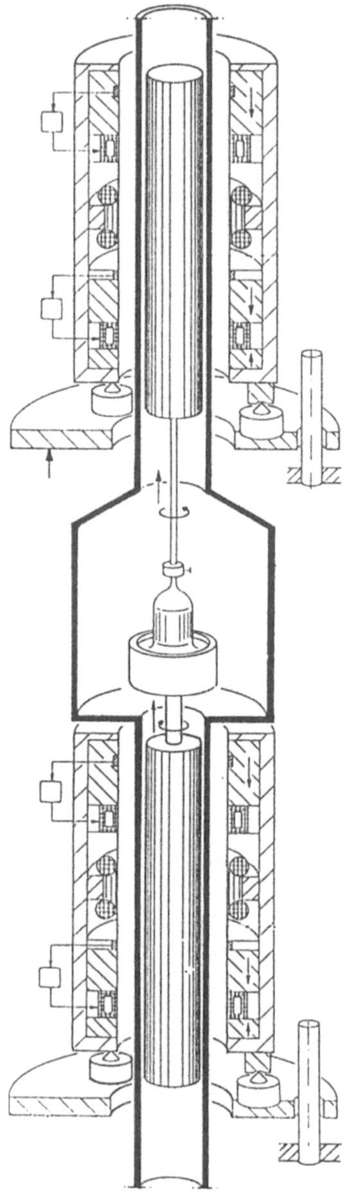

11

Ultra-Low-Friction, Zero-Power Magnetic Suspension System of the Spinning Rotor Vacuum Gauge

J.K. Fremerey and B.E. Lindenau

Institut für Grenzflächenforschung und Vakuumphysik,
KFA-Jülich GmbH, Postfach 1913, D-5170 Jülich
F.R.G.

Summary

Magnetic suspension systems have been developed at the University of Virginia (Holmes and Beams 1937) for the special purpose of spinning small bodies up to ultimate speed. The frictional torque acting on a rotating body when being suspended in a magnetic field with a highly symmetric flux distribution around the vertical axis was found to be many orders of magnitude lower than the torque exerted by gas friction at atmospheric pressure. Consequently, that type of suspension system was proposed and used lateron for pressure determination in rarefied gases. Drag torques down to 10^{-17} Nm have been resolved under optimum laboratory conditions. The present paper describes the magnetic suspension system of the "Spinning Rotor Vacuum Gauge" (SRG) which as a commercial unit resolves about 10^{-14} Nm under regular laboratory conditions. This figure corresponds to a pressure resolution of about 10^{-7} mbar, nitrogen equivalent, of the gauge. The SRG was introduced as a transfer standard in 1979 by BIPM and is commercially available since 1981. The gauge head functions will be discussed with special emphasis on minimization of the residual drag.

Introduction

The achievement of low bearing friction is a general aim in mechanical engineering. Ultra-low friction is demanded if rotational drag is to be determined for metrological purposes. The most advanced practiced system of this kind is the spinning rotor gauge (SRG), which allows pressure determination in rarefied gases by evaluation of the gaseous drag on a freely spinning rotor [1,2]. In a high vacuum, i.e. at pressures below about 10^{-2} mbar, reasonable pressure resolution can be achieved only in the absence of mechanical bearing friction. The development of a non-contacting magnetic rotor suspension by Holmes [3] in 1937 is to be regarded as the starting point of high-resolution friction metrology. The residual bearing friction in a magnetic rotor suspension results from eddy current losses in the rotor due to asymmetries of the magnetic field with respect to the rotor spin axis, as well as from eddy currents in conducting material in the vicinity of the rotor due to a non-uniform rotor magnetization. The inner losses (inside the rotor) are minimized by providing a highly symmetric suspension field while outer losses can be widely reduced by making the rotor from homogenious material and giving it a highly symmetric shape. The above conditions have been realized to a high degree by the type of permanent magnet suspension described in the present paper. Torque resolutions of the order 10^{-17} Nm have been achieved in laboratory units especially designed for basic research in the field of gravitational dynamics [4,5]. The above resolution corresponds to a deceleration time constant of several hundred years as observed on a 2.5 mm steel ball spinning at 200 Hz. Time constants of still several years can be resolved by the present commercial spinning rotor gauges.

Gauge head functions

A schematic of the gauge head is shown in Figure 1. The design of the stator package was guided by the need for a most symmetric arrangement of all suspension and operation components around the vertical spin axis of the rotor. The rotor is typically a 4.5 mm diam ball bearing. The ball is stabilized between two mutually attracting rear-earth-cobalt magnets supported and centered inside a mild steel case. The case provides magnetic flux return and effective shielding of the stator package against outer magnetic fields. The field symmetry in the center of the 35-mm magnetic gap is enhanced by circular pole discs made of Armco iron. The axial field strength is tuned by the number and location of the discs so that the magnetic forces acting on the ball just balance the gravitational forces when the ball is in the center of the stator package. At this position the ball is unstable along the vertical axis and natural stability is given in the horizontal plane, where the ball favours the location of highest field strength right on the symmetry axis of the system.

Only minimum active axial control is necessary for keeping the ball to the central equilibrium position. This is done by two coils coaxial with the magnets and equidistant above, respectively below the ball. The coils serve for sensing of the vertical rotor position (inductive pickup) and simulta-

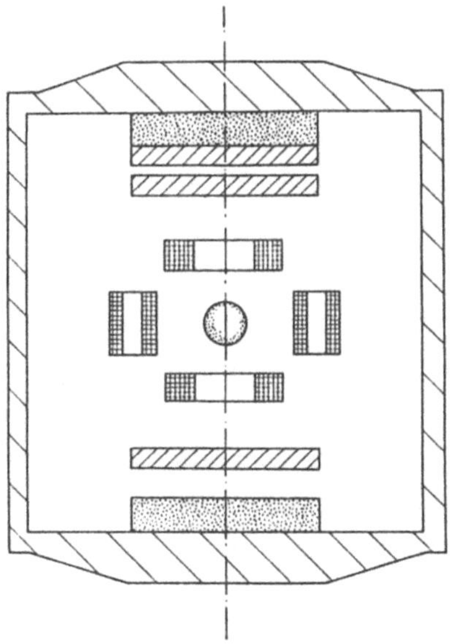

Figure 1. Single-axis permanent magnet suspension

neously for magnetically restoring and damping of the rotor along the vertical.

When the ball is at the magnetic equilibrium position the sensor output voltage and the control current become zero. A special section of the axial stabilization electronics serves for automatic tuning of the sensing-and-control amplifier to the zero-power condition. Thus, heat generation within the coils is reduced to a negligible amount associated with the inductive pickup.

In order to suppress or damp out lateral oscillations of the ball which would lead to disturbancies of the rotational field symmetry, the ball is surrounded by additional four coils in 90-deg separated, off-axis positions. In Figure 1 two of these coils are shown. The magnetized ball rotor induces a voltage proportional to its lateral velocity in one of these coils which is used for pickup while the opposite coil serves as the corresponding damping actuator.

The arrangement of further coils which are necessary for operation of a spinning rotor gauge are shown in the perspective view of the complete gauge head in Figure 2. Four drive coils with horizontal axes in connection with a two-phase generator drive the ball up to the operation frequency of about 410 Hz. After spin-up, the frequency of the freely rotating ball is picked up by two coils next to the rotor. An a.c. voltage is generated in these coils due to the horizontal component of the rotor magnetization. This rotating component mainly results from the asphericity of the ball. In view of a useful pickup amplitude, high-grade production ball bearings are too spherical, even though desirable for lowest eddy current drag. Low eddy currents and distortions of the magnetic field symmetry by the vacuum encapsulation is achieved by making this from an amagnetic steel (1.3952) tubing providing high electric resistivity and low permeability.

Vertical line-up (±0.5 deg) of the stator package is required for the rotor to be matched with the field symmetry axis.

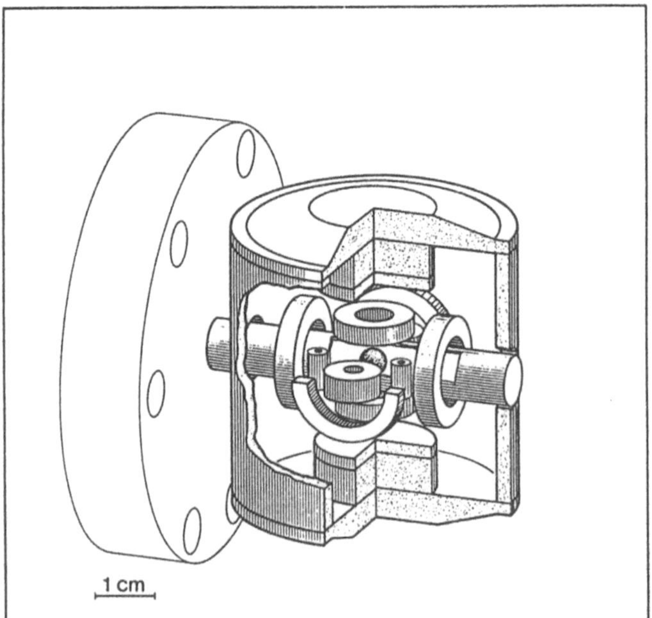

Figure 2. Spinning rotor vacuum gauge (SRG) head

Performance

The attributes of the ultra-low friction magnetic bearing structure with respect to resolution of low drag effects is visualized by Figure 3. The rotor frequency was monitored at a background gas pressure below 10^{-9} mbar over a period of three days.

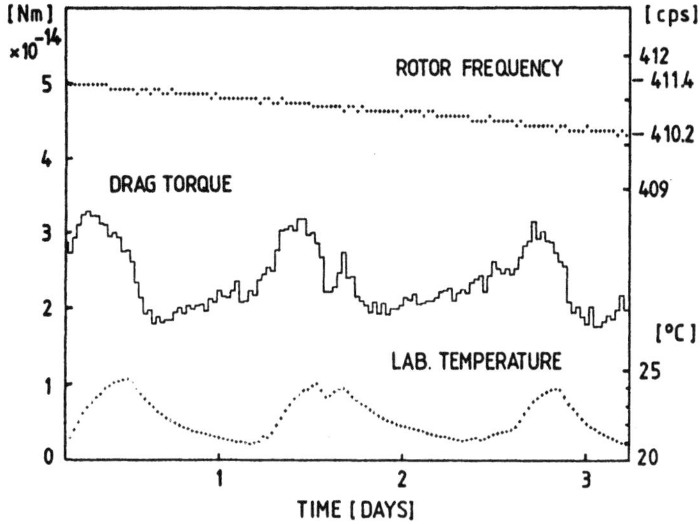

Figure 3. Frequency decay of SRG rotor and equivalent drag torque in correlation with laboratory temperature

The absolute frequency decay during this time was 1.2 cps. With a mean spin frequency of 410.8 cps this gives a decay ratio of 1.5×10^{-8} sec^{-1}, or a decay time constant of about two years.

With the gas friction eliminated, the decay is due to the eddy current drag. The mean value of the drag torque as obtained from the frequency decay is about 2×10^{-14} Nm and the resolution is of the order 10^{-15} Nm. In terms of gas pressures this resolution is of the order $\pm 1 \times 10^{-8}$ mbar.

The obvious fluctuations of the drag torque, which amount to about 1/3 of the overall drag are correlated with the drift of the laboratory temperature. The fluctuations are caused by the change of the rotor size with temperature and corresponding changes of the rotor inertia. Due to conservation of the angular momentum of the freely spinning rotor fluctuations of the rotational speed come into appearance. The example was chosen in order to demonstrate that no longer the bearing friction of the SRG suspension, but other effects determine the limit of torque resolution [6].

The temperature effect underlines the importance of a zero-power stabilization, which became feasible with the permanent magnet bearing concept.

A natural limitation of the residual drag torque acting on the ball is given by the Corriolis effect [7]. This is caused by the tendency of the ball to keep its spin axis fixed in space, while the support field turns with the earth. For central European latitudes the resulting drag torque is of the order 10^{-15} Nm. Thus, the drag actually observed in a commercial SRG is not far from the Corriolis limit.

References

1. J.W. Beams, J.L. Young, and J.W. Moore, J. Appl. Phys. 17, 886 (1946)
2. J.K. Fremerey, J. Vac. Sci. Technol. A3, 1715 (1985)
3. F.T. Holmes, Rev. Sci. Instrum. 8, 444 (1937)
4. J.K. Fremerey, Phys. Rev. Lett. 30, 753 (1973)
5. J.K. Fremerey, G.H. Comsa, and G. Comsa, Electrón. Fisc. Apli. 17, 193 (1974)
6. B.E. Lindenau, Proc. 1st European Vacuum Conf., Manchester, U.K., April 1988, Vacuum (in print)
7. J.K. Fremerey, Rev. Sci. Instrum. 43, 1413 (1972)

Centrifuge for Epitaxial Growth of Semiconductor Multilayers

E. BAUSER[1], G. SCHWEITZER[2], H. P. STRUNK[3], and A. TRAXLER[2]

[1]Max-Planck-Institut für Festkörperforschung, Stuttgart, FRG
[2]ETH, Zürich, Switzerland
[3]TU Hamburg-Harburg, Hamburg, FRG

Abstract

We describe a vacuum centrifuge which consists of a long vertical rotor that is at its upper end electromagnetically suspended inside a vacuum tank. The design is arranged such that it fulfills the requirement for a thoroughly clean vacuum recipient. Experiments are discussed which illustrate the application of the centrifuge to liquid phase epitaxy of silicon multilayers.

Introduction

Preparation for the production of high quality epitaxial semiconductor layers requires conditions of extreme purity. Impurities and foreign atoms, even when present in minute traces, can impede the uniform growth of new crystalline material on the substrate and thereby impair the required structural perfection of the multilayers. In addition, impurities may alter the electrical and optical properties of the epitaxial layers.
Liquid phase epitaxy (LPE) requires crystal growth systems that provide for brief dwell times for the liquid applied to the substrates. Transporting the solution by centrifugal force in a rotating crucible is, therefore, advantageous. **Short** dwell times of the solutions ensure growth of the desired **thin** crystalline layers on the substrates. In growth systems equipped with conventional bearings for rotary motions, however, the quality and uniformity of the opto-electronic material properties and the reproducibility of the growth- results suffer severely if contaminated by lubricants, or because of wear and abrasion.
Magnetic bearings provide a unique opportunity for avoiding

such difficulties [1,2]. The vacuum centrifuge described in the following text uses magnetic bearings and meets all requirements of liquid phase epitaxy. The description will be followed by a brief outline giving basic principles of LPE and explaining the features of operation of the rotating crucible within the centrifuge. Finally, a few examples of high quality silicon multilayers grown in the centrifugal system will serve to demonstrate the advantages which the centrifugal technique utilizing magnetic bearings possesses.

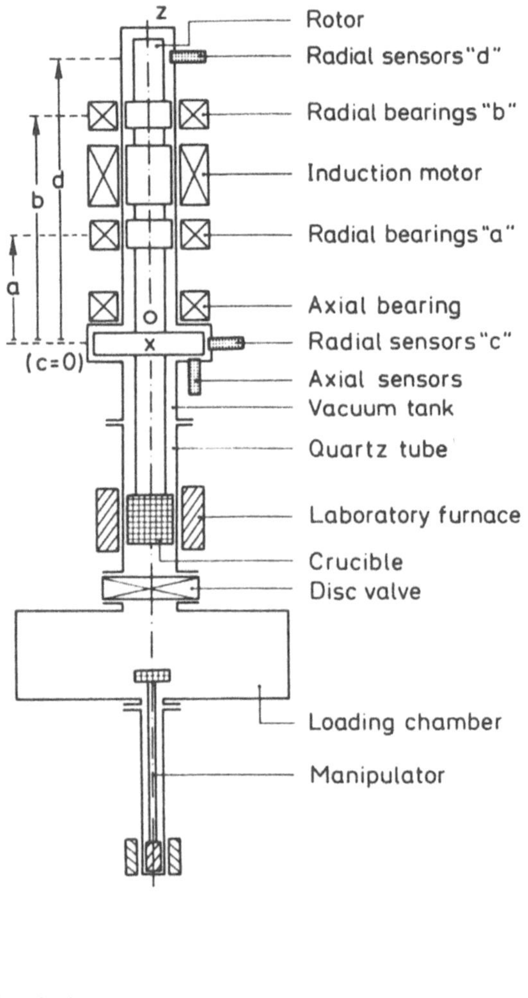

Fig.1. LPE centrifuge with contactless electromagnetic suspension of the rotor, schematic longitudinal section after [1-4]. The bearing and sensor plane positions a,b,c and d are measured from the centre of gravity at x(c=0). The planes are perpendicular to the z-axis (see also Table 1).

Vacuum epitaxy centrifuge

Figure 1 shows a schematical longitudinal section of the epitaxy centrifuge. The epitaxy crucible is mounted at the lower end of a vertical rotor. The rotor's upper end constitutes a part of a contactless electromagnetic suspension arrangement inside a thin-walled vacuum tank [2]. The bearing magnets and the rotor drive are outside the tank. The upper part of the tank surrounds the rotor and consists of a stainless steel tube. A flange connects the stainless steel tube with a quartz recipient. The lower end of the rotor and the crucible are surrounded by a quartz recipient. The crucible is heated from outside the quartz tube by a laboratory furnace. To reduce contamination of the system, the crucible is loaded and unloaded through a loading chamber. The loading chamber is flushed with pure hydrogen and is equipped with a magnetic feedthrough and a manipulator. Since temperatures of about 1000°C may be required in the crucible, the magnetic bearing and the rotor drive are water-cooled. The cooling jacket provides also for the mechanical rigidity of the system, Fig. 2.

Fig.2. Upper part of LPE growth system with centrifuge. Loading chamber and manipulator are removed.

Magnetic bearing surrounded by cooling jacket

Rotor and magnetic bearing

Figure 1 illustrates the arrangement of bearings and sensors. Figure 3 shows the rotor of the vacuum centrifuge. The various physical system parameters are indicated in Table 1. The electromagnet which generates the bearing force is part of an automatic control loop. The radial gap between rotor and bearing is 1.8 mm wide. More than half of the distance is filled by

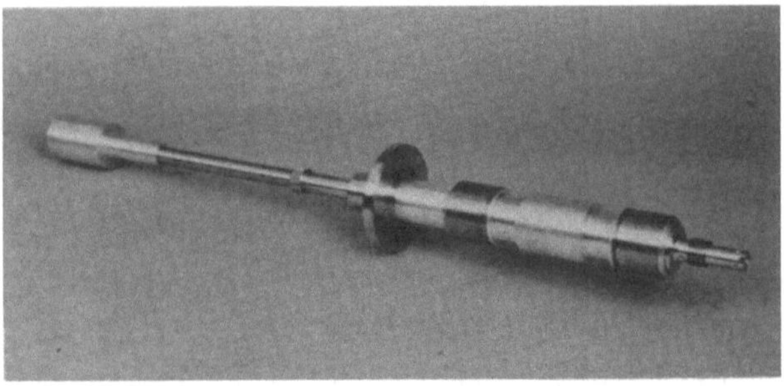

Fig.3. Rotor with two bearing rings for the radial bearing (right), support disk for the axial bearing (center) and payload crucible (left).

the 1 mm thick wall of the vacuum tank. As a consequence, rotor movements in the tank are restricted to 0.4 mm. In order to prevent tank damage owing to "overload" or sudden stoppage of the rotor, two graphite rings are fastened in the tank and serve as emergency bearings. The radial bearing has an 80 mm inner diameter and the comparatively large gap width mentioned above. The ratio of bearing force to control current is approximately 60 N/A. The gain factors are selected so as to achieve a maximum radial force of 60 N. The bearing's stiffness obviously depends on the dimensioning of the controller and is frequency-dependent. Bearing forces are fully effective up to a frequency of about 30 Hz. As shown in Fig. 4 they decrease with a further increase of the frequency. The system must be sufficiently vibration-free [5].

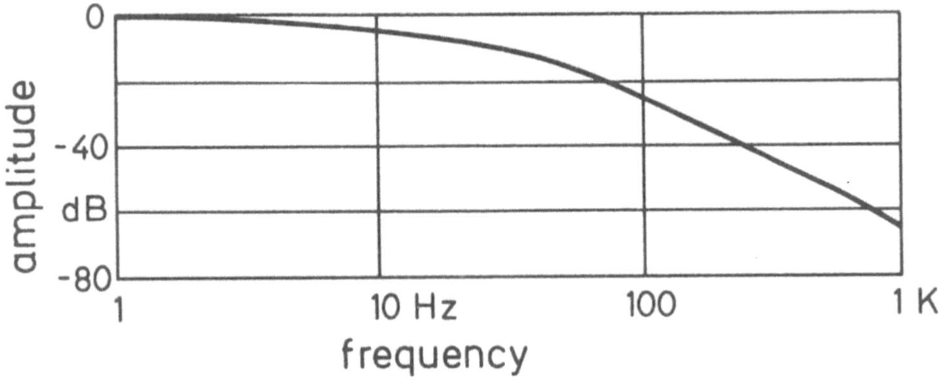

Fig.4. Amplitude-frequency relation of radial rotor motion.

Table 1. Physical System Parameters for the Centrifuge

Rotor length	L	= 1	m
Rotor mass	m	= 9	kg
Radial moment of Inertia	I_x	= .616	kg m^2
Axial moment of Inertia	I_z	= .0092	kg m^2
Rotational speed	Ω	= 1000	rpm
Force-current factor	k_i	= 66	N/amp
Force-displacement factor	k_s	= 80	N/mm
Bearing and sensor positions (Fig.1):			
Bearings "a"	a	= 118	mm
Bearings "b"	b	= 346	mm
Sensors "c"	c	= 0	mm
Sensors "d"	d	= 411	mm

Description of the crucible and epitaxial growth of multilayers

In LPE, single crystalline layers are deposited from saturated solutions on usually single crystalline substrates. The solvents applied for this purpose consist of molten metals saturated with the substance which is to be deposited epitaxially. The saturated solutions are brought into contact with the substrate, Fig. 5. Cooling of the system causes an epitaxial layer to grow on the substrate. When the solution is removed from the surface of the grown layer the epitaxial growth is terminated.

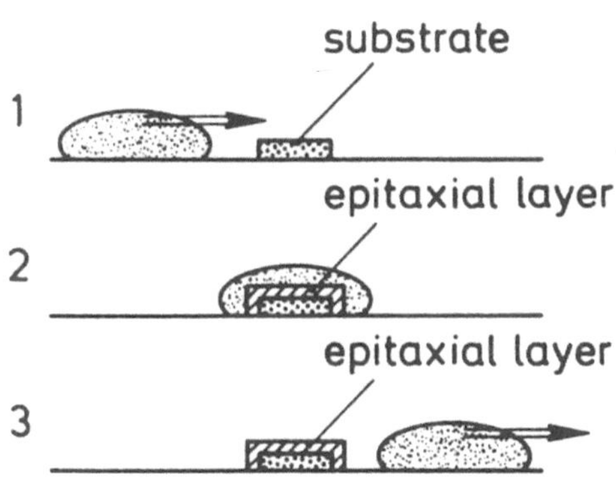

Fig.5. Liquid Phase Epitaxy (LPE). The three basic stages in layer growth procedure:

1 solution and substrate are brought into contact

2 cooling of the system: growth of the epitaxial layer

3 solution is removed from substrate.

One of our graphite crucibles of simple design which allows for the basic processes which are illustrated in Fig. 5 is schematically shown in Fig. 6. This crucible comprises four containers for solutions as indicated by the number "1" in Figs. 6a and b.

One way in which the growth procedure of the npn-multilayer can occur is the following [1]: Two of the containers opposite to each other, as for example those close to A and C, are filled with As- and Ga-doped, saturated solutions, respectively. Then two substrate wafers are placed in positions A and C, and two undoped wafers serving resaturation of the solutions are placed at positions B and D. When the crucible starts to rotate, the solutions flow from container position 1 to the growth positions by which they cover the substrates 2, whereby filling the gap-like channels above the substrates A and C. In this particular example, the crucible then rotates with 230 rpm.

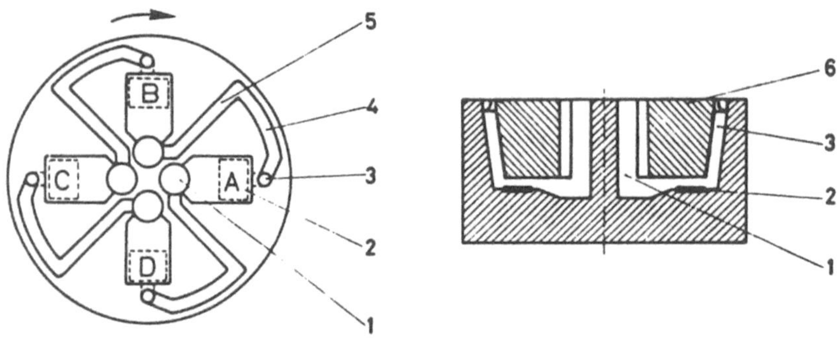

Fig.6. Crucible of the LPE centrifuge (a) cross section, (b) longitudinal section. The crucible itself contains no moving parts. The solutions are transported by centrifugal force. (1) containers for solution, (2) substrates, (3), (4) and (5) solution-flow channels, (6) substrate holder blocks.

If the temperature of the revolving crucible is reduced, growth of epitaxial layers on the substrates A and C results. Increasing the rotational frequency of the crucible causes the solutions to be withdrawn from the substrates. Growth is then terminated and the solutions move upwards through the flow-channels 3 into the storage-channels 4. The solutions remain there as long as the crucible rotates at higher speed, e.g. in this example 600 rpm. When the rotational frequency is reduced again, the solutions fall through the channels 5 (Fig. 6a) into the inner containers close to B and D. If the frequency of the boat is thereupon once more increased up to 230 rpm, the solutions creep into the gaps above the wafers B and D. These wafers being undoped, the solutions become, with an increase of the temperature of the crucible resaturated at the expense of the wafers B and D. During further increases, decreases and again increases of the rotational frequency of the boat, the solutions move on to substrates C and A. There, during times of temperature reduction, the next epitaxial layers grow. This growth cycle for pn- and np-structures can be repeated many times during one run. Both the operation of the epitaxy centrifuge and the epitaxial growth process are computer controlled. After charging the crucible the automated epitaxy apparatus produces single- or multilayers according to a chosen program. It needs no supervision.

Examples of grown multilayers

The repetitive and clean operation of the epitaxy centrifuge serves to grow high quality multilayers, such as pnpn-structures [4,5] or other sequences of layers. Fig. 7 shows a silicon pnpn-multilayer grown from two different saturated solutions of silicon in Indium, doped with Ga and As, respectively. The scanning electron micrograph of a photo-etched cleavage face of this sample shows that the thickness of the individual layers can be precisely controlled because of the automation of the system. An example which shows the defect-free growth of multilayers with even much thinner individual layers, as observed with the help of transmission electron microscopy, is given in Fig. 8, and the sequence of As- and Ga-doped layers is clearly visible. No crystalline defects have been formed during

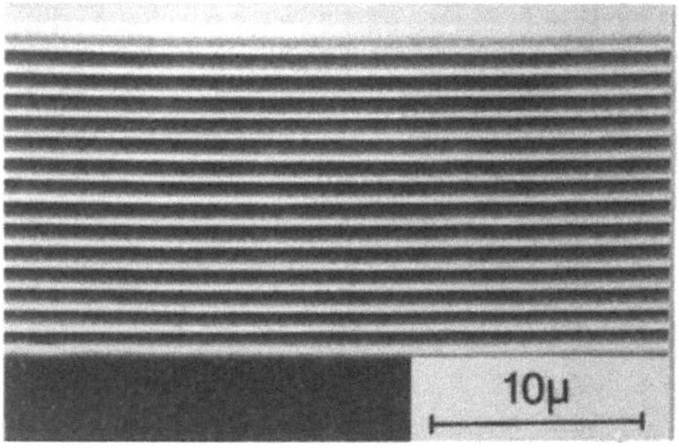

Fig.7. Silicon pnpn-multilayer grown in centrifuge. Photoetched cleavage face, SEM-micrograph. p-layer $10^{18}cm^{-3}$ Ga, n-layer $10^{18}cm^{-3}$As [7].

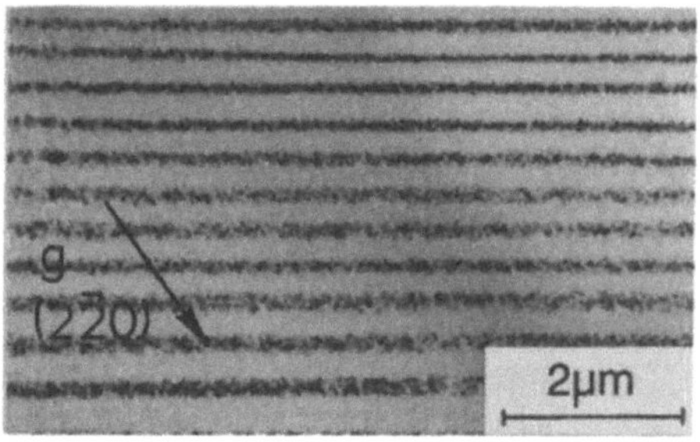

Fig.8. Transmission electron micrograph of pnpn-multilayer. Cross-section specimen, g diffraction vector. The multilayer is free of defects.

growth. This result is mainly due to the cleanliness which has to be observed during the whole growth process. It has to be noted that in the two cases illustrated in Figs. 7 and 8 the interfaces are remarkably planar which is a prerequisite for the technological application of multilayers.
Other examples of epitaxial growth, in particular the growth of multilayers on structured substrates, and of selective epitaxial growth yielding silicon on insulators, are given in the literature [1,4-6].

Conclusions

The vacuum centrifuge equipped with magnetic bearings has been successfully used in many experiments for the growth of semiconductor multilayers by liquid phase epitaxy. The operation of the automated system has proven to be reliable, economical and productive of multilayers having highest crystalline and electronic quality.

Acknowledgements The authors are grateful to Professor H.-J. Queisser for valuable contributions during discussions. The work was supported by the German Federal Ministry of Research and Technology (BMFT).

References

1. E. Bauser: The Preparation of Modulated Semiconductor Structures by Liquid Phase Epitaxy. From: Thin Film Growth Techniques for Low-Dimensional Structures. Farrow, R.F.C.; Parkin, S.S.P.; Dobson, P.J.; Neave, J.H.; and Arrott, A.S.; (eds.) Plenum Publishing Corporation 1987.
2. Schweitzer, G.; Traxler, A.; Bleuler, H.; Bauser, E.; and Koroknay, P.: Magnetische Lagerung einer Epitaxie-Zentrifuge bei Hochvakuumbedingungen, Vakuum-Technik 32 (1983) 70-74.
3. Bleuler, H.: Decentralized Control of Magnetic Rotor Bearing Systems. Thesis ETH 7573, Swiss Federal Institute of Technology, Zürich 1984.
4. Bauser, E. und Strunk, H.: Silizium Epitaxieschichten mittels Lösungstransport durch Fliehkraft und deren strukturelle und elektrische Charakterisierung überwiegend mit elektronenmikroskopischen Methoden. Forschungsbericht BMFT-FB-T 86-142 (1986).
5. Käss, D.; Warth, M.; Appel, W.; Strunk, H.P.; and Bauser, E.: Silicon Multilayers Grown by Liquid Phase Epitaxy. Bean, J.C. (ed.) Proc. First Int. Symp. on "Silicon Molecular Beam Epitaxy". Proc. Vol. 85-7, 250-258. The Electrochem. Soc., Pennington NJ, USA 1985.
6. Bauser, E.; Käss, D.; Warth, M.; and Strunk, H.P.: Silicon Layers Grown on Patterned Substrates by Liquid Phase Epitaxy. Mat. Res. Soc. Symp. Proc. Vol. 54. 1986 Materials Research Society.
7. H.P. Trah et al., to be published.

Magnetic Bearings for Precision Linear Slides

David Eisenhaure (SatCon), Alexander Slocum (MIT),
Richard Hockney (SatCon)

SatCon Technology Corporation
71 Rogers Street
Cambridge, MA 02142

Massachusetts Institute of Technology
77 Massachusetts Avenue
Cambridge, MA 02139

Abstract

Early uses of scanning tunneling microscopes indicate that many areas of science could greatly benefit if centimeter scale objects could be probed on an atomic scale. This could be accomplished by the development of a linear slide having one Angstrom resolution and a one-tenth meter range of travel. Magnetic suspension technology has been proposed as an ideal candidate for the suspension of the linear slide. This technology is attractive since it combines low noise, excellent positioning accuracy, zero wear, and requires no lubrication. The paper examines the feasibility of developing a magnetic suspension with a required resolution exceeding one Angstrom and establishes the requirements for this suspension. The achievable resolution is shown to be determined by position measurement accuracy, suspension gain, suspension bandwidth, and disturbance force levels. Expected disturbance force levels from ground motion, air currents, and acoustic effects are projected. Given the expected disturbance force levels, measurement accuracy, and required controller performance, a magnetic suspension could be developed which would provide resolution better than the one Angstrom required.

Introduction

Several Angstrom resolution devices currently exist. Scanning electron microscopes (STMs) are flexural linkage structures with Angstrom resolution and range of motion only on the order of one micron [1]. Because of its small range of motion, an STM can be made small enough to make it virtually immune to thermal and vibration problems that would plague a machine with centimeter range of motion. Researchers at the National Physical Laboratories (NPL) in England have built a linear stage with 1 A smoothness of motion over a range of centimeters. The stage uses Rulon™ pads as bearings that are arranged kinematically on a Vee way [2]. However, questions regarding long-term stability of the plastic bearings and controllability in the presence of sliding friction remain unanswered. Current designs of other precision machines such as wafer steppers and diamond turning machines [3] are principally performance limited by mechanical contact between moving parts, misalignment between actuators and bearings, bearing stability, and attainable temperature control [4,5,6]. To help overcome these problems, coarse-fine positioning systems, have evolved as shown in Figure 1. Although they can be effective, they are mechanically cumbersome and are intrinsically difficult to control. The design envisioned for the ARMM would address these issues by way of its kinematic magnetic bearing design.

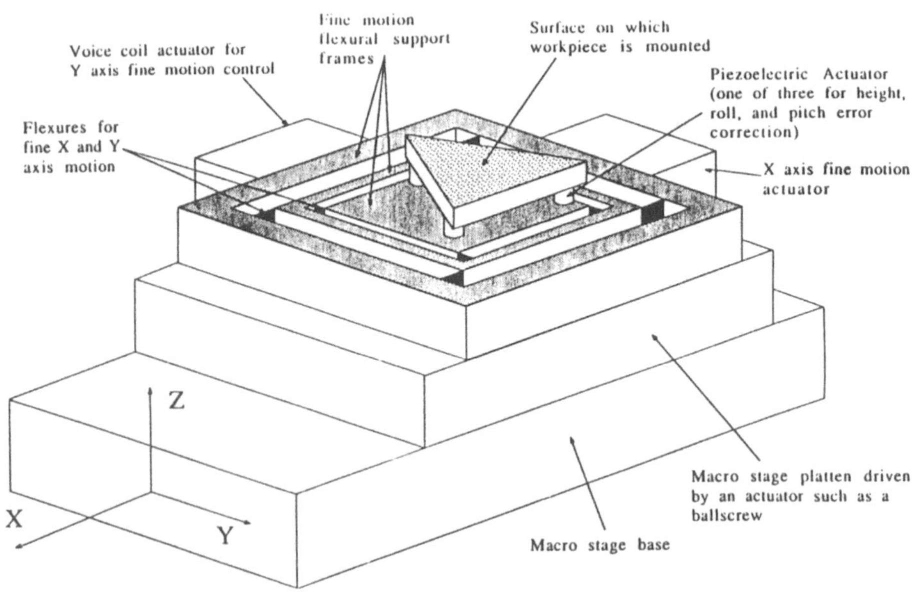

Fig.1. Example of a Coarse-Fine positioning system used to correct for slide errors caused by errors in slide geometry and forces caused by a misaligned actuator

The design of the ARMM evolved into the crossed axis design shown in Figures 2-4. A two-axis version only requires fabrication and installation of a "mirror image" of the first axis. The resultant two-axis machine would have a spherical or cubic shape to maximize structural efficiency with respect to stiffness and thermal stability. The axis' moving structural members would be mirror finished and be used as reflectors for the laser interferometers that provide position feedback signals to magnetic bearings (The axis would be designed so servo forces from the bearings and the weight of the part caused less than 0.1 A deformation.). The magnetic bearing actuators would be configured kinematically: five magnetic bearing actuators positioned to control five degrees of freedom.

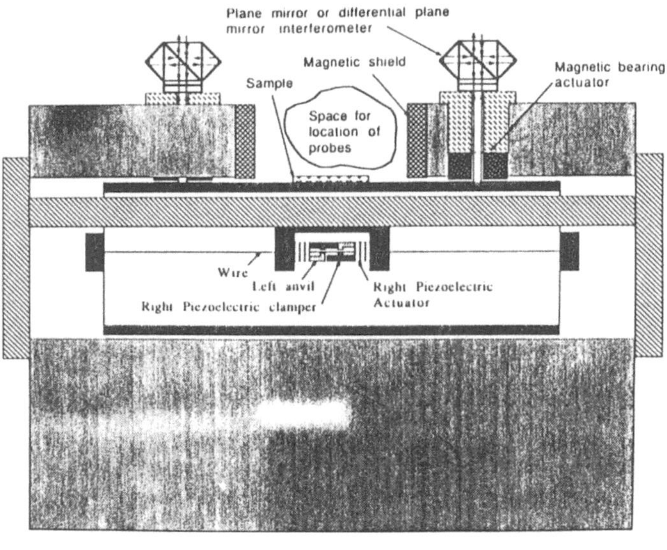

Fig.2. Cutaway side view of Atomic Resolution Measuring Machine (ARMM)

MIT/SatCon Magnetically Suspended Linear Slide

Current designs of precision machines such as wafer steppers and diamond turning machines are generally performance limited by mechanical contact between moving parts, misalignment between actuators and bearings, stability of air

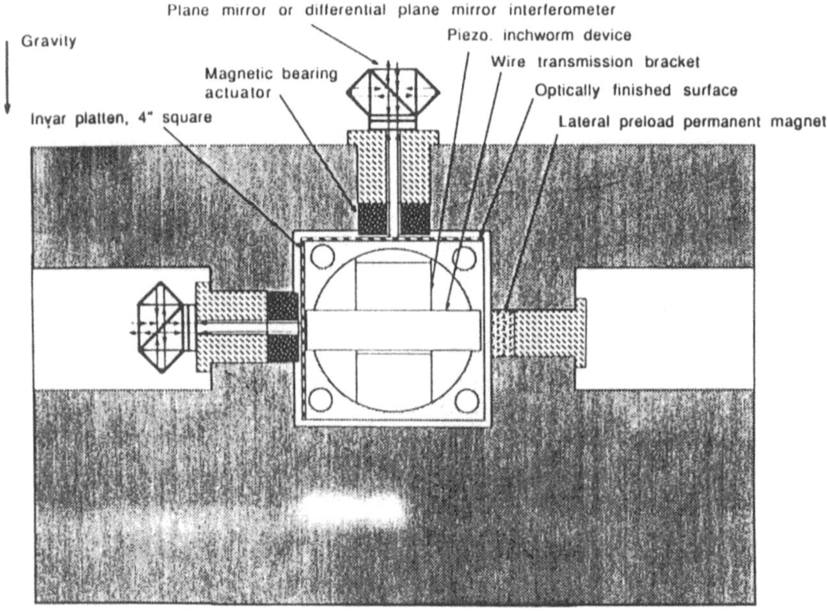

Fig.3. Cutaway right end view of Atomic Resolution Measuring Machine (ARMM)

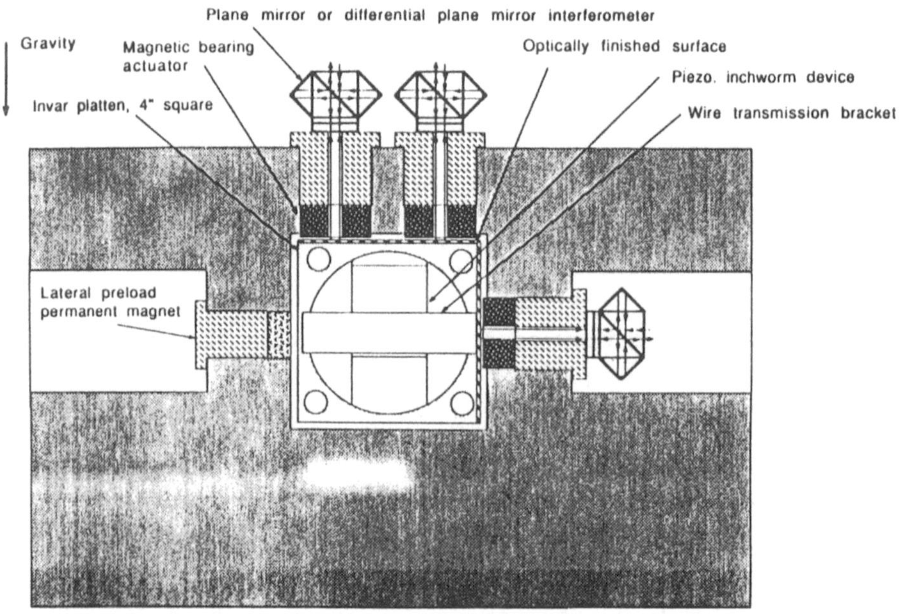

Fig.4. Cutaway left end view of Atomic Resolution Measuring Machine (ARMM)

bearings, and/or attainable temperature control. To help overcome some of these problems, coarse-fine positioning systems have evolved such as shown in Figure 1. Although effective, they are mechanically complex when compared to a magnetic bearing supported slide.

In order to maximize performance and minimize cost of magnetic bearing supported linear slides, kinematic design principals must be adhered to. With respect to spacing of components, in one extreme, the farther apart bearings are placed, the less chance of the system "walking". However, the farther apart the bearings are, the larger the structure needs to be in order to span the bearings, and thus the lower the system natural frequency.

Hence spacing and size of members should be kept in proportion: following the rule of the Golden Rectangle seems to provide for dynamically stable and structurally stiff systems. The Golden Rectangle was discovered by the Greeks and is seen in their ancient architecture quite frequently. A Golden Rectangle is a rectangle of dimensions a x b, whose proportion is such that when a square of size a x a is cut from the rectangle, the ratio of the sides of the remaining rectangle is the same as the ratio of the sides of the original rectangle. This ratio is (1 + (5)0.5)/2.

If a kinematic design is used, then a closed form solution for the effect of motion induced by one of the actuators on the motion of any point of the sample can be formulated. Similarly, if the position and orientation of a region of the slide is measured, the inverse kinematic solution allows for the precise quasi-independent servoing of each of the magnetic bearing actuators.

Even if direct measurement of the position and orientation of a region on the slide is made, precise position measurements of each bearing's distance from the surface of the slide must be made in order to maximize performance. This is due to the fact that a magnetic bearing's performance is sensitive to the speed of its controlling servo-loops and the accuracy of the gap measurement. If the inverse kinematic solution were used in lieu of a direct sensor measurement, the speed of the servo-loop may be five to ten times slower.

The baseline slide as currently envisioned would have envelope dimensions of 12 inches by 6 inches by 2 inches. This baseline slide would have a weight in Earth's gravity of ten pounds. In this paper the baseline slide will be used as a conceptual model to examine the resolution capabilities of a magnetic suspension system. The analysis assumes that the slide dynamics are beyond the bandwidth of the magnetic bearing and will not be a factor in the suspension's performance.

A baseline magnetic suspension block diagram is shown in Figure 5. The principle of operation of a magnetic suspension system is quite simple: the position of the suspended slide is measured and the control system regulates the current to an electromagnet by a gain the slide at the desired location. A simple control system might be characterized by a gain (stiffness) which relates errors in position to applied magnetic force and a bandwidth which indicates the frequency range over which the magnetic force may be applied. In addition to gain and bandwidth, the control system would require electronic compensation for stable operation. However, discussion of this is beyond the scope of this paper and has no impact on the conclusions. Magnetic suspensions generally have unstable plant characteristics which lead to a minimum required system bandwidth for stability of about 10 Hz for many suspensions. Maximum achievable bandwidths could range from 100 Hz for a simple attractive system to 40 KHz or higher for systems implemented with ferrite or voice coil actuators. As with all position control servos, accuracy is determined by position measurement accuracy, controller gain, controller bandwidth, and disturbance forces acting on the suspended slide.

Suspension Performance

At low frequencies suspension performance is determined almost completely by the ability of the controller to cancel disturbances. The primary control system parameter effecting disturbance cancellation is the controller gain which determines suspension stiffness. The higher the suspension stiffness, the greater the ability to reject force

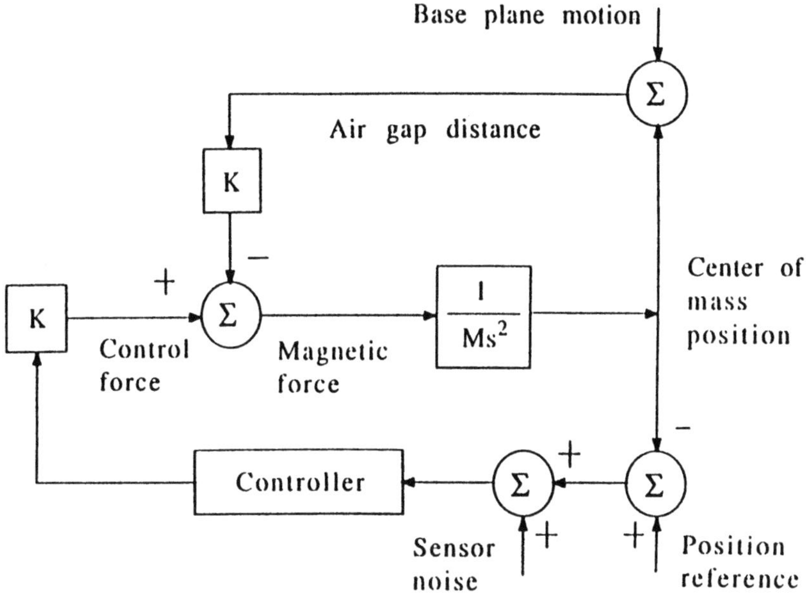

Fig.5. Basic magnetic suspension block diagram

disturbances. At high frequencies, however, all the disturbance force is absorbed by the slide inertia. The required bandwidth is the frequency at which the inertia of the slide alone could keep the slide motion below the required resolution. At frequencies below the required bandwidth, position measurement errors translate directly into resolution errors in the slide. At frequencies above the required bandwidth, resolution is independent of position measurement errors. Figures 6 and 7 show the achievable resolution at various disturbance force levels as a function of suspension stiffness and bandwidth, respectively. The disturbance force represented is modeled as a broad band disturbance over the entire frequency range of interest. For the simple control system considered, the bandwidth and stiffness requirements are related by bandwidth = $(\text{stiffness/mass})^{1/2}$ x 2Pi.

<u>Position Measurement</u>

There are four principal sensor systems that need to be designed for the ARMM:

(1) Environmental. Temperature, pressure, and humidity all need to be monitored to enable the environment to be

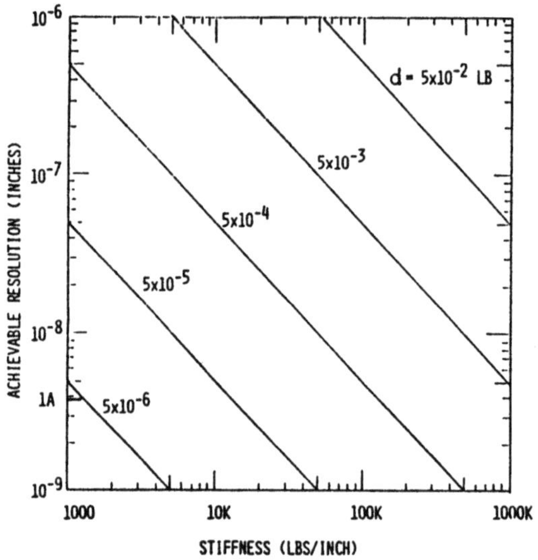

Fig.6. Achievable suspension resolution vs. suspension stiffness

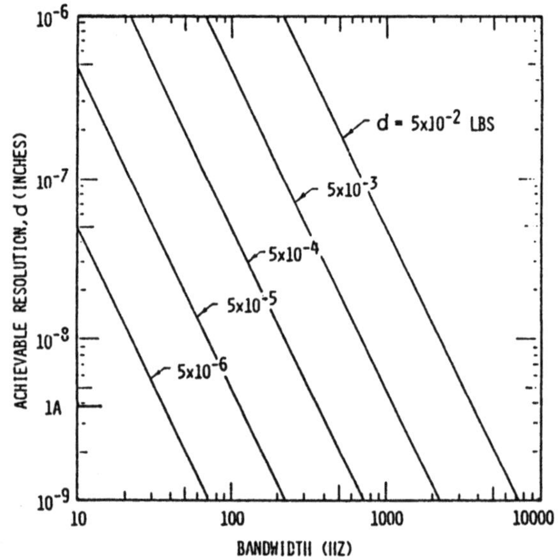

Fig.7. Achievable suspension resolution vs. suspension bandwidth

controlled to ensure accuracy of the machine. It is anticipated that temperature control good to 0.01-0.001°C/cm will be required for the ARMM. In order to achieve such high resolution and accuracy, a Mach-Zehnder interferometer will probably be used.

(2) Large range of motion, high resolution. For example, the axial position of the platen has to be measured with a resolution on the order of 1.0-0.1A over a range of motion on the order of 50 mm. Differential plane mirror interferometers will probably be used for this application. Existing technology allows for resolution to $\lambda/512$ (about 10 Angstroms).

(3) Small range of motion, ultra-high resolution. For example, the lateral position of the platen, the magnetic bearing gap, needs to be measured with a resolution of 0.1-0.01A over a range of motion on the order of 1μm. A Fabry-Perot interferometer with the capability to resolve to $\lambda/10^6$ could even be used for this type of application where one of the reference optics is moving orthogonal to the distance being measured.

(4) Measuring location of atoms. For example, scanning tunneling probes need to be adapted for use on the ARMM. Existing technology developed for the STM could be used; however, a single probe would take many thousands of years to map a 50 mm diameter specimen. Thus multiple probe techniques would be needed.

Capacitance probes or laser interferometers can be used to make ultra precision measurements of the bearing gap. Regardless of the system used, however, a through the bearing measurement method must be used. This maintains accuracy of the kinematic model and provides for checking measurement closure should direct measurement of the position and orientation of a region of the slide also be made.

If a laser interferometer system is used as shown in Figure 8, then the slide itself must be polished to optical quality. The cost of a laser interferometric measurement system is on the order of $10,000 per measurement axis. It is envisioned that in the near future the resolution of differential plane mirror interferometers will exceed the one Angstrom level. Thus a single magnetic bearing supported slide would require $50,000 worth of laser interferometers.

For substantially less cost, on the order of $2,500 per axis, capacitance probes, also shown in Figure 8, can be used for through the bearing sensor. They have the advantage of not

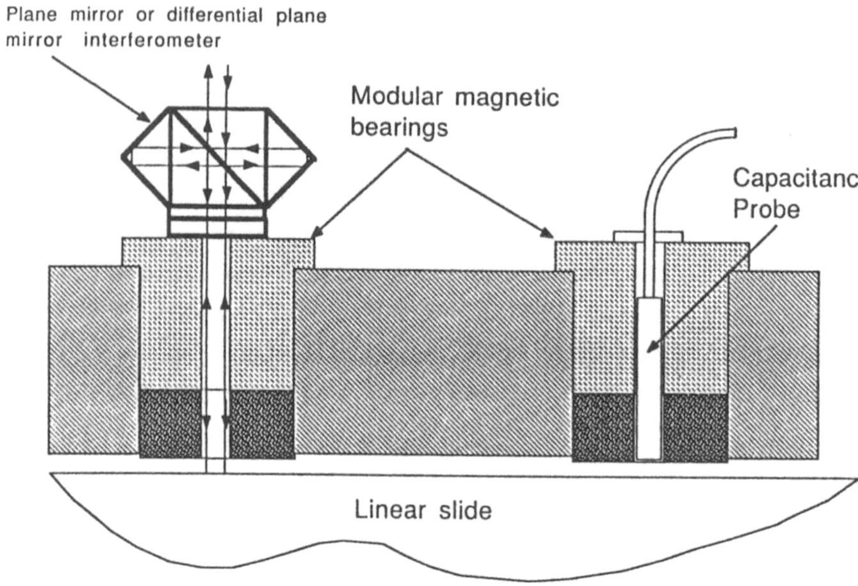

Fig.8. Sensor arrangements for through the bearing measurement

requiring the surface of the slide to be optically polished. The finite area of their measuring tips creates an averaging effect which reduces the effect of surface finish errors on the gap measurement. However, long-term drift problems may render the probes inadequate for continuous operation of the system over a period of days.

All measurement systems, as well as the machine itself, are very sensitive to thermal effects. Thus control of the overall temperature and of temperature gradients is extremely important. A detailed discussion of temperature control methods is beyond the scope of this paper, but some general comments on the effects of magnetic bearings on the overall system thermal budget are in order:

(1) Magnetic bearings with copper windings do dissipate heat; however, power dissipated can be minimized with a proper balance of permanent magnets to support 98% of the weight of the slide. Copper windings would then be used to provide servoing capability.

(2) Resistance heaters can be used to make the bearings constant power devices, thereby minimizing the effect of

transients on the system. The system could then be more easily operated at its thermal equilibrium.

(3) The availability of "warm" superconductors could dramatically change the way magnetic bearings are designed and used, and is thus a research area that required serious attention.

Disturbance Forces

The principle disturbance forces acting on the slide in a laboratory environment are due to air current, acoustic disturbances, and ground plane motion. The forces were modeled to have a broad frequency content over the range of interest. Figures 9 and 10 show the disturbance forces for impinging air flow and acoustic disturbances, respectively. Figure 11 shows the disturbance forces as a function of base motion for various suspension gaps. A comparison of Figures 9 through 11 shows that base motion is the largest single contributor to slide disturbance force. If necessary, the sensitivity to base motion could be reduced by one order of magnitude by incorporating magnetic flux feedback in the control loop and by two orders of magnitude by employing voice coil actuators. Even without these changes the total disturbance force level

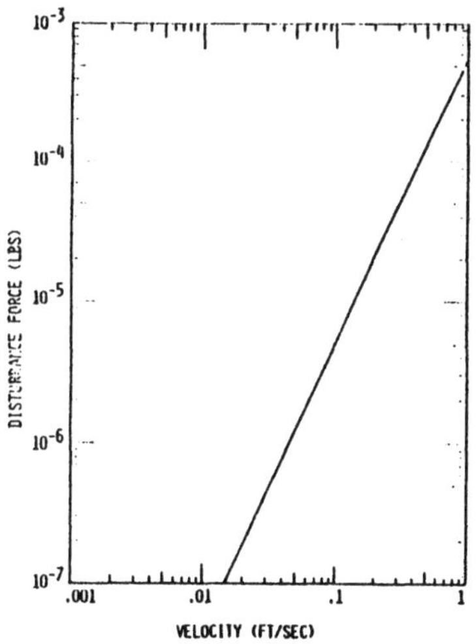

Fig.9. Slide disturbance force as a function of impinging air flow

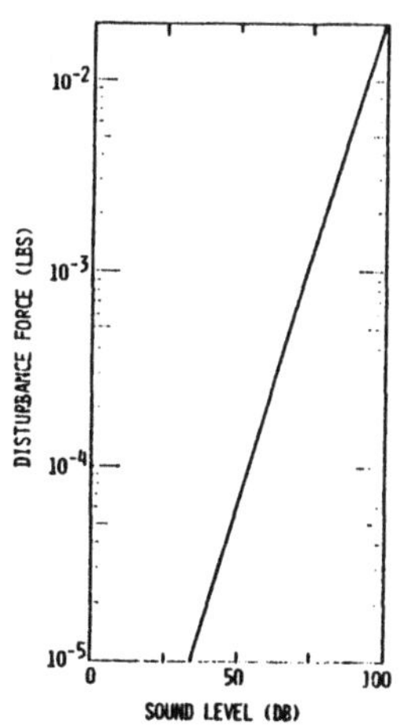

Fig.10. Slide disturbance force vs. acoustic noise level

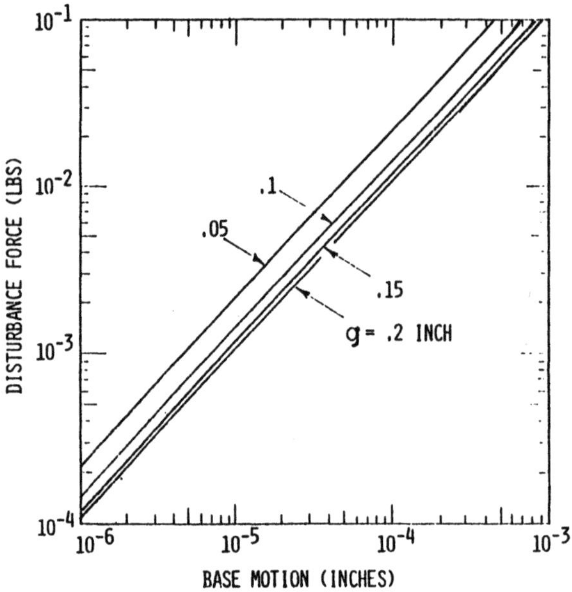

Fig.11. Slide disturbance force as a function of base plane motion

could be kept below 5×10^{-3} pounds. This is based on the MIT experience with expected base motions, air velocity, and acoustic effects in a well-controlled laboratory environment.

Conclusions

The factors that determine achievable resolution for a magnetically suspended slide are position measurement accuracy, controller bandwidth, controller gain, and disturbance forces. Laser interferometers are expected to be available in the near future with resolutions exceeding one Angstrom and adequate accuracy for a one Angstrom resolution slide suspension. At the expected disturbance force level of 5×10^{-3} pounds, the suspension would be required to have a bandwidth of 1,000 Hz and a stiffness of 1,000,000 pounds per inch to maintain a slide resolution of one Angstrom. These suspension characteristics are achievable with state-of-the-art magnetic and electronic components. Based on these considerations, the authors believe that a magnetically suspended slide having a resolution better than one Angstrom could be developed using state-of-the-art and near-state-of-the-art technology.

Bibliography

1. Binnig, G. and H. Rohrer: Scanning Electron Microscopy. Helvitica Physica Acta, Vol. 55 (1982) 726-735.

2. Lindsey, K. and P. Steuart: NPL Nanosurf 2: A Sub-nanometer Accuracy Stylus-based Surface Texture and Profile Measuring System with a Wide Range and Low Environmental Susceptibility. 4th Int. Precision Engr. Sem. Cransfield Inst. of Tech., U.K. (11-14 May 1987) 15.

3. Donaldson, R. and S. Patterson: Design and Construction of a Large Vertical-Axis Diamond Turning Machine. SPIE's 27th Annual International Technical Symposium and Instrument Display (21-26 August 1983).

4. Moore, W.: Foundations of Mechanical Accuracy. The Moore Special Tool Company. Bridgeport, Connecticut (1970) 24-28.

5. Biesterbos, J., et al: A Submicron I-Line Wafer Stepper. Solid State Technology (February 1987) 73-76.

6. Nuhn, M.; Yao, S.; Avrit, B.: I-Line Wafer Stepper Used for Low Volume Production of 0.5 Micrometer GaAs Integrated Circuits. Solid State Technology (February 1987) 81-84.

7. Berthold, J.; Jacobs, S.; Norton, M.: Dimensional Stability of Fused Silica, Invar, and Several Ultra-low Thermal Expansion Materials. Metrologia 13 (1977) 9-16.

Applications in Robotics

Applications of Magnetic Bearings in Robotics

TOSHIRO HIGUCHI

Institute of Industrial Science
University of Tokyo, Japan

Abstract

Magnetic bearings have the ability to support objects without mechanical contact. In addition they have the functions of controlling the stiffness and damping of levitation and controlling the position and attitude of the supported object with an accuracy of a micro meter or submicrons. By taking advantages of these characteristics of magnetic bearings, robots are expected to improve their abilities, to expand the field of activity in clean rooms and in vacuums, and to have good tools for the tasks which need precise manipulation. In this paper the following researches and attempts which are related to applications of magnetic bearings in robotics are introduced. These are: development of clean room robot of which joints are composed of magnetic bearings, development of super clean actuators which have the function of magnetic bearings and stepping motors in one construction, and development of magnetically supported intelligent hand as an end effector of robots to do precise assembling.

Clean Room Robot

Cleanliness of production environment is said to greatly affect the reliability and yield rate of semi-conductor circuits. Since the integration of semi-conductor circuit is increasing and the width of the circuit pattern reaching the order of sub-micron, prevention of dust and oil vapor generation in clean rooms is becoming an even greater necessity. Human operators being the greatest source of contaminant, it is desirable to replace them by robots and automated machines. However, when even greater degree of cleanliness of environment is needed, dust and oil generation from the machines can not be overlooked. So, the robots to be used in super clean rooms should not contaminate the air.

There are two ways to avoid the dust generation from robots. The first and conventional one is applying sealing technique to separate the inside of a robot where dusts and oil vapor are generated from motors, gears, ball screws, and guides etc. from the outside of the robot where cleanliness is required. Most of the commercialized clean room robots are made based on the method and also keep the inside pressure of robots lower than the pressure of clean room to prevent escape of dusts. The second and ideal one is to develop inherently clean robots by using magnetic bearings and magnetic levitation to eliminate mechanical contacts which are the worst source of dusts.

The clean room robot installing magnetic bearings in its joints is more expansive than the conventional sealed robots. But, the increasing demand for perfection of cleanliness will choose the robot with magnetic bearings. And since the robots without mechanical contacts need no lubricant, they are also suitable for applications in a vacuum and in space.

In order to make the clean room robot, it is also necessary to use clean actuators, which will be described later. Under gravitation, it is not easy to support all links of multi-link robot by installing a magnetic bearing to each joint. Therefore, as the first step, we [1] designed and constructed the parallel link robot of which upper two joints are composed of magnetic bearings as shown in Fig.1.

The construction of the robot manipulator is shown in Fig. 2. Links 1 and 2 are driven by DC motors at joints 1 and 2 respectively. Joint 3, 4, and 5 are passive joints without driving mechanism. Magnetic bearings are applied to joints 4 and 5. Joints 1, 2, and 3 are composed of usual ball bearings. As the air flow of clean rooms is made downward, joints 4 and 5 are the most important joints to be clean. Link 4 is completely levitated by the two magnetic bearings.

Links 1 and 3 have equal length of 300 mm. The length of link 2 and the distance between joint 4 and joint 5 are 170 mm.

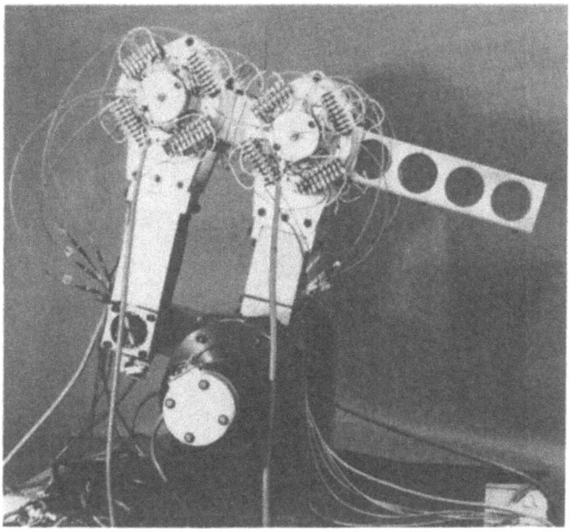

Fig. 1. Prototype of clean room robot with magnetic bearings.

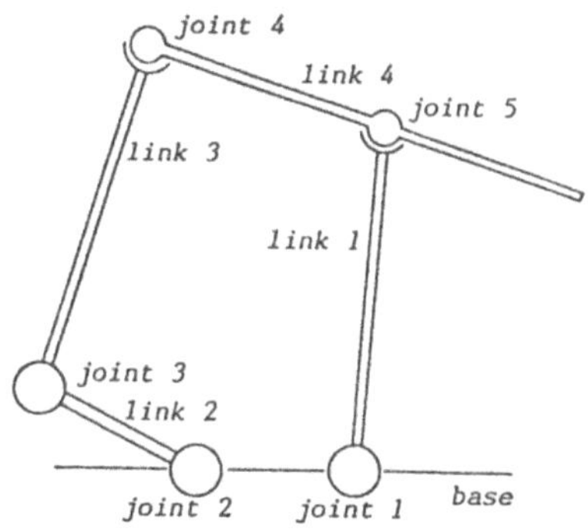

Fig. 2. Construction of the prototype of parallel link clean room robot.

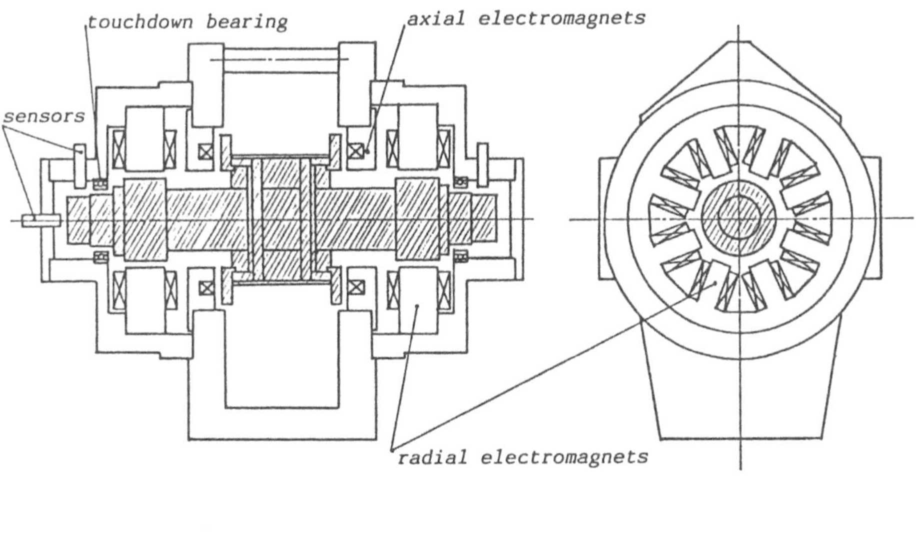

Fig. 3. Structure of the magnetic bearings of the joint.

The structure of the magnetic bearings of joints 4 and 5 is shown in Fig. 3. It is composed of two radial bearing units and one axial bearing unit. And the axial one is located between the radial ones. The rotors of the magnetic bearings of joint 4 and joint 5 are installed into the link 4 as shown in Fig. 4. The air gaps of the electromagnets are 0.7 mm and the rotor can move ±0.5 mm in radial direction and ±0.7 in the axial direction. Mass of a stator is 6 kg, that of a rotor is 1.3 kg, and the total mass of the link 4 is approximately 5 kg.

The gap sensors are eddy current type with resolution of 0.5μm. All of the five degrees of freedom of each magnetic bearing are controlled actively by means of 32 bit micro computer.

So far the levitation of the link 4 has been succeeded. The clean room robots using magnetic bearings are expected to have the advantages and functions beside the cleanliness as follows, so studies are continuing to investigate the functions by using the developed prototype.

1) Being free from troublesome problems of frictional forces.
2) Function of micro manipulation of work piece by active rotor position control of the magnetic bearings.
3) Function of controlling exerting forces and stiffnesses at the handling point (tip of link 4).
4) Function of sensing of forces acting on the robot.
5) Function of active vibration control of robot arm.

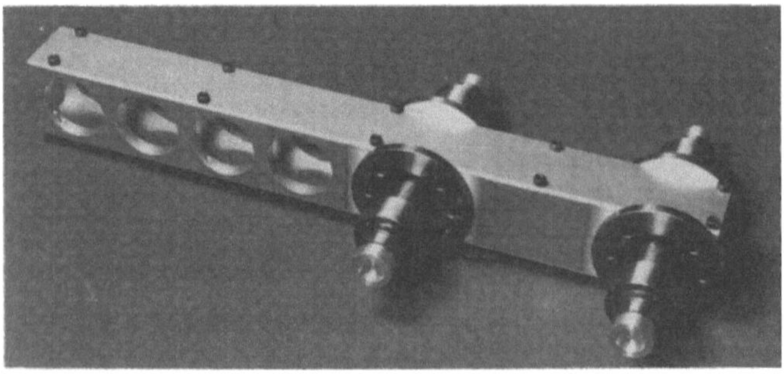

Fig. 4. Photo of the link 4 with magnetic bearing stators.

Development of Super Clean Actuators

In order to complete the clean room robots clean actuators are indispensable for driving and positioning. For robots and automated machines in clean rooms, vacuums, and space, we [2] developed new actuators which have the functions of magnetic bearings and the functions of stepping motors in one construction. The rotor of the motor is suspended exactly like usual a magnetic bearing and can rotate completely like a stepping motor. In this paper, the structure and the design of the actuators are presented with introduction of some prototypes.

Explanation of the idea and basic construction of the actuator will be done in the case of a linear motor. A conventional linear PM (Permanent Magnet) stepping motor is shown in Fig.5a. Propulsive force is gained by superposition of the magnetic flux generated by the currents of coils onto the magnetic flux generated by the permanent magnets. A strong attractive force, approximately ten times of the maximum propulsive force, acts between the mover and stator. Ball bearings, which are the source of dusts and oil vapor, are used to maintain a constant gap. Fig.5b shows a concept of an active magnetic suspension. The current of the electromagnet is controlled according to the output of gap sensor to keep the gap. Fig.5c shows the basic element of the super clean actuators. Its structure is quite similar to that of PM stepping motor except that the permanent magnets are replaced by electromagnets (main coils). By controlling the current of main coils based on gap sensor signal, mover is expected to levitate with a constant gap in spite of the toothed pole structure inherent to stepping motors. The developed super clean actuator may be regard as an amalgam of stepping motor and magnetic bearing.

The schema and structure of the element of the actuators are shown in Fig.6. Six coils are wound to the laminated core. A pair of lager coils (main coils) connected in series produce main magnetic flux. Coils A1 and A2 are connected in series and when one coil produces a magnetic flux in the same direc-

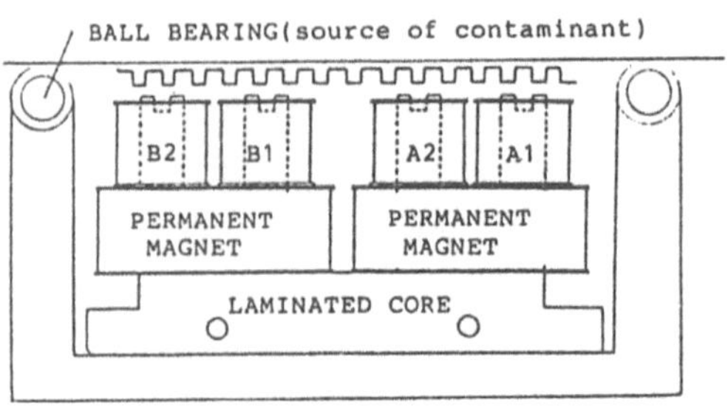

a) PM stepping motor

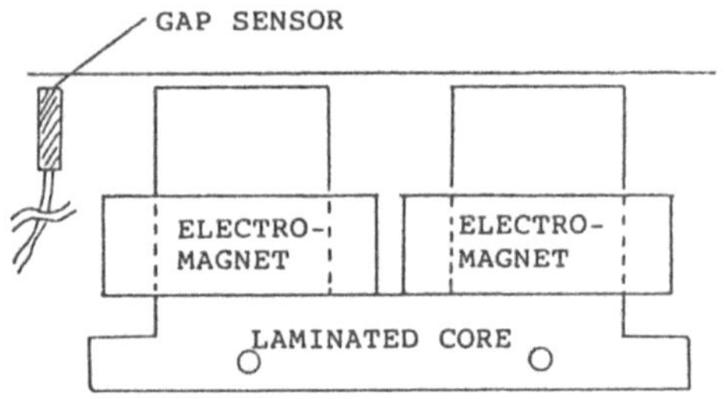

b) Active magnetic suspension

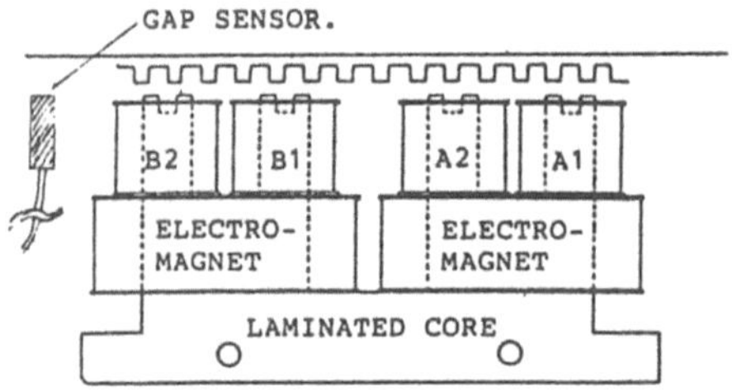

C) Basic element of super clean actuator

Fig. 5. Basic idea of super clean actuator (magnetically suspended stepping motor).

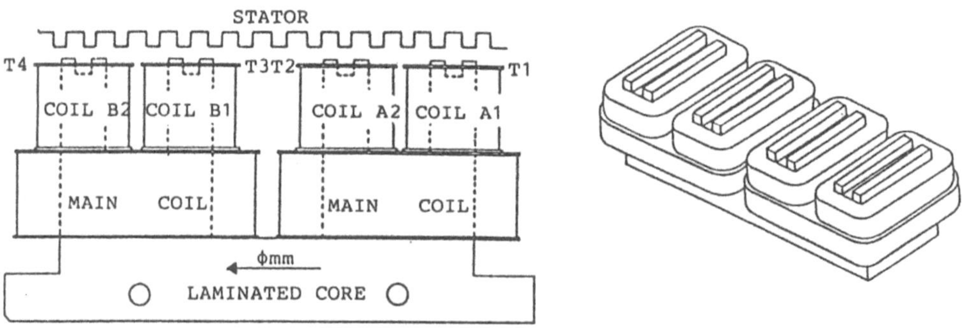

Fig. 6. Basic structure of super clean actuator element.

tion as the main flux, the other produces a magnetic flux against the main flux. The same applies for coils B1 and B2. The core has four toothed poles. Let one pitch of the teeth be expressed as 2pi radian, then the phase relation of the toothed poles are presented as T2-T1= 1 pi rad., T3-T1= $-\frac{1}{2}$ pi rad., and T4-T1= $\frac{1}{2}$ pi rad. The equivalent circuit of the magnetic circuit of the element is shown in Fig. 7.

Attractive force and propulsive force of the motor element can be calculated as the derivatives of the magnetic co-energy by H (gap direction) and X (traveling direction) respectively. The calculated value of the attractive force is shown in Fig.8. Due to the toothed pole structure, the attractive force acting on each toothed pole fluctuates severely as the line indicated by F1 along the direction of X. However, summation of the forces of four poles is rather constant and insensitive to change of position X. This comes from the phase relationship of the toothed poles. It was found from calculation that the fluctuation of the total attractive force takes minimum value when ratio of tooth width to pitch lies between 0.45 and 0.5. From the numerical analyses and experimental results related to the attractive force [2], it is confirmed that active control of magnetic suspension can be realized only by controlling current of the main coil independently of traveling position X.

Driving of the motor can be done in exactly the same manner as 2 phase PM stepping motor by using coil A (A1+A2) and coil B (B1+B2). The driving method commonly called mini-step drive or micro-step drive, where the currents are provided like sinusoidal forms, is preferable to this actuator. Rapid change of the currents of coil A and coil B may disturb the calm suspension. And micro-step drive is also desirable for precise positioning and smooth movement.

Since the element shown in Fig.6 has no function of attitude control in itself, totally contactless suspension of a rigid body requires several elements or co-use of electromagnets.

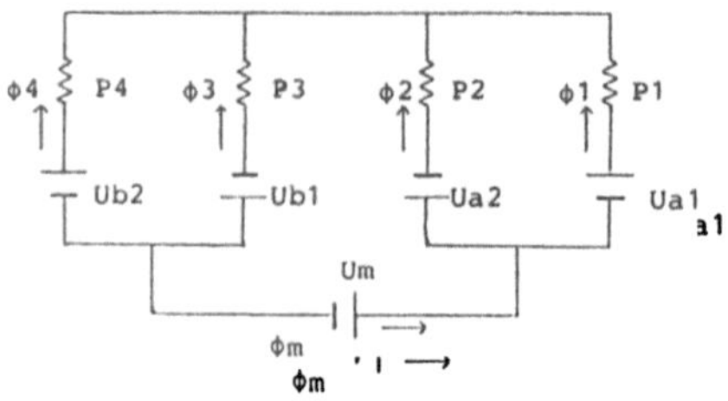

Fig. 7. Equivalent circuit of magnetic circuit of the basic element.

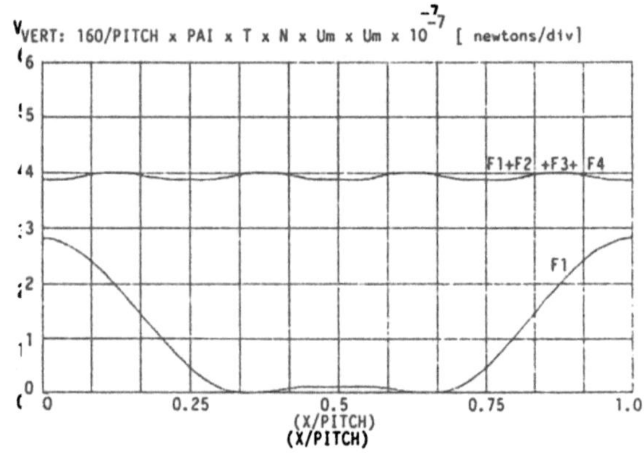

Fig. 8. Calculated value of attractive force.

Fig. 9. Prototype of super clean rotary motor.

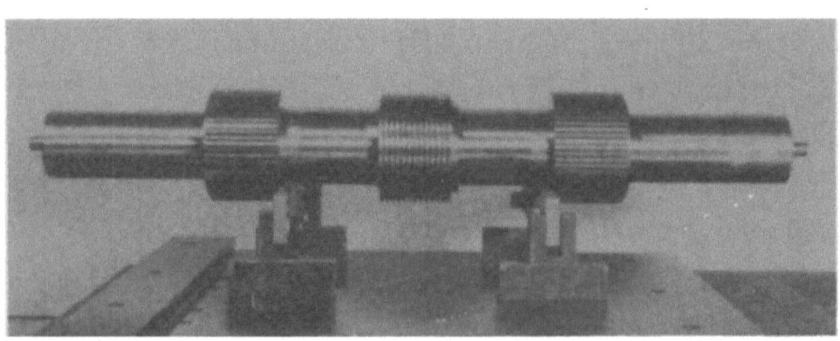

Fig. 10. Rotor of the prototype of super clean rotary motor.

By this time we have made and tested more than 6 types of super clean actuators based on the idea. Two of them are linear motion type and the others are rotary type. Here two typical examples will be explained.

The first prototype of super clean rotary motor is shown in Fig. 9. It consists of three stator blocks and a rotor shown in Fig. 10. The two outer side blocks are stepping motor units whose stator core is shown in Fig. 11. Each unit is composed of four basic elements shown in Fig. 6. The unit has the same function of radial bearing unit of common totally active magnetic bearing in addition to the function of stepping motor. And it has two capacitance type gap sensors for active magnetic suspension. The block at the center is used for passive thrust position control by means of circular grooves. Since it has also the function of a radial bearing unit, one of the outer side blocks is able to leave out. But experiments were done mainly by using the outer units as radial bearings.

When preset gap was 0.25 mm, gap fluctuation has been confirmed to be no greater than 4 µm p-p at an angular velocity of 0 to 40 pi rad/sec (velocity of rotor surface is 0 to 5 m/sec). Stiffness of suspension in the gap direction was 6 N/µm. Its maximum torque was about 1.0 Nm. And angular positioning accuracy was almost the same as usual stepping motors.

The other example is shown in Fig. 12. It was developed by NSK for presentation in robot show. The rotor is suspended magnetically like the first example. But the construction is outer rotor type. On the upper part of inner surface of the rotor cylinder teeth are made so as to control the rotary motion. And on the lower part of both the inner and outer surfaces of the rotor cylinder circular grooves are machined in order to control the height position (axial motion) of the arm. Four eddy current type gap sensors are used. Rotary motion and up and down motion of the rotor are controlled by means of the function of stepping motor. The other freedoms are controlled actively like magnetic bearings.

On the top of the rotor an arm is attached. Though it has only two degrees of freedom of large motion control, it can do simple tasks in clean rooms and vacuums. So it may be the simplest clean room robot by itself.

Since the super clean actuators are completely friction free, they have the possibility to be the ideal actuators for DD (Direct Drive) robots where frictions of bearings are obstacle to precise motion control.

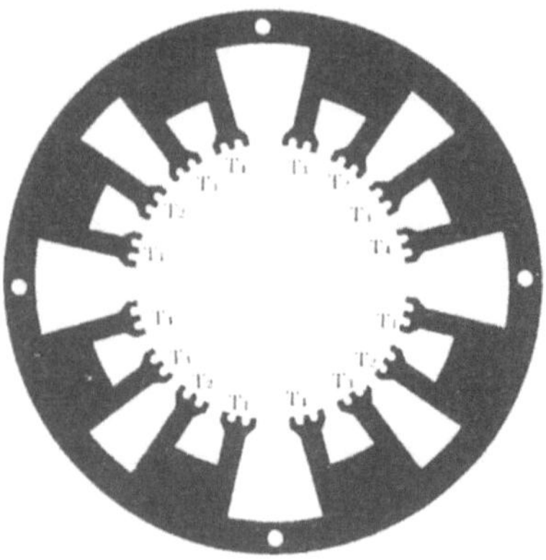

Fig. 11. Stator core.

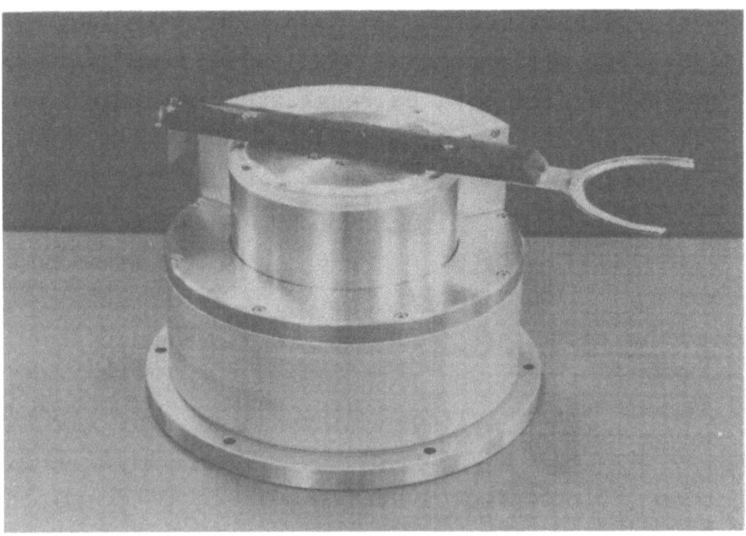

Fig. 12. Super clean actuator with both rotary and thrust motions in one body.

Magnetically Supported Intelligent Hand

Magnetically supported intelligent hand is a new end effector for robots to do precise assembling by using magnetic bearings. In the assembling process, usually it is very difficult to align a part to another part. So it is necessary to correct the attitude and position of the part. The developed hand can achieve very precise assembling tasks by correcting the the alignment errors automatically by means of both passive method and active method. For the passive method, the magnetic bearing supports one of the mates with appropriate compliances like RCC. And for the active correction, the magnetic bearing manipulates the part according to information of forces acting between the mates. The force can also detected by the currents of coils and air gaps of the magnetic bearing without using additional force sensors. In this paper conception of the hand will be described, and its structure, control system, functions, and the experimental system will be explained.

Insertion operation is a basic task of assembling and it is a kind of geometric problem. If a couple of mating parts could be aligned perfectly, insertion operation would always be done successfully. In actual cases, however, cost and technical limitations make the initial perfect alignment impractical. Successful insertion requires that the initial misalignment shown in Fig. 13 should be automatically corrected during the insertion process.

Many methods to correct the initial misalignment have been suggested. They can be classified into two methods: A) passive method using elasticity which absorbs the alignment errors passively, B) active method using sensory feedback to correct the errors actively. The Remote Center Compliance (RCC) is one of the devices estimated most highly for insertion operation due to its simplicity and performance. It is a mechanical elastic device of the passive method developed by Whitney and Nevins [3] [4]. Many similar mechanical elastic devices have been developed.

We [5] have developed the magnetically supported intelligent hand (MSIH) which has the function corresponding to that of RCC. Further, it has the advanced RCC function and the other useful functions explained later. It has the unique structure shown in Fig. 14 incorporating a magnetic bearing. It is composed of a hand shaft, two radial magnetic bearing units and one axial magnetic bearing unit. The hand shaft which hold a work piece with its gripper is supported by the magnetic bearing. Five gap sensors are installed and five degrees of freedom of motions of the shaft are controlled actively.

In most cases of the passive insertion method, a part is supported compliantly and it is tilted and translated by the mating forces and moments during the insertion process to correct misalignment between parts to mate. If certain symmetry conditions are satisfied, then support system can be represented mathematically by a compliance center and compliances at and around the compliance center. In the RCC theory, it is suggested that the optimum location of the compliance center is a point close to the tip of the supported part and it is concluded that the RCC has the ability to reduce mating forces and the chance of jamming by the function shown in Fig. 15.

The MSIH can posses a function corresponding to that of the RCC. In addition to that, it can easily and intentionally move the compliance center and change the support stiffness by altering the control parameters of the magnetic bearing. Whereas the mechanical RCC devices have a fixed compliance center and stiffnesses, the MSIH has the flexibility to be adapted to many kinds of parts.

The MSIH can realize more effective passive insertion by controlling the compliance during an insertion operation. And it is posible to support a work piece stiffly during transportation by the robot, whose arm is connected to the MSIH, to insertion spot, but flexibly during insertion in order to save the settling time of vibration of the hand shaft caused by the rapid robot motion.

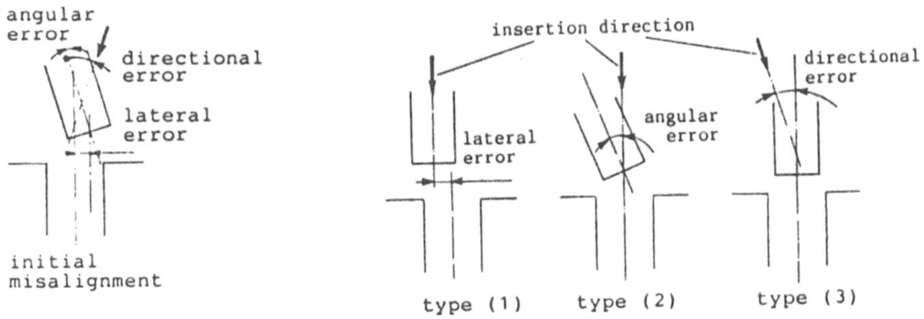

Fig. 13. Errors of misalignment and three types of errors.

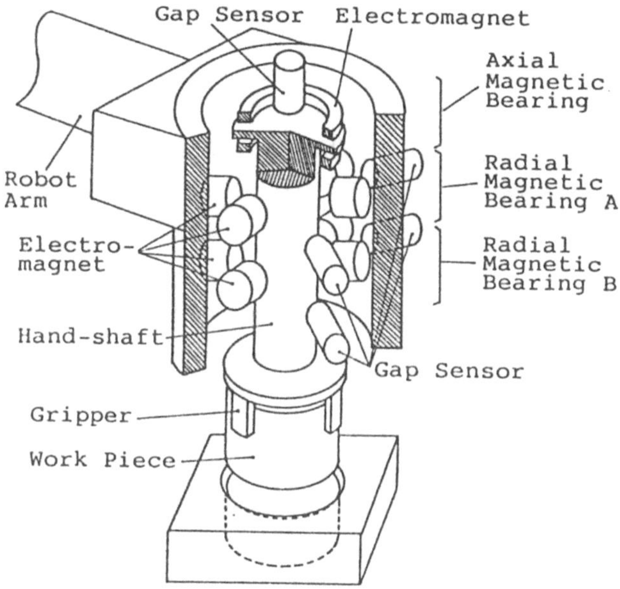

Fig. 14. Structure of magnetically supported intelligent hand.

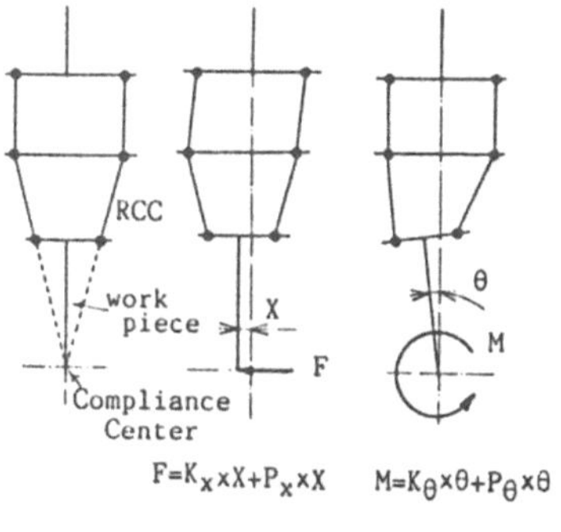

Fig. 15. RCC function.

The MSIH can not give large movements to the work piece, since the air gaps of the magnetic bearing limit the movable range. It can, however, translate and tilt the work piece with a resolving power almost equal to that of gap sensors. The prototype of MSIH has positioning accuracy of 1 µm. This high resolution enables MSIH to do active correction of small errors during insertion. The active insertion method is expected to do such difficult tasks as insertion with very close clearance, mating of parts with large diameters, mating of fragile parts, and press fit insertion.

In order to realize the active insertion, it is necessary to acquire information of misalignment and forces during the process. Position and attitude of the work piece are calculated from the outputs of the gap sensors. The forces and moments acting between parts to mate can be calculated by solving the static force balance equations where magnetic forces of the electromagnets are calculated from the coil currents and the air gaps. Therefore MSIH has the ability to infer the direction and degree of the initial error and to estimate the situation during the insertion process.

The experimental setup is shown in Fig. 16. It is composed of the MSIH and the xyz-positioning system and the control sytem. The cross section of the MSIH is shown in Fig. 17. The average gap of the radial bearing A is 2 mm, that of radial bearing B is 1.1 mm, and that of the axial bearing is 1.6 mm. The mass of the hand shaft is 1.4 kg. On the end of the hand shaft a peg is attached directly for experimental study of insertion.

The xy-table positions a mating part with a hole with resolving power of 1 µm. It is used to set various initial errors between the parts (the peg and the hole). The z positioning device moves the MSIH in the z direction to execute the insertion.

The control system was designed to utilize digital control as much as possible. It is composed of a Digital Signal Processor (DSP) circuit, a personal computer, A/D converter circuits, D/A converter circuits, coil current controller, and xyz-posi-

Fig. 16. Experimental setup.

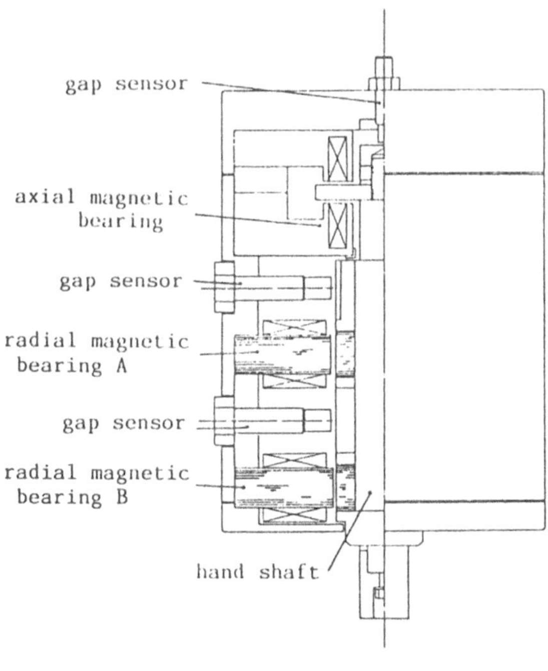

Fig. 17. Cross section of MSIH.

tion controller. All five degrees of freedom of the magnetic bearing is controlled by DSP. Sampling period of DSP for suspension is 60 μsec. Schematic of the control system is shown in Fig. 18. The feedback matrices K and P which determine the compliance matrices and the damping matrices of the suspension is given from the personal computer. Detail of the control system is described in the paper [5].

The hand shaft can be supported magnetically with the accuracy of about ±2 μm by digital control. By using the RCC function of the MSIH, we have succeeded in automated insertion with the experimental system under any following conditions: the diameter of the peg is 19.995 mm, the diameters of the holes are 20.003 mm, 20.013 mm, 20.024 mm, and 20.052 mm, the lateral initial errors are 0.2 mm, 0.35 mm, 0.4 mm, and 0.5 mm.

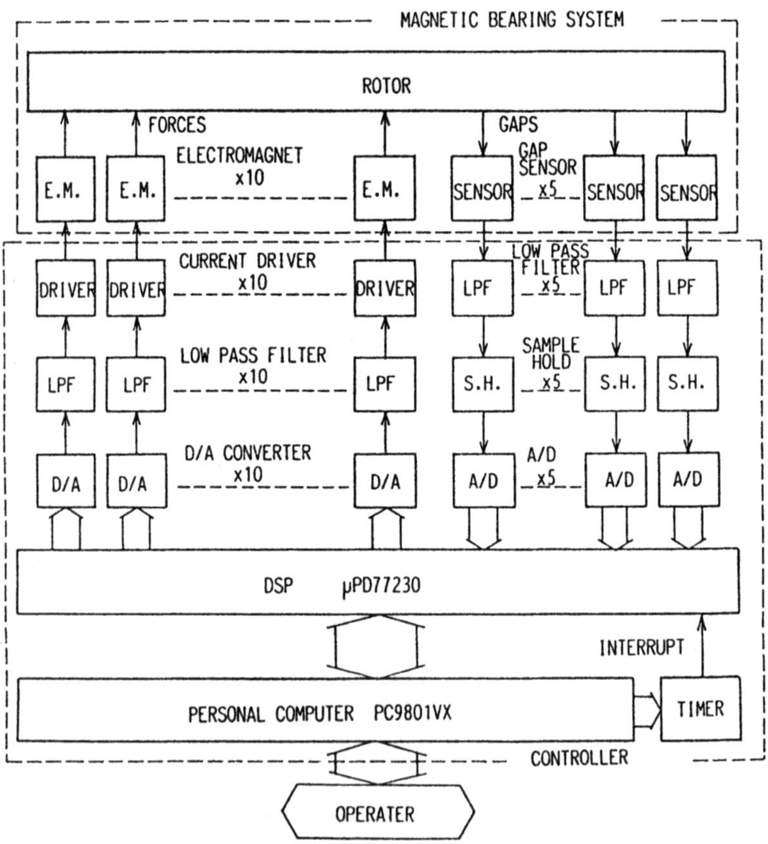

Fig. 18. Control system of MSIH.

Conclusion

The ability and the performance of robots should be progressed by using magnetic bearings for their joints and their end effectors. The examples presented in this paper are only the beginning of the applications of magnetic bearings and magnetic levitation. Further studies and developments are demanded to increase the supportable load, stiffness, and accuracy of positioning of magnetic bearings and to decrease the weight and size of magnetic bearings.

References

1. Higuchi, T.; Oka, K.; Sugawara, H: Development of clean room robot with contactless joints using magnetic bearings. Proc. of USA-Japan symposium on flexible automation (1988)

2. Higuchi, T.; Kawakatu,H.: Super-clean actuator for machines and robots. Proc. of IECON'87 (1987) 303-310.

3. Whitney, D.E.; Nevins, J.L.: What is the remote center compliance and what can it do? Proc. of 9th ISIR (1979) 135-152.

4. Whitney, D.E.:Quasi-static assembly of compliantly supported rigid parts. ASME, J. DSMC, 104 (1982) 65-77.

5. Higuchi, T.; Tsuda, M.; Fujiwara, S.: Magnetic supported intelligent hand for automated precise assembly. Proc. of IECON'87 (1987) 926-933.

Special Bearings

Lift Force Analysis in a Controlled - PM LMS Maglev Carrier

K. Yoshida, E. Zen and H. Inoguchi, S. Sonoda, T. Nakao
Kyushu University, Japan Yaskawa Electric MFG. Co. Ltd., Japan

Abstract

The lift force in the controlled permanent-magnet (PM) linear synchronous motor (LSM) Maglev carrier is theoretically evaluated by using the analytical formulas previously proposed by one of the authors and numerical finite-element method (FEM). The proposed theory and the FEM analysis are verified from comparing the calculated results with measured values in the experimental carrier designed on a basis of our analytical formulas.

Controlled-PM LSM Maglev carrier

A long-stator type of LSM with controlled-PM excitation has the integrated functions of LSM propulsion and attractive-mode levitation. Modern PMs with high energy products, such as SmCo or NdFeB material, are applied for basic

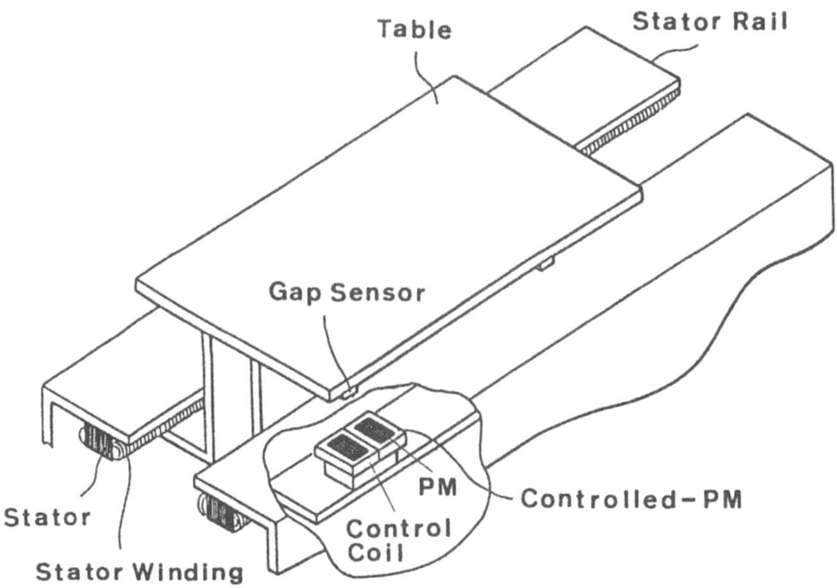

Fig. 1 Configuration of a controlled-PM LSM Maglev carrier

excitation. This machine can be used as contactless actuators or carrier for factory automation (FA) and office automation (OA).

Figure 1 shows configuration of a controlled-$SmCo_5$PM LSM Maglev carrier which has a 2-pole controlled-PM at each of the corners. The experimental carrier is studied here theoretically as well as experimentally.

A 2-pole LSM with controlled-PM excitation shown in Fig. 2 is theoretically treated using an analytical method and numerical FEM in the following.

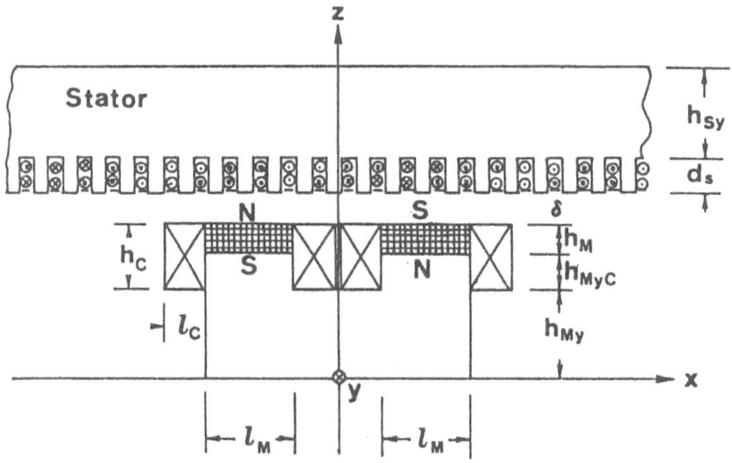

Fig. 2 Cross-section of 2-pole LSM with controlled-PM excitation

Analytical Approach

The translator in Fig. 3 is made of an array of PMs with alternating poarity fixed on a magnet yoke and the control coils wound directly around them. The field analysis of the machine is carried out analytically using a simple PM model [1], which is available for SmCo and NdFeB magnets. The PM modelling is based on the conception of an equivalent non-magnetic layer region which has an impressed volume-current density equivalent to the coercive force and geometries of PM. An impressed volume-current density of the control coils is superposed to that of the PMs in the non-magnetic layer model. The boundary-value field problem is solved analytically using the transfer-matrix method [2]. The formulas for thrust and lift forces are simultaneously derived in the analytical form [3].

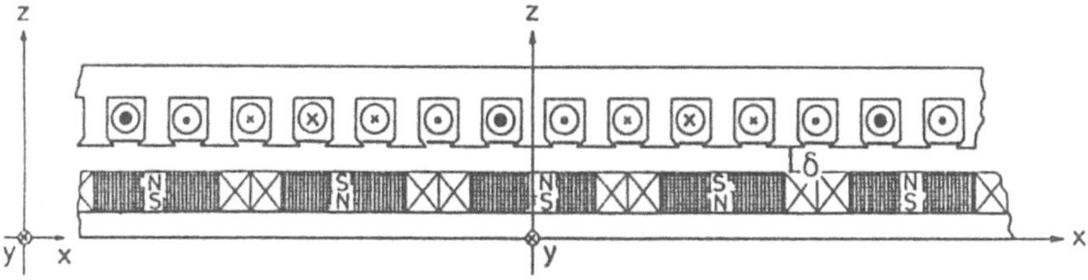

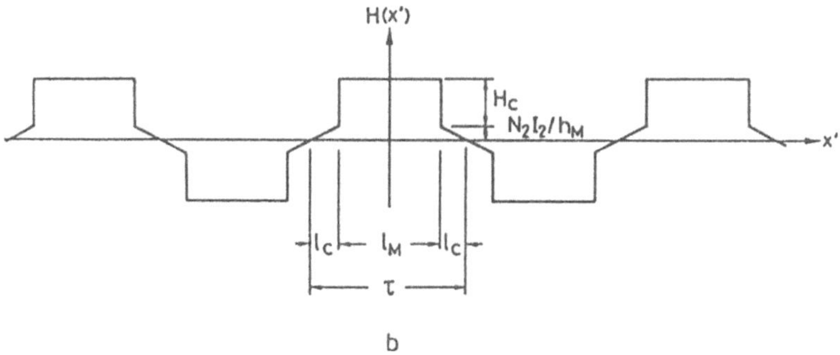

Fig. 3 (a) Cross-section of the LSM with many controlled-PMs mounted directly on magnet yoke

(b) Distribution of the resultant MMF per unit length along the PM height

The lift force consists of three components. The first one is an attractive force of the controlled-PMs attracting magnetically the stator laminated iron. This is the main levitation force and is given as follws :

$$F_{zM}(I_2,\delta_e) = H_c^2 \sum_{n=1,\,3,\ldots} K_{zMn}(\delta_e)\left(1 + \frac{1}{\pi}\frac{N_2 I_2}{h_M H_c}\frac{\tau}{\ell_C}\, n \cot k_n \frac{\ell_M}{2}\right)^2 \tag{1}$$

$$K_{zMn}(\delta_e) = \frac{16}{\pi^2}\mu_0 p b \tau \frac{1}{n^2}\sin^2 k_n \frac{\ell_M}{2}\,\frac{\sinh^2 k_n \frac{h_M}{2}\cosh^2 k_n\left(\frac{h_M}{2} + h'_{Myn}\right)}{\sinh^2 k_n(\delta_e + h_M + h'_{Myn} + h'_{Syn})} \tag{2}$$

where

$$\tanh k_n h'_i = \frac{\mu_0}{\mu_i}\,\frac{\dfrac{\mu_i}{\mu_0}\dfrac{|k_n|}{k_n} + \tanh k_n h_i}{1 + \dfrac{\mu_i}{\mu_0}\dfrac{|k_n|}{k_n}\tanh k_n h_i} \qquad i = My,\ Sy \tag{3}$$

$$\delta_e = k_c\delta \quad ; \quad k_c = \text{Carter's coefficient} \tag{4}$$

When Eqs.(1) - (4) are applied to the calculation of lift force in the 2-pole machine in Fig.2, attention is paied to the difference between Figs. 2 and 3. The longitudinal end effect due to leakage flux and the effect of iron saliency between PM and magnet yoke can not be taken into account in the analytical model in Fig. 3.

Numerical Analysis via FEM

Numerical analysis method is suitable for mathematical model of some complex geometry as shown in Fig. 2. FEM is used, via which the non-linearity of magnet and stator yokes are taken into considerations.
When PM is magnetized constantly along the PM height with a magnitude of coercive force H_c, the equivalent volume-current density is given by [4]

$$i_M(\ell_C, z_0) = H_c\delta(\ell_C, z_0) \;,\quad i_M(\ell_C+\ell_M, z_0) = -H_c\delta(\ell_C+\ell_M, z_0)$$
$$(h_{My}+h_{MyC} \leq z_0 \leq h_{My}+h_{MyC}+h_M) \tag{5}$$

where $\delta(\ell_C, z_0)$ and $\delta(\ell_C + \ell_M, z_0)$ are δ function.
This Eq.(5) means that PM can be replaced by equivalent current sheet on the side surface of the PM. Ampere-turns $H_c h_M$ and $-H_c h_M$ are equally distributed at the nodes of each side surface of the PM.
Current-carrying control coils are treated similarly by distributing equally ampere-turns N_2I_2 or $-N_2I_2$ at the nodes in each cross-section of the control coils.

Numerical Calculations and Experiments

The flux pattern, and the space distribution of the x-and z-components of airgap flux density and attractive force are calculated using FEM and are shown in Fig. 4. It is found,especially from symmetricity of attractive force per pole with respect to each pole axis, that the end effects are almost negligible. In case of carrier with very small airgap length, airgap flux is concentrated strongly within the magnet length, due to very high coercive force of $SmCo_5$ magnet.

Figures 5 and 6 show a comparison between calculated results from Eqs.(1)-(4) and via FEM and measured values, regarding the stator without slots.

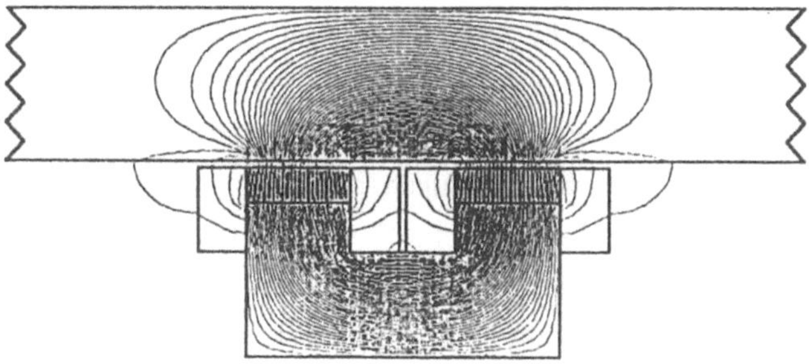

(a) Magnet field contours

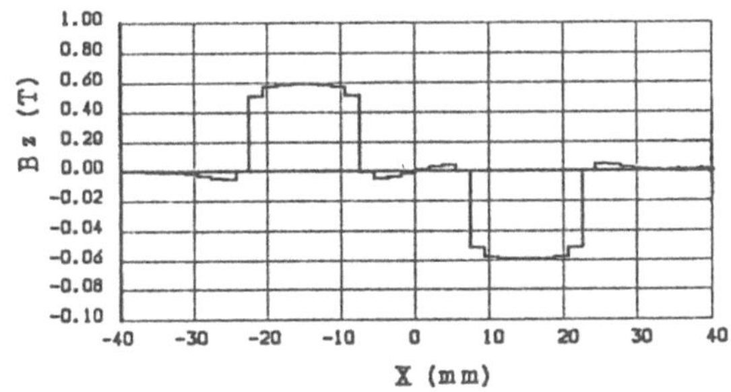

(b) The z-component of the airgap flux density at the stator surface

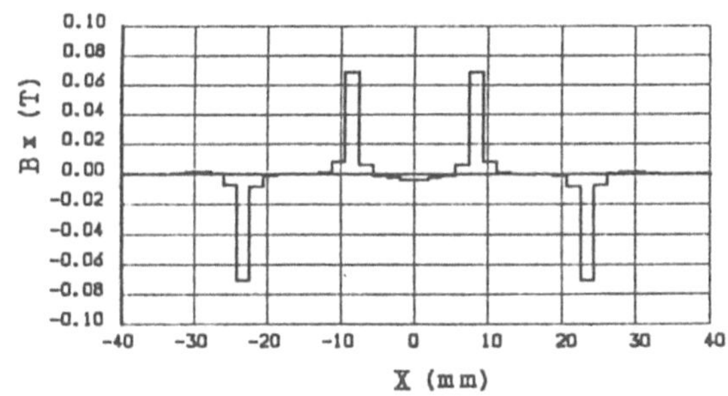

(c) The x-component of the airgap flux density at the stator surface

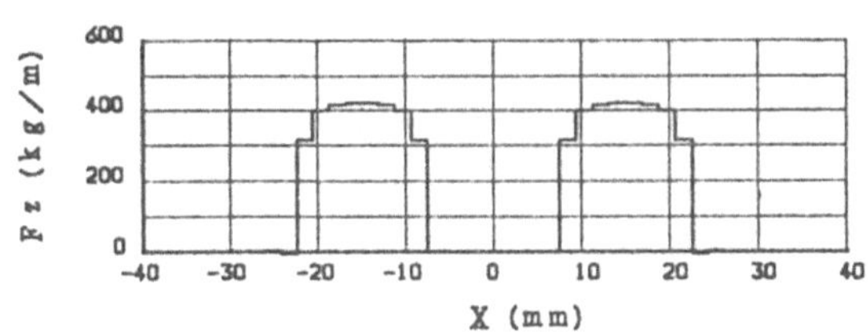

(d) The attractive force at the stator surface

Fig. 4 Spacae distribution of parameters at $\delta = 1mm$
under control current I_2=-15A

The measured values are obtained by adding the four values measured independently for each controlled-PM of carrier. Figures 7 and 8 show a comparison between the calculated results and measured values, regarding the stator with slots. The measured values are obtained when the experimental carrier is levitated stationary by the controller. In Fig. 7 the control current I_2 is controlled to be almost zero, by regulating sensitively the airgap length of the four controlled-PM. The net weight of the carrier is about 26kg, which corresponds to the airgap length of about 3.5mm as shown in Fig. 7.

The theoretical values are somewhat larger than the experimental ones. But they generally show a good agreement. Though FEM treats more exact model, it has tendency to overestimate the lift force. Analytical method explains more precisely the experimental results. The analytical formulas propsed are verified experimentally and theoretically to be effective in the design of 2-pole controlled-PM LSM with iron saliency, which is used for Maglev carrier.

Acknowledgements

Thanks are due to Mr. N. Teshima for carrying out the lift force measurements and to Mr. H. Kawanami for preparation of this paper.

References

[1] Yoshida, K. and Weh, H. : A Method of Modeling Permanent Magnets for Analytical Approach to Electrical Machinery, Archiv f. Elektrotechnik, No. 4 Vol. 68 (1985) 229-239

[2] Yoshida, K. : New Transfer-Matrix Theory of Linear Induction Machines, taking into account Longitudinal and Transverse Ferromagnetic End effects, IEE Proc. 128 Pt. B (1981) 225-236

[3] Yoshida, K. and Weh, H : Analytical Approach to Linear Synchronous Motors Excited Using Flat Type Controlled Permanent-Magnets, Proc. of Int. Conf. on Elect. Machines, Part 1 Lausanne Spt. 1984

[4] Yoshida, K and Zen, E. : Theoretical and Experimental Study of Lift Force in Small Type LSM with Controlled-PM Excitation, Report-88-18 of the Studying Committee on Rotating Machinery

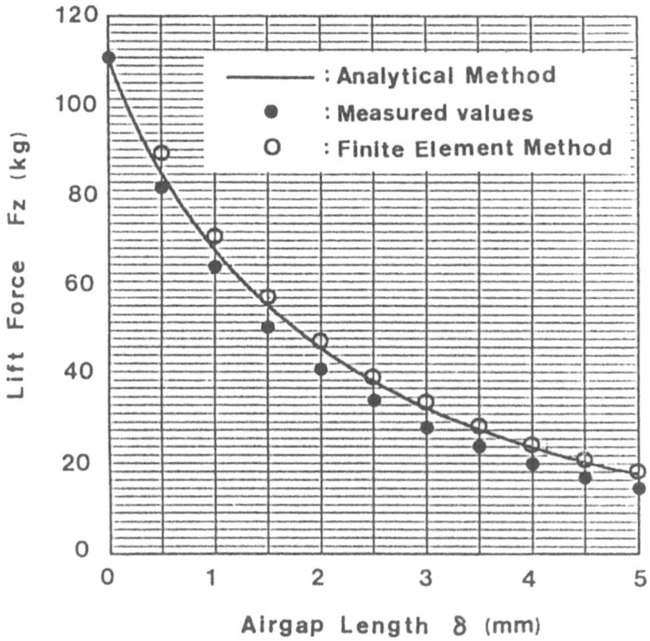

Fig. 5 Lift force vs. airgap length characteristics for slotless stator-laminated-iron under the control current $I_2 = 0$

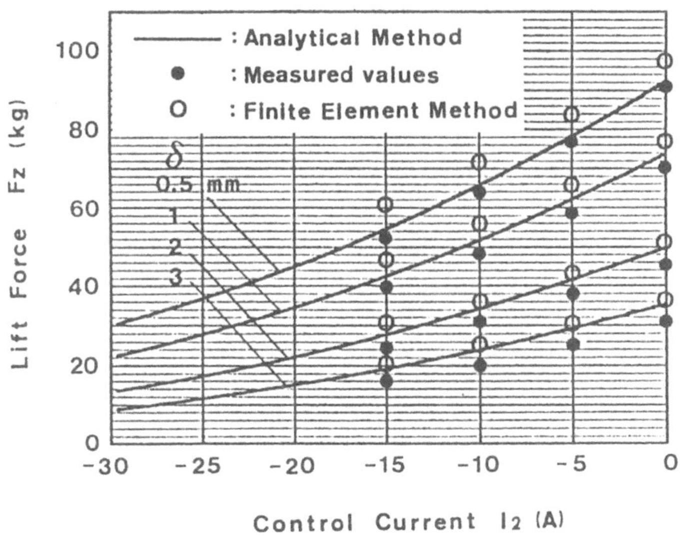

Fig. 6 Lift force vs. control current characteristics for slotless stator-laminated-iron under the airgap length δ = 0.5mm, 1mm, 2mm and 3mm

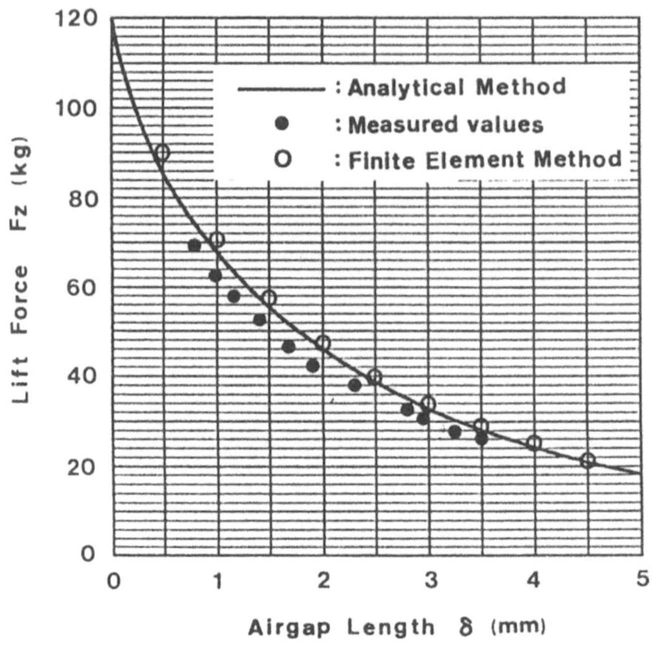

Fig. 7 Lift force vs. airgap length characteristics for slotted stator-laminated-iron under the control current $I_2 = 0$

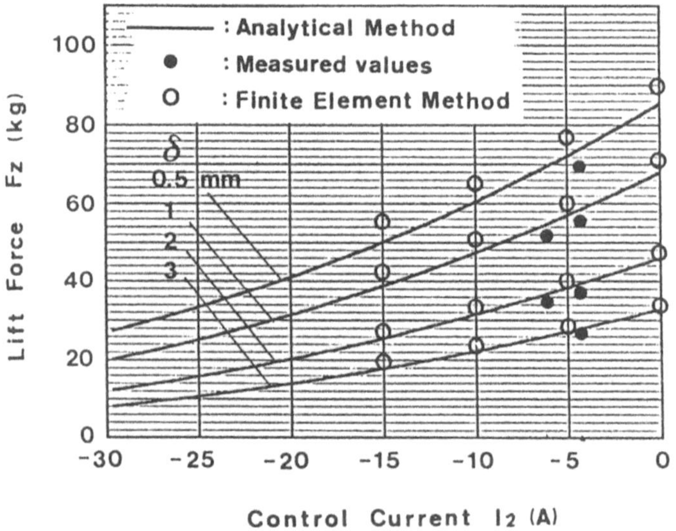

Fig. 8 Lift force vs. control current characteristics for slotted stator-laminated-iron under the airgap length δ = 0.5mm, 1mm, 2mm and 3mm

Experimental Investigation of an Eddy-Current Bearing

JORGEN L. NIKOLAJSEN

Texas A&M University
Mechanical Engineering Department
College Station, Texas

Summary

An AC-electromagnetic bearing based on eddy-current repulsion has been built and tested. The bearing was found to be inherently stable and able to provide support in five degrees of freedom, i.e. a short rotor was fully supported by a single bearing without any feedback control. However, with the current design configuration, the load carrying capacity and the stiffness and damping are lower than for the conventional DC-electromagnetic bearing and the power consumption is higher. Experimental confirmation of the inherent bearing stability and the support capabilities was reported in an earlier paper. The current paper reports on the experimental investigation of two additional capabilities, namely motoring and emergency shutdown without catcher bearings. In addition, quantitative experimental results are presented by which the goodness of each of the bearing capabilities can be judged. Based on these results, design modifications are proposed for improving the bearing capabilities.

Introduction

The magnetic bearing described in this paper is based on the so-called Electromagnetic River suspension for high speed vehicles which was proposed and demonstrated by Eastham & Laithwaite [1] in 1974. The Magnetic River was turned into a journal bearing by bending it into a circular shape (see Figure 1). The resulting Eddy-Current Bearing was designed and built on the basis of the Magnetic River behavior reported in the literature (see, for example, references quoted in [2] and [3]). No analytical work has yet been conducted. The objective was to determine whether the bearing would inherit the basic desirable characteristics of the Magnetic River such as inherent stability, support capability in five degrees of freedom, motoring capability, and emergency shutdown capability. This paper reports on the experimental findings to date. Preliminary results were reported in [4]. Additional background can be found in [5]. Work on other types of AC-electromagnetic bearings has been reported in [6] through [9].

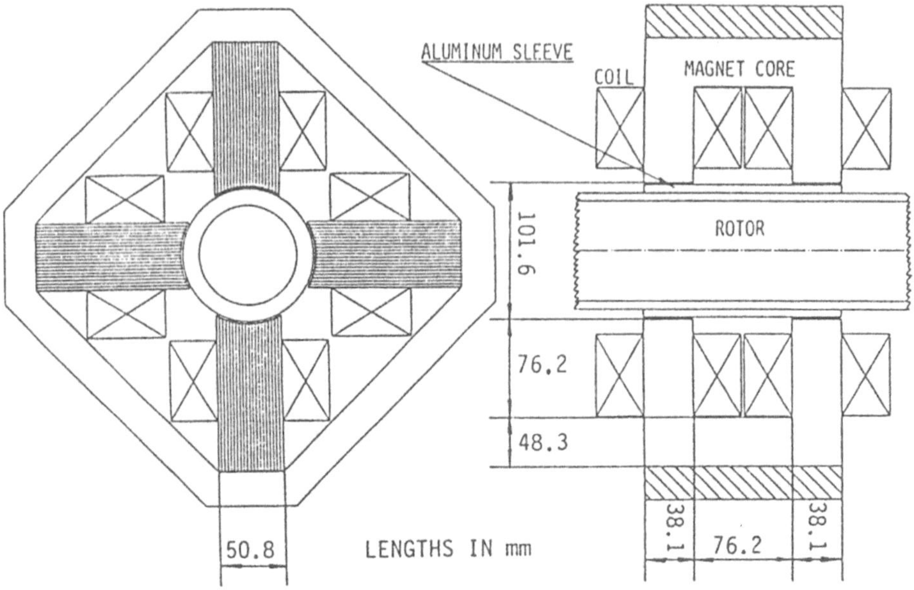

Figure 1 The Eddy-Current Bearing (Cross Section)

Figure 2 The Eddy-Current Bearing (Prototype)

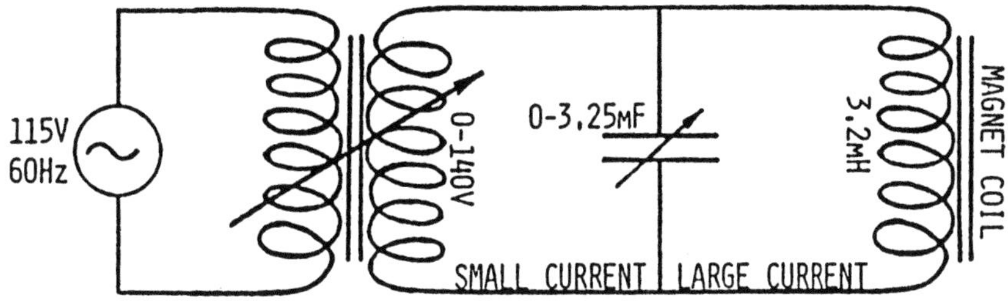

Figure 3 Electric Circuit for Each Magnet

The Experimental Apparatus

Four U-shaped electromagnets are spaced 90° apart in a star formation to form a 10.16 cm diameter bearing as shown in Figures 1 and 2. The magnets are mounted in a non-magnetic stainless steel housing. The magnet cores are made of grain-oriented 0.356 mm laminations with a saturation flux density of about 2 Tesla. The coils have 58 turns each. They are wound with two parallel flat copper wires with a total cross section of approximately 6.35 mm x 4.32 mm. The inductance of two coils mounted on one core was measured to be about 3.2 mH.

Each of the four magnets has an electric circuit as shown in Figure 3. Power is supplied from a 115V 60 Hz single phase outlet. The power is adjusted by means of the variable transformer. The variable capacitor is used to adjust the power factor (or tune the circuit) such that a large current will circulate between the capacitor and the coil while only enough current is drawn from the supply to cover the I^2R loss in the coils and in the rotor. The variable capacitor consists of a bank of oil-filled capacitors in parallel. Each capacitor bank has 13 100μF capacitors and 13 150μF capacitors connected such that the capacitance can be varied in steps of 100μF from zero to 3250μF. Fine tuning of the power factor is therefore not possible but the reactive power can be reduced sufficiently to permit experimentation with the available equipment. It was generally found that 11 to 13 capacitors needed to be switched in to minimize the supply current.

Five rotors made of construction aluminum were available for levitation. Three of these were 15.2 cm long with an inner diameter of 7.62 cm and outer diameters of 9.50 cm, 9.73 cm, and 9.91 cm respectively. They were used to study the effect of bearing clearance. A 15.2 cm long solid iron cylinder with 7.62 cm diameter was made to fit snugly inside the aluminum sleeves to provide extra weight. Two additional aluminum sleeves with lengths 12.7 cm and 17.8 cm were used to study the effect of sleeve length. A pure copper sleeve with length 15.2 cm and inner and outer diameters of 7.37 cm and 9.50 cm was also made to study the effect of improved conductivity. The masses of the three 15.2 cm aluminum sleeves were 1.038 kg, 1.176 kg, and 1.288 kg. The iron core mass was 4.883 kg and the copper sleeve mass was 3.859 kg.

Results

Levitation and Stability

All three 15.2 cm aluminum sleeves were successfully levitated as shown in Figure 2, confirming the inherent stability of the bearing and the five degree-of-freedom support capability. Contrary to expectations, slightly larger currents in all four magnets were required to achieve stable levitation of the large diameter sleeve than for the medium and small diameter sleeves (75A [82V] versus 60A [65V] in the bottom magnet). The larger current was not required to lift the sleeve but to provide sufficient support stiffness to prevent excitation of radial vibrations of the sleeve by the 60 Hz magnetic flux pulsations. Such excitation otherwise led to rattling of the sleeve against the pole faces.

With the iron cylinder inserted in the aluminum sleeves, there was insufficient power available to energize all four magnets for full levitation. With the bottom magnet excited only, the 6 kg sleeve/core combination could be lifted free of the bottom with 130A [128V] and 140A [133V] for the large and medium diameter sleeves respectively while the small diameter sleeve was unable to fully lift off at 160A [142V] which was the maximum power available. For comparison, 40A [44V] to the bottom magnet was required to lift each of the three sleeves without the iron core. Thus, a maximum lift capacity of 59N has been demonstrated so far.

Although the copper sleeve geometry is almost identical to the small diameter aluminum sleeve, it suffered axial instability and tried to exit the bearing when lifted. The long (17.8 cm) aluminum sleeve behaved similarly suggesting that the copper sleeve might be stabilized by reducing its length. The length of the copper sleeve was reduced to 13.8 cm which made it stable in the axial direction. However, pitching oscillations would very slowly build up and eventually the sleeve would start colliding with the pole faces. It is not known whether this was a true self excited instability or a subharmonic response to the 60 Hz nonlinear magnetic flux pulsations. In any case, it confirms that the inherent damping is very low.

Although the long (17.8 cm) aluminum sleeve suffered from axial instability, it had a significantly better radial support capability than the 15.2 cm sleeves. On the contrary, the short (12.7 cm) aluminum sleeve had less radial support capability than the 15.2 cm sleeves but considerably better thrust capability. This suggests, as expected, that it will be difficult to optimize the bearing with respect to all its capabilities simultaneously.

The magnetic support stiffnesses, with the three 15.2 cm aluminum sleeves levitated, were estimated by impacting the sleeves and counting vibration periods with a stopwatch. The average radial, axial and angular stiffnesses were found to be in the neighborhood of 80 N/m, 60 N/m, and 0.3 Nm/rad. The damping was again found to be very low, as evidenced by the long time taken for the vibrations to die out. These preliminary results showed, as expected, that the axial stiffness increased as the bearing clearance decreased. However, contrary to expectations, both the radial and the angular stiffnesses were found to increase with increasing clearance.

Motoring

The motoring capability was confirmed. With single phase current supply, the three 15.2 cm aluminum sleeves were levitated and would rotate in the bearing when the magnet currents were adjusted to position the sleeve eccentrically within the bearing clearance. They would stop rotating when brought back to the concentric position and would rotate in the opposite direction when brought to a diametrically opposite eccentric position. The larger the eccentricity, the higher the speed. The maximum speed recorded with single phase current was about 50 rpm. With 3-phase current, the rotational speed could be increased to about 750 rpm with the sleeves supported mechanically. It was not possible to fully levitate the sleeves with 3-phase current, apparently because the torque at zero speed would rotate the sleeves before metal contact could be broken, thus initiating a backward whirl instability.

Emergency Shutdown

It was predicted [5] that the rotor could be decelerated safely after a power failure by switching to battery operated DC-power. The operating principle would change to so-called electrodynamic levitation with the eddy-currents generated in the rotor by the high speed motion of the rotor surface past a 'row' of DC-magnets. Both levitation and braking forces would be generated. This is the principle used in levitation of high speed trains by superconducting magnets [10].

It has not yet been possible to fully demonstrate the electrodynamic levitation with the available equipment. The 9.50 cm diameter, 15.2 cm long aluminum sleeve was mounted on a 350 VA electric hand drill, inserted in the bearing, and accelerated to 1190 rpm. The top and side magnets were disconnected and the bottom magnet was connected to a large DC-power supply. A strong braking torque was developed as the DC-power was increased. At 8V, 250A, the braking torque almost stopped the drill but no appreciable lift force was detected. Apparently, the surface speed of the sleeve (6 m/s) was too low [11].

Efficiency

After less than five minutes of operation, the aluminum sleeves got too hot to be hand held whereas the magnets remained cool with only a modest temperature increase to be felt. Sleeve temperatures over 110°C were measured with corresponding magnet temperatures of about 35°C. The copper sleeve ran slightly cooler than the aluminum sleeves, attesting to its greater conductivity. An increase in supply current to one circuit would result in a similar current in the other circuits, indicating strong mutual inductance between the magnet coils.

Power Frequency

Finally, the effect of power supply frequency was studied using a 60 Hz magnet and a 400 Hz magnet. The 60 Hz magnet was identical to the bearing magnets except that the pole faces were flat. The 400 Hz magnet had identical geometry but was made of 0.1 mm laminations, and the number of turns was 15 per coil to achieve the same magnetic flux density as with the 60 Hz magnet but using a 150V, 400 Hz power supply. According to the analysis of [8], a frequency increase should result in a significant improvement of the lift capacity. This did not occur. The main effect was that thinner aluminum plates could be lifted with 400 Hz power supply, presumably because the eddy-current penetration depth decreases with increasing frequency.

Conclusions

The experiments have confirmed that the Eddy-Current Bearing retains the basic advantages and disadvantages of the Magnetic River suspension:

1. A single eddy-current bearing is sufficient to fully support a short rotor in all five degrees of freedom simultaneously.
2. The bearing provides inherently stable rotor support. No feedback control is needed.
3. The bearing will act as a motor and as a support simultaneously.
4. The lift capacity and the stiffness and damping achieved so far are well below those achieved by DC-electromagnetic bearings.
5. The efficiency of the bearing is low due to a high I^2R loss in the rotor. This is considered to be the most serious problem which must be overcome before the bearing finds practical application. The current thinking is to move the design closer to a conventional induction motor to gain motoring efficiency. The bearing could then be used in space-based applications such as flywheel energy storage systems where it could support the flywheel and also act as the motor/generator.

Acknowledgment

This work was supported by the TAMU Turbomachinery Research Consortium. Rig hardware was donated by IMC Corporation and by Magnetic Bearings, Inc.

References

1. Eastham, J. F. and Laithwaite, E. R., "Linear Induction Motors as Electromagnetic Rivers," Proceedings IEE, 121, 10, (1974), 1099-1108.

2. Jayawant, B. V., "Electromagnetic Levitation and Suspension Techniques," Edward Arnold, London, 1981.

3. Jayawant, B. V., "Electromagnetic Suspension and Levitation," IEE Proceedings, 129, A, 8, (1982), 549-581.

4. Nikolajsen, J. L., "An AC-Electromagnetic Bearing for Flywheel Energy Storage in Space," to be published in NASA CP, Magnetic Suspension Workshop, NASA Langley, Feburary 1988.

5. Nikolajsen, J. L., "A Magnetic Bearing Based on Eddy-Current Repulsion," NASA CP 2443, Rotordynamic Instability Problems In High Performance Turbomachinery, (1986), 460-465.

6. Bolton, H., "An Electromagnetic Bearing," IEE CP 120, Linear Electric Machines, London, (1974), 45-50.

7. Iskierka, S., "Influence of Irregularities of Exciting Current Density Distribution on Induction Bearing Work," Acta Technica Csav, 5, (1984), 572-582.

8. Iskierka, S., "Analysis of an Induction Bearing by the Finite Element Method," Archiv fur Electrotechnik, 67, (1984), 375-380.

9. Connor, K. A. and Tichy, J. A., "Analysis of an Eddy-Current Journal Bearing," ASME Paper 87-Trib-10, (1987).

10. Rhodes, R. G. and Mulhall, B. E., "Magnetic Levitation for Rail Transport," Clarendon Press, Oxford, 1981.

11. Foner, S. and Schwartz, B. B. (Editors), "Superconducting Machines and Devices," Chapter 6: Iwasa, Y., "High Speed Magnetically Levitated and Propelled Mass Ground Transportation," Plenum Press, New York, (1973), 347-399.

Stabilization of Rotor Motion

Analysis on the Operating Stability of a Magnetic Bearing

Zhao Hongbin, Tsinghua University, Beijing, PR China,
Zhang Zuming, Beijing management Institute of Machinery Industry, Beijing, PR China.

ABSTRACT

In this paper, the production mechanism of self-excited vibration of a magnetic bearing-rotor system is explained theoretically, and the related operating stability is investigated. A clan of the stable boundary curve derived theoretically is basically in agreement with the experiments. Those provide a foundation to choose the parameters of a rotor system, such as the parameters of rotor, damper and magnetic bearing. In this paper, a prototype device could be run up to a high speed(30,000r.p.m.) stably and repeatedly.This device possesses a rotor which weighs 10 kilogram and is suspended with a controlled magnetic bearing.

I. INTRODUCTION

The rotor controlled with a magnetic bearing is easily produced self-excited vibration which interrupts the operating stability and which comes from the states of the controlling circuit and the hysteresis of the material, when it turns with a high speed. Therefore, the technical problem of how to eliminate self-excited vibration must be solved. Early in 1946, J.W.Beams[1] had possessed a specific device to avoid self-excited vibration in a ultra-centrifuge, and then other scientists also studied the similar devices. G.Schweitzer[2] studied another kind of magnetic bearing, possessing radial rigidity and damping.

The device described in this paper was developed early before, and its construction is very simple, but its mechanical performance was satisfied. A simple description of this device is shown in Fig.1. The rotor is suspended by an attraction of iron core which is immersed into damping oil and is hung at the axis of the hole coil through a steel string. Then we could have the fairly good result to avoid self-excited vibration if we change the parameters of the diameter of steel string and viscosity of the damping oil. The philosophy of choosing all these parameters will be given theoretically later.

II. THE MECHANICAL-MODAL OF MAGNETIC BEARING-ROTOR SYSTEM

When the theoretical analysis is undertaking, a suitable mechanical-model should be built by simplify of the mechanical device. The basic hypothesis will be given as follows.

(1)Both rotor and iron core are regarded as particles, neglecting their moments of inertia.

(2)In equilibrium state, the mass centre of the electromagnetic iron core, the mass centre of the measured point down side are all in a same vertical line.

(3)Only the horizontal movements of the rotor and iron core are considered, when rotor is turning around.

(4)Being not in its equilibrium place, the rotor will be acted by a restored force which is proportional to the relative displacement between the rotor and the iron core.

(5)The damping force acting on the rotor is viscous when rotor moves in the oil. The symbol applied in this paper is given in the following:

m_1---The mass of the electromagnetic iron core (kg).

m_2---The mass of the rotor (kg).

f_0---The restored force of the rotor (N).

k_1---Equivalent stiffness of steel string (N/m).

k_0---Equivalent stiffness of magnetic force (N/m).

d_1---External damping coefficient from the viscosity of the oil (N/s.m).

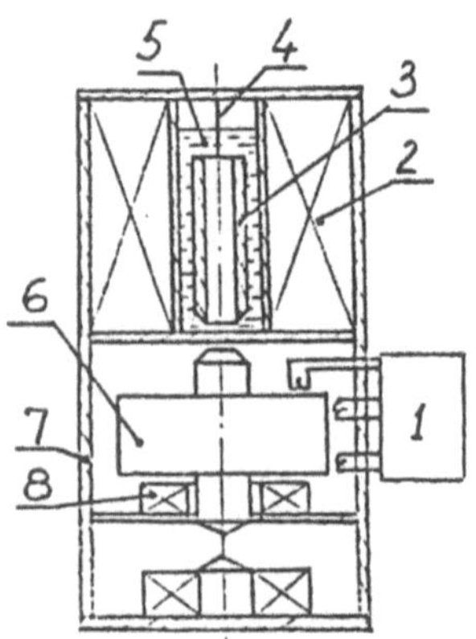

1.vibration meter
controlling box(circuit)
speed meter
2.controlling coil
3.iron core
4.steel string
5.damping oil
6.rotor
7.vacuum casing
8.motor

Fig.1. Experimental rig

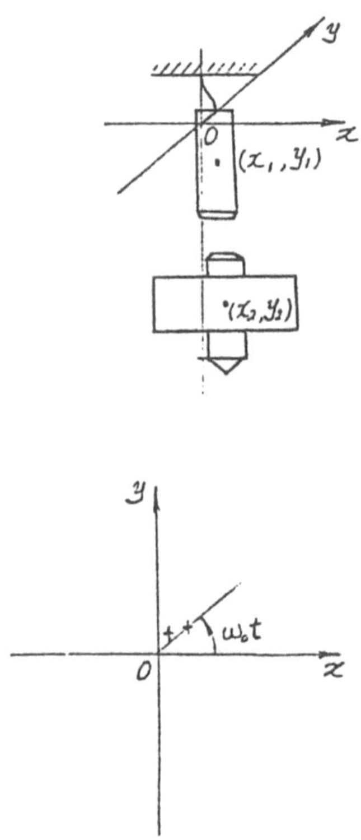

Fig.2. Coordinate system

d_o---Internal damping coefficient of the magnetic damping (N/s.m).

ω_o---Angular speed of the rotor (rad/s).

$\bar{m}$---The dimensionless mass ($\bar{m}=m_1/m_o$).

$\bar{d}$---The dimensionless damping ($\bar{d}=d_1/d_o$).

$\bar{k}$---The dimensionless stiffness ($\bar{k}=k_1/k_o$).

III. EQUATION OF MOTION

As shown in Fig.2, a coordinate system XOY is established fixedly on the foundation, "x_1,y_1" represents the coordinates of the mass centre of the electromagnetic iron core,"x_2,y_2" represents the ones of the rotor. By these, we obtain equation of motion of the iron core and rotor:

$$m_1\ddot{z}_1 + d_1\dot{z}_1 + k_1 z_1 - f_o = 0$$
$$m_2\ddot{z}_2 + f_o = 0 \tag{1}$$

Where z_1, z_2 are the following complex expression:

$$z_1 = x_1 + iy_1 = \zeta_1 e^{i\omega_o t}$$
$$z_2 = x_2 + iy_2 = \zeta_2 e^{i\omega_o t} \tag{2}$$

From that we conclude

$$\dot{z}_k = \dot{\zeta}_k + i\omega_o\zeta_k e^{i\omega_o t}$$
$$\dot{\zeta}_k = \dot{z}_k - i\omega_o z_k e^{i\omega_o t} \qquad k=1,2 \tag{3}$$

Assuming the rotor and the iron core move corresponded, then obtain:

$$f_o = [k_o(\zeta_2 - \zeta_1) + d_o(\dot{\zeta}_2 - \dot{\zeta}_1)]e^{i\omega_o t}$$
$$= k_o(z_2 - z_1) + d_o(\dot{z}_2 - \dot{z}_1) - i\omega_o d_o(z_2 - z_1) \tag{4}$$

Considering Eq.(2,4), Expression(1) becomes:

$$m_1\ddot{z}_1 + (k_1 + k_o)z_1 - k_o z_2 + (d_1 + d_o)\dot{z}_1 - d_o\dot{z}_2 - i\omega_o d_o(z_2 - z_1) = 0$$
$$m_2\ddot{z}_2 + k_2 z_2 - k_o z_1 + d_o\dot{z}_2 - d_o\dot{z}_1 - i\omega_o d_o(z_2 - z_1) = 0 \tag{5}$$

Introducing the symbols :

$$\Omega_1^2 = k_1/m_1 \ , \quad \Omega_o^2 = k_o/m_1 \ , \quad \Omega_2^2 = k_o/m_2$$
$$\varepsilon n_1 = d_1/m_1 \ , \quad \varepsilon n_o = d_o/m_1 \ , \quad \varepsilon n_2 = d_o/m_2 \tag{6}$$

Assuming that the damping is light, the quantity ε is a small one, the Eqs.(5) are transformed as

$$\ddot{z}_1 + (\Omega_1^2 + \Omega_o^2)z_1 - \Omega_o^2 z_2 = \varepsilon[n_o\dot{z}_2 - (n_1 + n_o)\dot{z}_1 + i\omega_o n_o(z_1 - z_2)]$$
$$\ddot{z}_2 + \Omega_2 z_2^2 - \Omega_2 z_1^2 = \varepsilon[n_o\dot{z}_1 - n_2\dot{z}_2 + i\omega_o n_2(z_2 - z_1)]$$

and the related homogeneous form is

$$\ddot{z}_1 + (\Omega_1^2 + \Omega_0^2)z_1 - \Omega_0^2 z_2 = 0$$
$$\ddot{z}_2 + \Omega_2^2 z_2 - \Omega_2^2 z_1 = 0 \tag{8}$$

Assuming the solution of Eqs.(8) possesses a form

$$z_1 = z_{1k} e^{i\omega t}$$
$$z_2 = z_{2k} e^{i\omega t} \tag{9}$$

Substituting Eqs.(9) into Eqs.(8), then obtain

$$(-\omega^2 + \Omega_1^2 + \Omega_0^2)z_{1k} - \Omega_0^2 z_{2k} = 0$$
$$(-\omega^2 + \Omega_2^2)z_{2k} - \Omega_2^2 z_{1k} = 0 \tag{10}$$

If the solution for Eqs.(9) is nontrivial, then an eigen-value equation must be satisfied

$$\begin{vmatrix} -\omega^2 + \Omega_1^2 + \Omega_0^2 & -\Omega_0^2 \\ -\Omega^2 & -\omega^2 + \Omega^2 \end{vmatrix} = 0$$

Then eigen-values are

$$\omega_k^2 = [(\Omega_1^2 + \Omega_0^2 + \Omega_2^2) \mp \sqrt{(\Omega_1^2 + \Omega_0^2 - \Omega_2^2) + 4\Omega_0^2\Omega_2^2}\,]/2 \quad _{k=1,2} \tag{11}$$

Noticing Eqs.(6), these expression can be simplified

$$(\omega_k/\Omega_2)^2 = [(1 + \bar{k} + \bar{m}) \mp \sqrt{(1 + \bar{k} - \bar{m})^2 + 4\bar{m}}\,]/(2\bar{m}) \tag{12}$$

Substituting Eq.(11) into first one of Eqs.(10), then obtain

$$\bar{z}_k = (z_{2k}/z_{1k})\big|_{\omega_k} = [(1 + \bar{k} - \bar{m}) \pm \sqrt{(1 + \bar{k} - \bar{m})^2 + 4\bar{m}}\,]/2$$
$$(\omega_k/\Omega_2)^2 = (1 + \bar{k} - \bar{z}_k)/\bar{m} \qquad k=1,2 \tag{13}$$

Rewriting these eigen-solutions into matrix

$$V = \begin{bmatrix} 1 & 1 \\ \bar{z}_1 & \bar{z}_2 \end{bmatrix} \tag{14}$$

Introducing a variable transform

$$\begin{Bmatrix} z_1 \\ z_2 \end{Bmatrix} = V \begin{Bmatrix} z_{m1} \\ z_{m2} \end{Bmatrix}$$

and multiplying Eqs.(7) with V^H which is a conjugate transpose matrix of V, then obtain

$$z_{m1} + \omega_1^2 z_{m1} = (\bar{z}_2 f_1(\varepsilon) - f_2(\varepsilon))/(\bar{z}_2 - \bar{z}_1)$$
$$z_{m2} + \omega_2^2 z_{m2} = (\bar{z}_1 f_1(\varepsilon) - f_2(\varepsilon))/(\bar{z}_1 - \bar{z}_2) \tag{15}$$

where $f_1(\varepsilon)$, $_2f(\varepsilon)$ represent the right of Eqs.(5). These equations are equations of motion in standard form.

IV. BUILDING THE STABILITY CRITERION BY THE METHOD[3,4] OF PERTURBATION

For quantity ε is small, we can use perturbation method to extract the stable boundary curve obtained in Eq.(15). Eq.(15) is a linear differential equation with constant coefficient, so its solution can be expressed as follows:

$$z_{m1} = (\sum_0^\infty A_{1k}\varepsilon^k)\exp(i\omega_1 + \sum_1^\infty \delta_{1k}\varepsilon^k) + (\sum_0^\infty B_{1k}\varepsilon^k)\exp(i\omega_2 + \sum_1^\infty \delta_{2k}\varepsilon^k)$$

$$z_{m2} = (\sum_0^\infty A_{2k}\varepsilon^k)\exp(i\omega_1 + \sum_1^\infty \delta_{1k}\varepsilon^k) + (\sum_0^\infty B_{2k}\varepsilon^k)\exp(i\omega_2 + \sum_1^\infty \delta_{2k}\varepsilon^k) \quad (16)$$

Substituting Eq.(16) into Eq.(15), then taking the zero order term ε^0, we can obtain a solution shown in Eqs.(11),(13). If taking the first order term ε^1, and equalizing the corresponding coefficients, we obtain the quantities δ_{11}, δ_{21}, for simplification, the details of derivation are omitted. The δ_{11}, δ_{21} are called modal damping which show the stability of a system, The system are stable, if $\delta_{11}<0$, $\delta_{21}<0$. We put them in following expression

$$\omega_0/\Omega_2 \le [1 + \bar{d}/(1 - \bar{z}_k)^2]\sqrt{(1 + \bar{k} - \bar{z}_k)/\bar{m}} \quad (17)$$

choosing the sign of equality in Eqs.(17), we obtain the stable boundary curves shown in Fig.3. The curve I is determined by k=1, corresponding the lower eigen-frequency, and the curve II is determined by k=2, cresponding the higher one. They are all expressed in non-dimensional forms. A clan of curve shown in Fig.3 indicates the influence on the stable area when the mass ratio $\bar{m}$ changes. The curves shown in Fig.4 expresses the influence of the damping ratio $\bar{d}$ to the stable area.

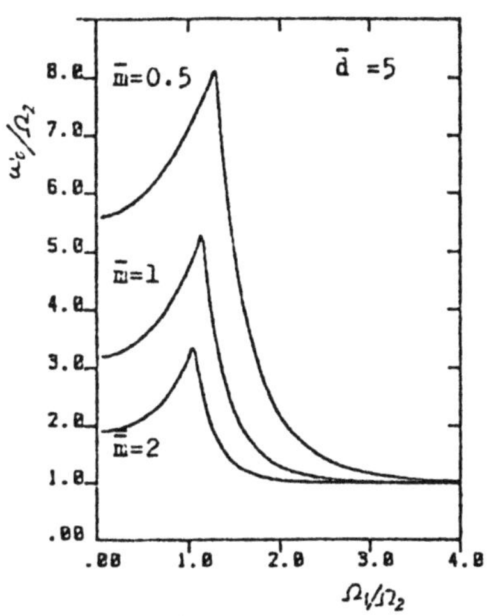

Fig.3 Stable speed ω_0 increase when $\bar{m}$ decrease.

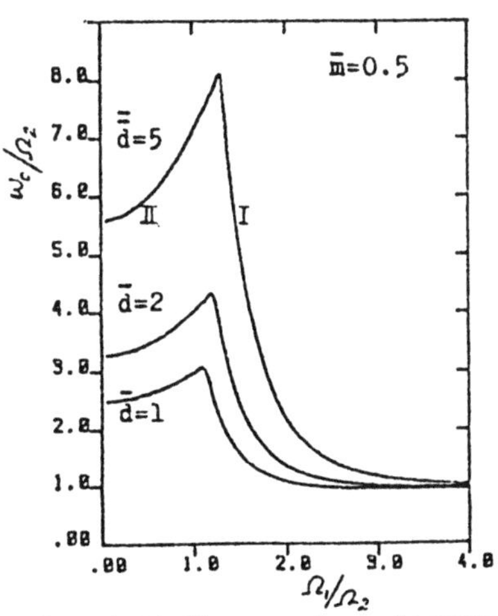

Fig.4 Stable speed ω_0 increase when $\bar{d}$ increase.

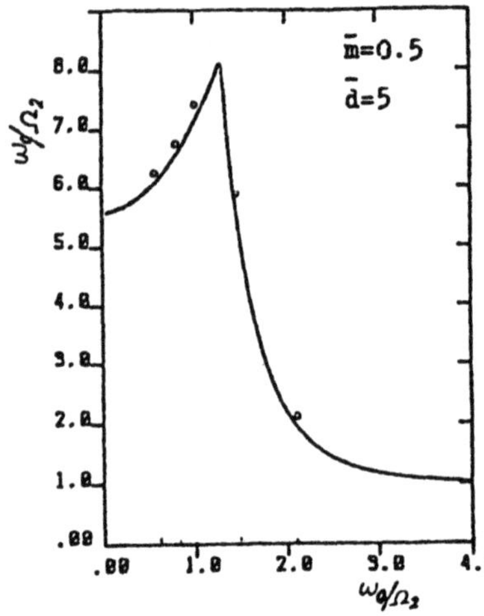

Fig.5 Comparison between theoretical curve and experimental points(∘).

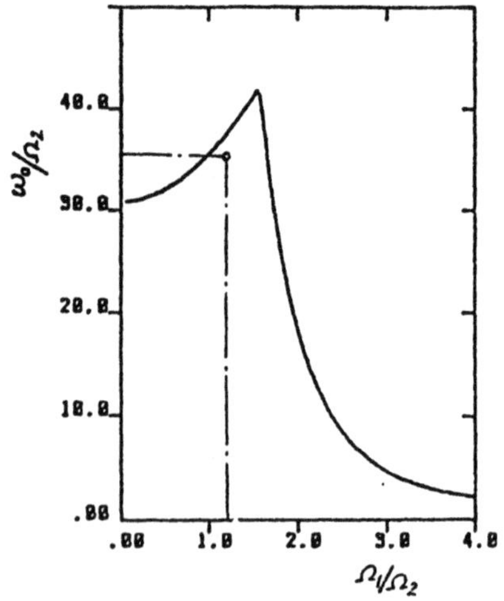

Fig.6 The work state of prototype(point ∘)

V. BUILDING THE STABILITY CRITERION BY THE METHOD OF ENERGY EQUILIBRIUM

In the paragraph IV, the perturbation method is used for searching the stable boundary curves. More accuracy results can be derived, if the higher orders of small quantity ε are put in the calculation. Of course, much more tedious calculation must be dealt with. However, another method of energy equilibrium is simpler for this purpose[2]. We consider that in eigen-state k=1,2 the damping force from the oil is shown as

$$F_{1k} = d_1 \dot{z}_{1k} = i\omega_k d_1 z_{1k}, \qquad k=1,2$$

When the rotor and the core vibrate, the force dissipates the energy in oil, the corresponding power is indicated by

$$P_{1k} = \dot{z}^H_{1k} F_{1k} = d_1 \omega_k^2 z_{1k}^2 \qquad k=1,2 \tag{18}$$

The internal damping force from hysteresis is shown as

$$F_{2k} = d_o(\dot{z}_{2k} - \dot{z}_{1k}) - i\omega_o d_o(z_{2k} - z_{1k}),$$

and the force introduces the energy into the system from motor, the corresponding power is as

$$P_{2k} = (\dot{z}_{2k} - \dot{z}_{1k})^H F_{2k} = d_o \omega_k(\omega_k - \omega_o)(z_{2k} - z_{1k})^2 \tag{19}$$

If the system is stable, the dissipated energy by d_1 must be larger or equal the energy introduced by d_o, i.e.

$$P_{1k} \geq P_{2k}, \qquad k=1,2 \tag{20}$$

Putting Eqs.(18),(19) into Eqs.(20), a criterion of energy for stable boundary can be transformed as

$$\omega_o/\omega_k \leq 1 + \bar{d}/(1-\bar{z}_k)^2 \qquad k=1,2 \tag{21}$$

$$\bar{z}_k = z_{1k}/z_{2k} \qquad k=1,2$$

Considering Eqs.(13 rewrite Eqs.(21) as

$$\omega_o/\Omega_2 \le [1 + \bar{d}/(1 - \bar{z}_k)^2]\sqrt{(1 + \bar{k} - \bar{z}_k)/\bar{m}} \qquad k=1,2$$

which has a same form as Eqs.(17). Therefore, the same results can be obtained from both methods, perturbation method and energy method.

VI. OPERATION EXPERIMENT

With the axial position controlled by controlling box, and driven by the high frequency motor, the rotor is run inside the vacuum-protect casing. Using the vibration meter to check the radial amplitude and using the speed meter to cheek the rational speed of the rotor.

The rotor is made from aluminum alloy CL4 with high strength, and the attraction top of which is made from high quality low-carbon steel, the iron core from pure iron. With the safety in mind, the experiment to verify the stable boundary curve is carried on the parameters of the low-speed states (choosing $\bar{d}=5$), the parameters of the experimental device is given in Tab.2.I. The eigen-frequency ω_1 of the electromagnetic iron core is changed by changing the diameter of the steel string. The values of ω_1 with respect to the steel string with length 100mm and varies diameters are given in Tab.1. Measuring five actual experimental points, on which the values is fairly approximate to the theoretical ones (just a little larger, ref. Fig.5), it is sufficient to verify that the preceding theoretical analysis is accuracy enough.

The actual prototypes requires to run in a high speed, namely a maximum rotational speed, 30,000r.p.m. is asked, (which is the limited speed of the material CL4 ϕ240mm), i.e.,the maximum rational speed ratio ω_o/Ω_2 is demanded to reach 35.7 nearly. To reach such a high rotational speed, the achievements in the above theoretical analysis should be used to choose the suitable parameters of a rotor system, such as the parameters of rotor, damper and iron core. The iron corn is designed to be hollow (Fig.1) so as to decrease the mass ratio $\bar{m}$ and to increase the damping ratio $\bar{d}$. The parameters of the prototype is given in Tab.2.II.

The work states of the prototype is shown in Fig.6; the rotor was run up to a high speed of 30,000rpm stably and repeatedly,

Tab.1. Ω_2 and the correspongding diameter of steel string

Diameter(mm)	1.5	2.0	2.5	3.5	5.0
Ω_1 (rad/s)	8.8	11.7	14.5	20.5	29.0
Ω_2 (rad/s)	0.6	0.8	1.0	1.5	2.1

Tab.2. The parameters of experimental rig

Items		Mass(kg)	Material	Dimension	Freq.(rad/s)	Note
Rotor		10.0	CL4	φ240x82	Ω_2=14.0	
I	Iron core	5.0	pure iron	φ60x230	Ω_1(Tab.1)	$\bar{d}$=5.0
II	Iron core	2.5	pure iron	φ50-25x220	Ω_1=17.0	$\bar{d}$=20.

nothing is found unusual, and the rotor which weighs 10 kilogram is suspended by a controlled magnetic bearing.

VII .CONCLUSION

(1) A damping device must be used to eliminate the self-excited vibration when the magnetic bearing-rotor system turns with a much higher speed than the eigen-frequency which is about 15 Hz (900 r.p.m).

From Fig.4, we know that if the external damping $\bar{d}$ is very light, only when the ratio ω_0/ω_k is a bit larger than 1,the system do not produce self-excited vibration. Namely, the rotor is stable only when the speed is about 15Hz (i.e. 900 r.p.m.),but up to a little higher speed,(for example,more than 900 r.p.m.),it will lost its stability.

(2) It is convenient to expand the stable area that the mass ratio $\bar{m}$ of the rotor to the iron core is decreased (Fig.3). But to decrease the ratio $\bar{m}$ is limited by the magnetism saturation of the iron.

(3) It is also very convenient to extend the stable area to increase the damping ratio $\bar{d}$. We should try to increase the external damping d_1. If d_1 is large enough, the rotor can be run up to a high speed.

(4) When the two eigen-frequencies of the rotoı and the iron core Ω_1, Ω_2 are matched each other, the best damping effect could be obtained.

REFERENCE

1. Beams, J. W.; Young, J. L; Moore, J. W., The Production of High Centrifugal Fields, J. Appl. Phys., 17(1946), p886.
2. Schweitzer, G.; Stabilization of self-excited Vibration by Active Damper, Dynamics of Rotors, IUTAM Symposium, Lungby, Denmark, 1974, Springer-Verlag, Berlin, 1975, p472.
3. Tondl, A, Some Problems of Rotor Dynamics. London, Chapman and Hall, 1965.
4 Nayfeh, A. H., Introduction to Pertubation Techniques. New York, Wiley-Interscience, 1981.

Design of a Controller for Magnetic Levitation

Pierri, P. S., Pereira da Silva, P. S. and F. Betti

COPESP II - Av. Prof. Lineu Prestes, 2242 - Cep 05508
USP - São Paulo, SP - BRASIL

SUMMARY

In this work we will apply a linear quadratic regulator for the controller of the magnetic levitation of a rigid body. A successfull implementation with a diferentiator to estimate the velocity is showed, using a linear and not noisy displacement sensor. Another implementation with a pragmatic control policy that worked well with a noisy and strongly nonlinear sensor is presented.

1.0) INTRODUCTION

Robustness is a well-known property of optimal linear quadratic control (LQR). Guaranteed phase and gain margins of respectively, 60 degrees and -6 dB to infinity is a measure of robustness with this control policy [1], [2]. For a system

$$\dot{x}=Ax+Bu, \qquad x(0)=x_o, \qquad v=Cx \tag{1.1}$$

in which we attempt to minimize the functional

$$J=\int_0^{+\infty}(x'Qx + u'Ru)dt \tag{1.2}$$

with Q, R symmetric and respectively, semi-definite and positive definite matrices, the steady state optimal solution doesn't depend on the initial condition x_o and it is of the state feedback form:

$$u(t)=Fx(t), \qquad F=-R^{-1}B'P \tag{1.3}$$

where P is the unique solution of the Riccati equation [3]. Furthermore, controllability of (A,B) ensures stability of the closed-loop system $\dot{x}=(A+BF)x$.

In this work we will apply a LQR for the controller of the magnetic levitation of a rigid body, [4], [5], for two cases. In the first one we will use a linear displacement sensor, and in the second one we will use a noisy and nonlinear sensor.

2.0) PLANT IDENTIFICATION

The state variable system representation will depend on the knowledge about the magnetic actuator, sensors, power amplifier and body weight. In this section, the plant identification and its linearization are described.

2.1) Magnetic actuator

The following law will be considered for the magnetic

actuator force:

$$f=C(I/X)^2 \tag{2.1}$$

where:
- f: force, (N)
- I: coil current, (A)
- X: displacement (m) and
- C: magnetic actuator constant.

The magnetic actuator constant C can be found through experimental methods, [6], analytical methods (in some cases) or by finite element approach. We used an experimental method, in which the current I is obtained for a constant f, at each gap X. Thus, we obtained $C=2.223E-5\ Nm^2/A^2$.

2.2) Displacement sensors

We used an eddy current sensor and an inductive sensor. The first one presents a good linearity while the second one is highly nonlinear. In the test rig we got the calibration curves (fig. 1):

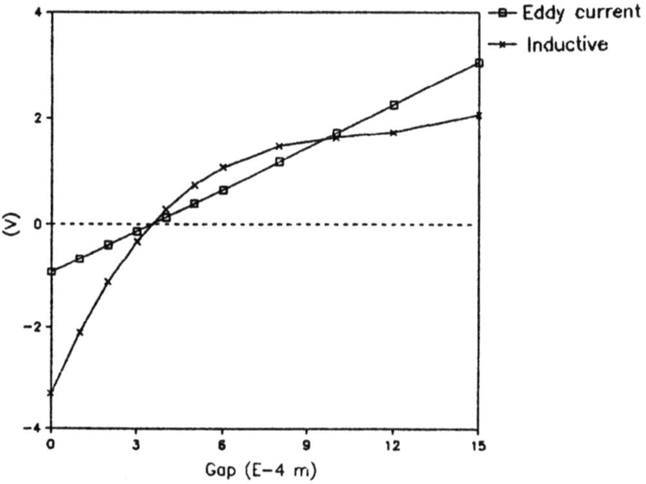

Fig. 1 - Sensors calibration curves

Eddy current sensor: $v=2625X-0.92,$ (2.2.1)

Inductive sensor: $v=-3.32+1.33E4X-1.22E7X^2+3.87E9X^3$ (2.2.2)

2.3) Power amplifier

The power amplifier used is linear and has a very flat frequency response. Then, we can describe it as a constant gain $K_a=0.42$ A/V.

2.4) State variable system representation

The linearized force around levitation gap is:

$$f = f(Io,Xo) + \frac{df}{dX}\Big|_{\substack{X=Xo,\\ I=Io}}(X-Xo) + \frac{df}{dI}\Big|_{\substack{X=Xo,\\ I=Io}}(I-Io) + \;\ldots \tag{2.4}$$

where Xo and Io are respectively the levitation gap and levitation current. If we write x=X-Xo and i=I-Io, and use (2.1), we get:

$$f \simeq CK_k{}^2 - K_x x + K_m i \tag{2.4.1}$$

where: $K_x = 2CIo^2/Xo^3$; $K_m = 2CIo/Xo^2$; $K_k = Io/Xo$.

Thus, the differencial equation for the system linearized around levitation gap is $M\ddot{x} = K_x x - K_m i$ (M: mass of body). In matricial representation (1.1):

$$\begin{bmatrix}\dot{x}\\ \ddot{x}\end{bmatrix} = \begin{bmatrix}0 & 1\\ K_x/M & 0\end{bmatrix}\begin{bmatrix}x\\ \dot{x}\end{bmatrix} + \begin{bmatrix}0\\ -K_m/M\end{bmatrix} i \tag{2.4.2}$$

We can change the mechanical state variables by the eletrical state variables:

$$\begin{bmatrix}\dot{v}\\ \ddot{v}\end{bmatrix} = \begin{bmatrix}0 & 1\\ K_x/M & 0\end{bmatrix}\begin{bmatrix}v\\ \dot{v}\end{bmatrix} + \begin{bmatrix}0\\ -K_m K_d K_a/M\end{bmatrix} Va \tag{2.4.3}$$

with $K_d = \frac{dv}{dX}\Big|_{X=Xo}$ and Va: control voltage output.

In this way, when evaluating the state feedback to get a LQR, we have to calculate the gains for displacement and velocity signals.

3.0) CONTROLLER DESIGN: EDDY CURRENT SENSOR CASE

Choosing Xo=0.35 mm for the levitation gap we obtained the following matrices for the system:

$$A = \begin{bmatrix}0 & 1\\ 5.604E+4 & 0\end{bmatrix},\quad B = \begin{bmatrix}0\\ -2.007E+5\end{bmatrix},\quad C = \begin{bmatrix}1\\ 0\end{bmatrix}' \tag{3.1}$$

It is easy to demonstrate that the system above is controlable and observable, [5].

We evaluated the LQR feedback matrices F for different choices of the ponderation matrices Q and R. Through computer simulation of the nonlinear system model, we found the best gains.

$$Q = \begin{bmatrix}1 & 0\\ 0 & 3.0E-4\end{bmatrix},\quad R = [\,1\,],\quad F = [-1.32 \quad -1.77E-2] \tag{3.2}$$

We got a good performance with this choice of F (Fig. 2, curve 1). For that ponderation the levitation was fast and without overshoot. The simulation results were very close to those ones we got in the test rig, where we used a differentiator to produce a velocity estimation. The implemented transfer function and its approximation for low frequencies were:

$$G(s) = 1.32 + \frac{2.7966E8s}{s^2 + 11.3E3s + 1.58E10} \simeq 1.32 + 1.77E-2s \tag{3.3}$$

3.1) Set-point regulator (SPR)

In order to obtain greater bearing stiffness around the levitation gap, we tried to increase the closed-loop gains using another F, but the levitation transient became worse due to growth of the overshoot (Fig. 2, curve 2). Overshoot appears because the magnetic actuator can't produce repulsive forces to brake the body.

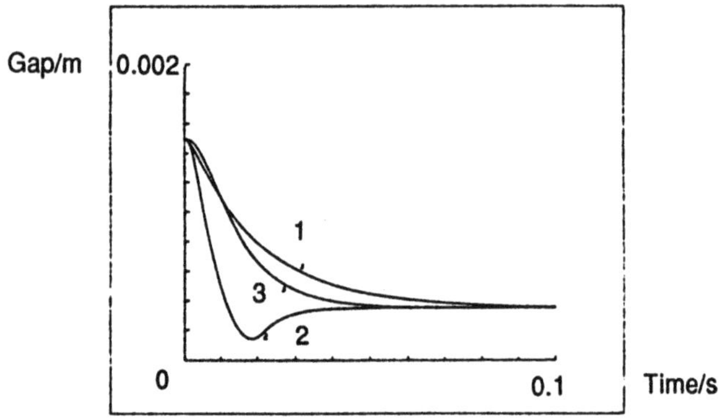

Fig. 2 - Levitation Transient: Gap(m) X Time (s)
1) F11=1.32; 2) F11=3.42 3) F11=3.42 with SPR

High closed-loop gains usually produce high currents with abrupt variation in the levitation transient. When magnetic material is submited to that flux variation, it becomes magnetized (magnetic remnant). This will produce greater overshoot in the test rig than that one obtained by computer simulation (where the magnetic remnant model is not available).

A solution found for this problem was to get slow levitation transients with high closed-loop gains. This was possible due to robustness of the LQR during transient levitation and by a slowly variable signal which was added to the set-point D.C. voltage for gap Xo. This corresponds to the following block diagram:

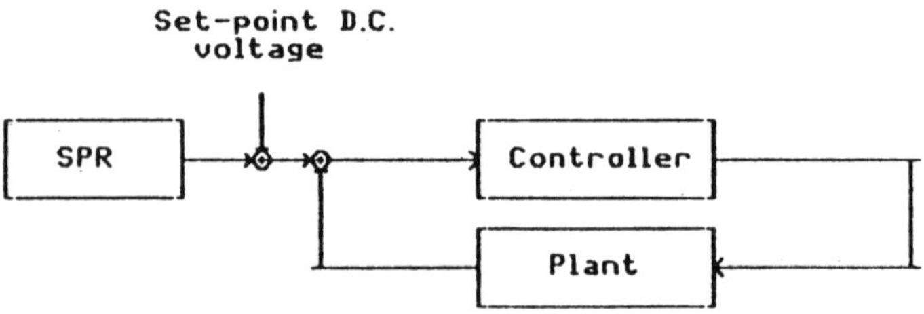

Fig. 3 - Set-point regulator (SPR) control block diagram

The SPR has been used in order to keep the control error signal very small during all the levitation transient and consequently to produce small body acceleration and coil current, even with high closed-loop gains (Fig. 2, curve 3).

4.0) CONTROLLER DESIGN: INDUCTIVE SENSOR CASE

In section 3 we presented a controller with a differentiator to produce a velocity estimation. This worked well because the Eddy current sensor is linear and not noisy. However, the same control policy presented a bad performance when using an inductive sensor. This is due to the noisy nature of that sensor. Furthermore, the inductive sensor is strongly nonlinear (Fig 1).

The controller for this case must be robust (specially due to the non-linearity of the sensor) and the closed-loop system must have good noise rejection properties.

The first possible approach is the design of an observer-based compensator, considering the closed-loop robustness. Transfer function computations will be necessary to determine the noise and mechanical disturbance rejection. For final verification, the sensor non-linearity will be taken in account in computer simulation.

Considering the differential equations of the system and a observer-controller topology [7] with a full state observer, we have:

$$\begin{bmatrix} \dot{x} \\ \dot{w} \end{bmatrix} = \begin{bmatrix} A & BF \\ -GC & A+GC+BF \end{bmatrix} \begin{bmatrix} x \\ w \end{bmatrix} + \begin{bmatrix} B \\ B \end{bmatrix} v(t) + \begin{bmatrix} B \\ 0 \end{bmatrix} q(t) + \begin{bmatrix} 0 \\ -G \end{bmatrix} r(t),$$

$$y = [C \quad 0] \begin{bmatrix} x \\ w \end{bmatrix}. \tag{4.1}$$

where r(t) is the sensor noise, v(t) is a known input and q(t) is an unknown mechanical disturbance. We can use the similarity transformation in order to compute the transfer functions:

$$\bar{A}_c = T^{-1} A_c T =$$

$$\begin{bmatrix} 0 & I \\ I & -I \end{bmatrix} \begin{bmatrix} A & BF \\ -GC & A+GC+BF \end{bmatrix} \begin{bmatrix} I & I \\ I & 0 \end{bmatrix} = \begin{bmatrix} A+BF & GC \\ 0 & A+GC \end{bmatrix}. \tag{4.2}$$

From matrix $\bar{A}_c$ (4.2) the so called deterministic separation property is clear. Indeed, the tringular-block form of (4.2) implies that the poles of $\bar{A}_c$ are the poles of (A+BF) and (A+GC). The controlability of (A,B) and the observability of (C,A) ensures arbitrary allocation of the poles of (A+GC) and (A+BF), respectively. For instance, we can use LQR algorithm to determine the optimal feedback F and use the dual system to determine a G' so that (A'+C'G') has adequate poles. Unfortunately, this ensures stability but there is no guarantee of robustness with observer-based controllers [8], [9]. In the new base defined by the similarity transformation (4.2) is very easy to compute all the transfer functions from v, q and r to x. Moreover, it's clear that the controler's transfer function is $L(s)=-F(sI-(A+BF+GC))^{-1}G$, and the plant's transfer function is $H(s)=C(sI-A)^{-1}B$. So it's not difficult to determine the open-loop L(s)H(s) (so the gain and phase margins) and noise and disturbance rejection. Then, we can proceed in the design with several choices of F and G, comparing the results of each one.

For our particular problem, we observed that the robustness and noise rejection of the closed-loop system led to relatively low gains of the controller. On the other hand we required that the controller must levitate the body from its start position, which is 1.15 mm distant from the levitation gap 0.35 mm. Due to the non-linearity of the sensor and actuator, the linearized open-loop gain around the start position is many times lower than that one around the levitation gap. In practice, the gain adjustment of the controller became almost impossible, because it must be high enough to produce a current in the magnetic actuator in order to levitate the body, but it has to be low enough because of the noise and stability restrictions. We solved this problem by adding an integrator to the system, and then proceeding with a LQR policy. The plant to be controlled (2.4.3) was a little bit different from the one in the section 3, and is given by:

$$A1=\begin{bmatrix} 0 & 1 \\ 5.6E+4 & 0 \end{bmatrix},\quad B1=\begin{bmatrix} 0 \\ -1.3E+5 \end{bmatrix},\quad C1=\begin{bmatrix} 1 \\ 0 \end{bmatrix}'. \qquad (4.3)$$

We added an integrator in the way that a new input was defined. The complete new plant will be:

$$A=\begin{bmatrix} 0 & 1 & 0 \\ 5.6E+4 & 0 & -1.3E+5 \\ K & 0 & 0 \end{bmatrix},\quad B=\begin{bmatrix} 0 & 0 \\ -1.3E+5 & 0 \\ 0 & -1 \end{bmatrix}. \qquad (4.4)$$

This corresponds to the following block diagram:

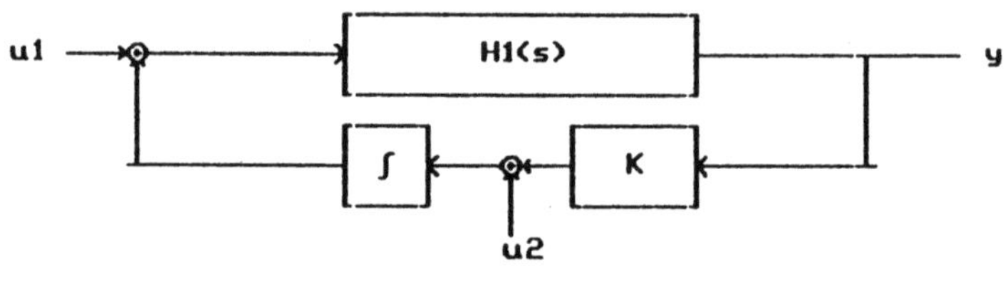

Fig. 4 - Integrator system block diagram - $H1(s)=C1(sI-A1)^{-1}B1$

We added an integrator and a gain K to the original system, defining a new input u2. The exact value of K was determined in a trial and error fashion. To this new system defined by (4.4) we evaluate the LQR feedback matrices with differents choices of Q, R and K in functional (1.2) and in the system (4.4). The main idea was to have a strong ponderation of the first input u1 compared with the second input u2. This was done in order to have lower gains in the proportional and derivative parts of the controller, but great gain in the integral part, which will produce high D.C. gain. With the choices:

$$Q=\begin{bmatrix} 400 & 0 & 0 \\ 0 & 0.05 & 0 \\ 0 & 0 & 0.5 \end{bmatrix}, \qquad R=\begin{bmatrix} 200 & 0 \\ 0 & 1 \end{bmatrix}, \qquad K=50. \tag{4.5}$$

the optimal feedback matrix was found to be

$$Fo=\begin{bmatrix} 2.31 & 1.69E-2 & 0.187 \\ -2.74 & 2.88E-4 & 8.27 \end{bmatrix}. \tag{4.6}$$

The reader can verify that the matrix Fo with $Fo_{22}=0$ will produce the same closed-loop poles. The controller transfer function would be $L(s)=1.69E-2s-2.31+38.42/(s+8.27)$.
The implemented transfer function was:

$$L(s)=L_o(s)+L_i(s) \tag{4.7}$$

where: $L_o(s)=\dfrac{-2.08E8(1.69E-2s-2.31)}{s^2+5.02E4s+2.08E8}$ and $L_i(s)=\dfrac{38.42}{s+8.27}$

The $L_o(s)$ produces an estimation of the proportional and derivative part of the control. Its poles were selected by trial and error, in order to have robustness and good dynamic response, dictated by the closed-loop poles. The $L_i(s)$ is the integral part of the controller.

In order to check the closed-loop system behavior, several computer simulations were performed. We check the controller performance considering sensor and actuator non-linearities, mechanical disturbances and sensor noise. In this way the controller (4.7) was found to be a good choice. As an example of simulation, we show in the figure 5 a typical levitation curve with the set-point regulator. These results were very close to the experiment results .

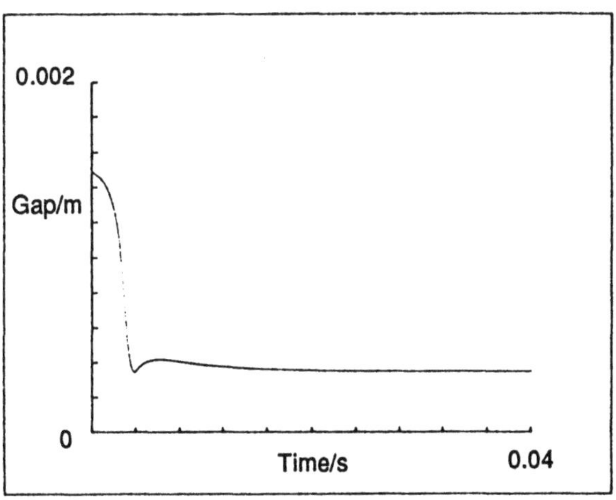

Fig. 5 - Levitation Transient: System with integrator

5.0) CONCLUSION

In section 3 we showed that, in some cases, the differentiator is a simple manner to produce a velocity estimation and an optimal control can be implemented with good closed-loop performace. In section 4, we showed that, in the case of a noisy and non-linear sensor, a simple design method was able to produce a robust controller. In this case an integrator was added and its pole was determined by an LQR algorithm. So we can call the controller of section 4 as an "optimal PID control".

We showed the great influence in controller design of the displacement sensors.

References:

1. Safonov, M. G. and M. Athans, "Gain and phase margins for Multiloop LQG regulator", IEEE Trans. Automat. Contr., Apr. 1977.
2. Doyle, J. C., "Guaranteed margins for LQG regulators", IEEE Trans. Automat. Contr., vol. Ac-23, pp. 756-757, Aug. 1978.
3. Anderson, B. D. O. and J. B. Moore, "Linear optimal control", Prentice Hall, Englewood Cliffs, N. J., 1971.
4. Schweitzer, G. and H. Ulbrich, "Magnetic Bearings - A novel type of suspension", Sec. Intern. Conf. on vibration in rotating machinery, Institution of Mechanical Engineers, London, Cambridge, 1980.
5. Deboni, J. E. Z. e W. R. Mello, "Estudos de controle otimo para levitação magnética", COPESP, internal report, 1986.
6. Traxler, A. and G. Schweitzer, "Measurement of the force characterist contactless eletromagnetic rotor bearing", IMEKO Intern Measurem. Conf, Symposium on Measurement and Estimation, Brixen, Italy, May, 1984, pp. 187-191.
7. Kailath, T., "Linear systems", Prentice Hall, Englewood Cliffs, N. J., 1980.
8. Okada, T., Kihara, M. and H. Furihata, "Robust control systems with observers", Int. J. Contr., vol. 41, num. 5, 1987, pp. 1207-1219.
9. Tsui, C. C., "On preserving the robustness of an optimal control system with observers", IEEE Trans. Automat. Contr., vol Ac-32, pp. 823-826, Sep. 1987.

Magnetic Bearing Design for a High Speed Rotor

E. H. MASLEN, P. E. ALLAIRE, M. A. SCOTT

Department of Mechanical and Aerospace Engineering,
University of Virginia, Charlottesville, VA, USA

P. HERMANN

Sundstrand Corporation
Rockford, IL, USA

Summary

The design of an example set of magnetic bearings for a small high speed compressor is presented. The bearings were designed to replace the fluid film bearings of an existing compressor design. Reduction of parasitic power losses due to the bearings is projected to be from the current 2 kW down to approximately 200 W.

Introduction

Magnetic bearings are beginning to see application to a broad spectrum of industrial machinery. The motives for using magnetic bearings are as varied as the machines in which they are found. In large turbomachinery, these bearings have proved attractive because they can eliminate complex lubrication systems [1]. In the petrochemical industry, canned pumps can benefit from the longevity of magnetic bearings in harsh chemical environments [2]. Rotary anode X-ray tubes have taken advantage of the ability of these bearings to operate in a vacuum [3]. For the application discussed here, the primary motive for using magnetic bearings is to reduce parasitic power losses in the bearings.

The application of magnetic bearings to small, high speed rotating machinery accentuates constraints in the design of the magnetic components which may be considered of secondary importance in larger machinery. First, the magnetic components generally must fit into a very small space. This tends to discourage the use of designs requiring complex pole structures such as is the case with many of the hybrid electro-permanent magnet designs. Further, minimum coil volume dictates high coil current densities so that providing adequate cooling for the coils may be difficult. Second, the mass added to the shaft must be minimized in order to avoid adverse effects to the shaft dynamics. As a percentage of total rotor mass, the added journal mass for magnetic bearings will typically be higher in small equipment than in large. Finally, any rotating components should, if possible, be permanently mounted to the shaft so as to minimize the eccentricity (and resulting unbalance) associated with slip fits. This is especially significant in machinery intended to operate at very high speed.

Partial funding for this work was provided by Sundstrand Corporation, the Army Research Office, and the Center for Innovative Technology of the Commonwealth of Virginia.

The purpose of this paper is to describe some of the factors which should be taken into account in the bearing analysis. No attempt is made here to investigate the detailed finite element analysis which should ultimately conclude any design project of this sort.

The example bearing design to be discussed here exemplifies this type of highly constrained problem. The machine for which the bearings were designed is a small, electrically driven compressor. It has an operating speed range of 30,000 to 70,000 RPM. The motor armature is integral with the compressor shaft and is centered between the two radial bearings whose span is 20 cm. The shaft mass is approximately 1.6 kg. Design of these bearings was somewhat complicated by the requirement that they be retrofit to an existing design, so that they had to fit within the physical envelope originally provided for fluid film bearings and the associated seals.

The actual performance specifications and design figures presented here are intended only to illustrate the considerations governing the design process. They do not precisely correspond to any bearing design actually being constructed either by the authors or their sponsors.

Nomenclature

A_g	pole face area
α	temperature coefficient of wire electrical conductivity
B_{sat}	magnetic saturation flux density
β	geometry factor: $\partial G / \partial x$
c_g	ratio of total conductor volume to total coil volume
δ	characteristic coil radius
ε	bearing derating factor
f	bearing force
G	nominal air gap between bearing stator and rotor
h_c	convection coefficient
I_b	nominal, or bias coil current: time invariant
i_p	coil perturbation current: varies with time
J	coil current density
k	coil bulk thermal conductivity
k_s	spool thermal conductivity
μ_o	magnetic permeability of free space
N	total wire turns linked by a given magnetic circuit
r_s	coil spool effective inner radius = perimeter length / 2π
r_i	coil effective inner radius
r_o	coil effective outer radius
σ_o	wire electrical conductivity at T_o
T_o	reference temperature at which wire electrical conductivity is measured
T_p	pole structure fixed temperature
T_∞	coolant free stream temperature

Design Overview

The integrated compressor/bearing design is shown in Figure 1. From a mechanical point of view, several features are of immediate interest.

The bearings are constructed in such a manner that all rotating components, i.e.: the laminated stators and the thrust disk, are permanently mounted to the shaft. Disassembly of the

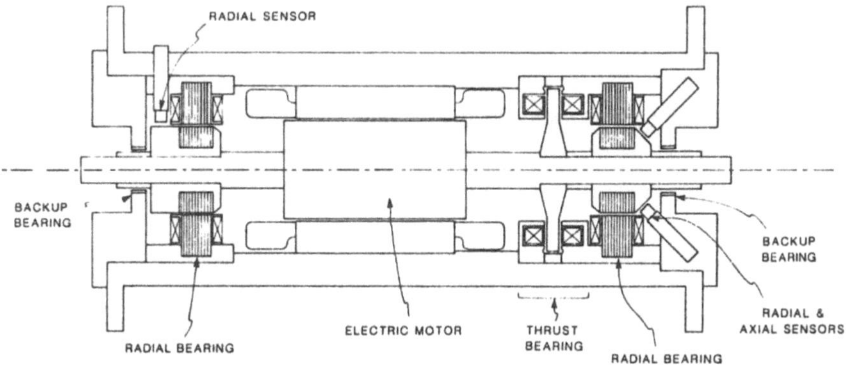

Figure 1 - Diagram of Compressor/Bearing Design

compressor does not require removal of any of these components, which permits the shaft to be balanced as an assembly outside of the compressor housing. The shaft assembly with laminated journals and thrust disk is shown in Figure 2.

Axial motion sensing can not be accomplished by simply directing a proximitor at the shaft end due to the impellers fitted to both ends of the shaft. Instead, both axial and radial motions are detected by a set of three probes at the right end of the compressor (see Figure 1.) These probes are located at ninety degree intervals around the shaft and are oriented at a forty five degree angle relative to the shaft axis. The axial, horizontal, and vertical motion of the shaft can readily be extracted from these probe signals by simple coordinate transformation. This scheme will inevitably lead to some crosstalk between the three axes due to calibration error and drift in the probes, but these effects are expected to be relatively minor.

The top shaft speed of 70,000 RPM dictates that hoop stresses in the various rotating components generated by the high centripetal forces cannot be neglected. Particularly affected by this is the thrust disk because it has the largest diameter of any of the rotating parts and also because it is constructed entirely of magnet iron, which has a substantially lower yield stress than do high strength steels. The design solution to this problem involves providing the thrust

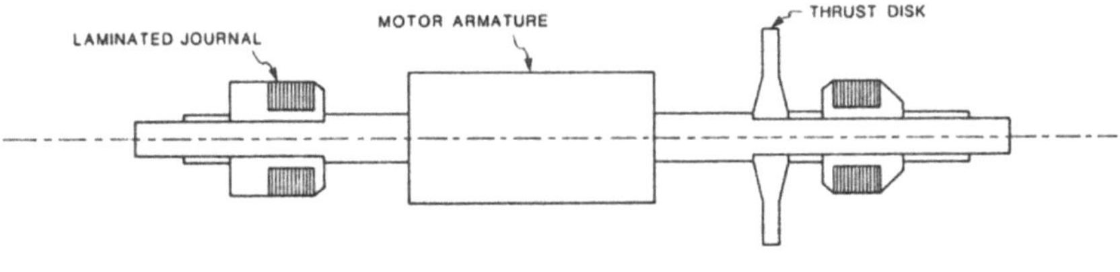

Figure 2 - Shaft with Laminated Journals and Thrust Disk

disk with a thick hub and tapering the disk to a thin section at the outer radius.

Backup bearings are provided at each shaft end in the event that the bearings should fail or an external load exceeding the capacity of the bearings should be applied. These are simple overbored bushings and are not intended for continuous operation. An axial backup bearing is provided by coating the faces of the thrust bearing with an anti-wear coating.

Performance Specification

The design of the magnetic components was guided by three major concerns: load capacity, thermal performance, and size. The radial and thrust actuators must be able to generate the magnetic forces required to meet the specified load requirements. At the same time, resistive heating in the coils must not exceed the capacity of a realistic cooling system to prevent overheating and failure of the coil insulation. Above all, any usable design must fit into the allotted space.

For the thrust bearing, the anticipated static axial force due to the pressure differential between the two stages is approximately 35 N. An additional sinusoidal component due to fluid forces generated by off-design operation of the impeller- volute interaction is expected to be on the order of ten percent of the static load. Excess load capacity of three times the total shaft weight, or 48 N is required for transient loads.

The radial bearings must support the static weight of the shaft (about 8 N per bearing) plus unbalance forces and transient loads. Impeller blade pass and hydrodynamic unbalance forces are expected to be substantially smaller than mass unbalance effects. The largest synchronous harmonic forces to be generated by the magnetic bearings depend both on the worst case unbalance distribution and the control algorithm employed. For the initial design work, it was assumed that these forces would be at most 4 N. The transient loads are expected to be no more than three times the rotor weight, or 24 N.

Cooling of the coils is provided by two primary mechanisms: conduction to the pole structure and convection to a cooling fluid. The thrust bearing coil is entirely enclosed by the pole structure and the thrust disk, so that only conduction to the pole structure is significant for this coil. For the radial bearings, however, the coils are exposed to the working fluid, so they can be convectively cooled as well. As design limitations for these two mechanisms, it was assumed that the temperature of the pole structure would not exceed 60°C and that the maximum allowable coolant flow rate is about 800 cc/sec. The coolant fluid is air at a maximum supply temperature of 50°C.

Space restrictions are the least flexible contraint in the problem. The outer diameter of the motor armature sets the minimum inner radius of the thrust bearing at 1.9 cm. If the thrust bearing pole pieces are of a smaller inner radius then disassembly of the compressor will require removal of the thrust disk. The casing dimensions fix the maximum outer diameter of both the thrust and radial bearings at 7.0 cm. Finally, the axial length of the thrust and radial bearing set on the right end of Figure 1 could not exceed 6.5 cm.

Magnetic Model

For each of the five axes of control, an axially opposed pair of electro- magnet pole structures as shown in Figure 3 provides a reversible control force. The magnet coils are energized with a fixed bias current of equal magnitude on each side. The coil currents are then symmetrically perturbed to produce the dynamic control force. The fixed bias current provides a limited linearization of the bearing force - perturbation current relationship, as well as permitting rapid changes in bearing force (high slew rate) even when the nominal bearing force is zero [4].

Simple analytic studies of this arrangement have been widely reported in the literature [5,6]. The analysis yields an analytic expression relating the bearing force to the various design parameters but neglects nonideal behavior of the magnetic circuit such as leakage and fringing. Allaire, et al. [2] have proposed a fairly simple scheme for computing a correction factor, ε, which derates the bearing performance based on an approximate calculation of the permeabilities of the leakage and fringing paths. The derating factor is fairly insensitive to small changes in design geometry so that various designs may be compared using the same derating factor. A factor of 1.1 was used for both the radial and thrust bearings at the design stage.

The resulting governing equation is:

$$f = \frac{\mu_o \beta N^2 A_g}{4\varepsilon^2}\left[\frac{i_1^2}{(G-\beta x)^2} - \frac{i_2^2}{(G+\beta x)^2}\right]$$

where the currents i_1 and i_2 are governed and restricted by

$$i_1 = I_b + i_p, \quad 0 \le i_1 \le i_{sat} \qquad i_2 = I_b - i_p, \quad 0 \le i_2 \le i_{sat}$$

and i_{sat} is the coil current which produces magnetic saturation in the pole structure:

$$i_{sat} = 2GB_{sat} / \mu_o N\varepsilon$$

Thermal Model

The ultimate goal of thermal models in magnetic bearing design is to determine the maximum wire temperature within the coil as a function of the coil current. This maximum temperature must be less than the manufacturer's thermal rating for the wire insulation.

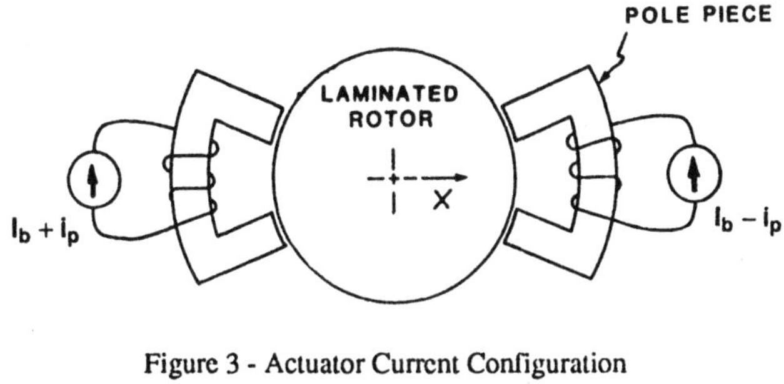

Figure 3 - Actuator Current Configuration

In this study, an analytic thermal model of the coils was used to predict the maximum wire temperature. This model is based on the following approximations:

1. the coil thermal conductivity can be modeled as a homogeneous bulk property
2. the coil geometry is adequately modeled as cylindrical
3. thermal gradients are purely radial
4. convection from the coil outer surface is uniform
5. the coil is wound on a spool of finite conductivity
6. the spool is in intimate contact with a pole piece at fixed temperature at its inner radius (a low spool conductivity can be used to simulate high contact resistance)
7. the electrical conductivity of the coil wire varies with temperature according to $\sigma = \sigma_o/(1+\alpha[T-T_o])$, where σ_o is the nominal conductivity measured at the temperature T_o: $\alpha = 0.00393/°C$ for copper wire, $\sigma(T{=}20°C) = 5.8\times10^6$ mho/cm

Given these approximations, the temperature in the coil can be represented by

$$T(r) = \tau c_1 J_o(r/\delta) + \tau c_2 Y_o(r/\delta) + T_o - 1/\alpha$$

where $\delta \equiv [(\sigma_o k c_g)/\alpha]^{1/2}/J$, $\tau \equiv T_\infty - 20 + 1/\alpha$, $\tau_p \equiv (T_p - 20 + 1/\alpha)/\tau$, J_o and Y_o are zero order Bessel's functions of the first and second kind, and the coefficients c_1 and c_2 are found from the matrix equation

$$\begin{bmatrix} \dot{J}_o(r_o/\delta) + h_c\delta J_o(r_o/\delta)/k & \dot{Y}_o(r_o/\delta) + h_c\delta Y_o(r_o/\delta)/k & 0 \\ \dot{J}_o(r_i/\delta) & \dot{Y}_o(r_i/\delta) & -\delta k_s/kr_i \\ J_o(r_i/\delta) & Y_o(r_i/\delta) & \ln(r_s/r_i) \end{bmatrix} \begin{bmatrix} c_1 \\ c_2 \\ c_3 \end{bmatrix} = \begin{bmatrix} h_c\delta/k \\ 0 \\ \tau_p \end{bmatrix}$$

Smith [7] provides a table originally developed by General Electric which correlates a conductivity enhancement factor (to be applied to the conductivity of the coil potting compound) with the percent metal by volume of the coil. For a typical wire packing factor of 0.72, the coil is about 57% metal by volume. Assuming full impregnation of the coil with polyester potting compound, the bulk conductivity of the coils should be about 0.012 W/cm/°C.

Given the specified limit on coolant flow rate and an estimate of the free space between the coils, the coolant free stream velocity around the radial bearing coils is expected to be on the order of 4 m/sec. The convection coefficient can be computed using standard expressions for multitube heat exchangers [8]. The predicted convection coefficient is 0.008 W/cm^2 /°C.

Assumption 1 produces a model which does not indicate the local thermal gradients surrounding the individual conductors, while assumptions 3 and 4 ignore stagnation zones in the coolant flow. Further, the thrust bearing coil is in contact with the pole structure both on its inner radial surface and on one of its axial faces (see Figure 1.) This additional conduction cooling is ignored by assumptions 3 and 6. Because of these three significant errors, the model can be expected to predict peak temperatures which are low for the radial coils and probably a little high for the thrust coils.

With the exception of coolant stagnation, these errors can be avoided by using a finite element model to compute the temperature distribution. The model must be quite detailed, however, in order to properly predict the local gradients surrounding the conductors.

The approach actually adopted in this study employed various engineering judgements in applying the coil current density limits determined by the analytic model. These judgments are discussed in the next section. The current densities arrived at in this manner appeared to be consistent with industrial experience in the design of electric motors. For both bearings, the current limits predicted by the thermal model were compared to the RMS current required to generate the stated static and harmonic load capacities.

Bearing Design

Given the close spatial limits for the bearings, the design approach was to first find the axially shortest thrust bearing design which would meet the performance requirements. The analytically determined thermal limits were followed fairly closely, with the modification that the peak coil temperature used to determine the maximum allowable current density was 40°C lower than the wire manufacturer's limit. This provided a safety margin which should accomodate localized gradients.

Once the thrust bearing design was determined, the radial bearing was designed to fill the remaining space. In this way, a thermally conservative design was obtained for the radial bearing for which the thermal model is expected to be the least certain.

Choice of the magnetic air gap length for the two bearings was somewhat arbitrary. For both bearings, a gap of 0.5 mm was used, which is twice the tolerable shaft excursion in each direction. This figure reflects a desire to limit actuator nonlinearities due to shaft displacement: a shorter gap would produce a strongly nonlinear bearing, whereas an excessively long gap produces a reduced current-to-force gain resulting in high power losses.

The resulting bearing designs are summarized below.

Thrust Bearing

Rated RMS load	37.5 N
Peak load capacity	273.8 N
Coil current density at RMS load	465 A/cm^2
Maximum safe current density (thermal model)	640 A/cm^2
Coil resistance	3.9 Ω
Coil inductance	6.4 mH
Wire turns per coil	323
Wire gage	26
Bias coil current	0.55 A
Nominal air gap	0.05 cm
Coil thickness (axial)	0.96 cm
Coil thickness (radial)	0.76 cm
Spool Thickness	0.08 cm
Total package length (2 bearings + thrust disk)	3.57 cm
Linearized open loop stiffness	-69600 N/m
Linearized actuator gain	77.8 N/A

Radial Bearing

Rated RMS load	10.8 N

Peak load capacity	45.1 N
Coil current density at RMS load	576 A/cm^2
Maximum safe current density (thermal model)	1270 A/cm^2
Coil resistance (per coil)	1.62 Ω
Coil inductance (per coil)	4.5 mH
Wire turns per coil	189
Wire gage	26
Bias coil current	1.30 A
Nominal air gap	0.05 cm
Journal length	1.00 cm
Journal diameter	3.50 cm
Coil length	1.18 cm
Coil thickness (radial)	0.41 cm
Spool Thickness	0.08 cm
Bearing axial length	1.80 cm
Linearized open loop stiffness	-83600 N/m
Linearized actuator gain	35.7 N/A

The axial length occupied by these two bearings is 5.4 cm out of the total 6.5 cm available. This leaves 1.1 cm for probes, coolant passages, and clearance around the motor windings in the right end bearing envelope.

Power Loss Estimates

There are three main sources of power dissipation which must be considered for magnetic bearings. These are resistive losses in the coils, inefficiency of the coil power supply and power amplifier, and combined eddy current and hysteretic losses in the magnet iron. The first are readily estimated from the expected loads, the actuator gain, and the coil resistances. For the radial bearings operating at rated RMS load, the anticipated power loss would be a total of 49.3 watts. The thrust bearing should dissipate approximately 3.9 watts at rated load.

The inefficiencies of the supporting power electronics provide a multiplicative factor for the coil current losses. For a conventional transformer-capacitor DC power supply and a pulse width modulated power amplifier, the total efficiency should be on the order of 75 percent. This raises the electrical power losses associated with each of the bearings by a factor of 1.33, yielding 65.7 watts for the radial bearings and 5.3 watts for the thrust bearing.

The hysteresis and eddy current losses are difficult to estimate accurately, but an upper bound can be determined by computing a worst case flux variation in the stator and rotor, assume that the fluctuations permeate the magnet iron uniformly, and use manufacturer's data to determine the resulting losses. For the radial bearing journal, the worst case operation would involve complete reversal of the saturation flux density in the rotor at a frequency of four times the highest running speed. This corresponds to a sinusoidal flux variation of amplitude 1.2 Tesla at a frequency of 4800 Hz. For the stator, the maximum amplitude of flux variation should be one half of the saturation density at a frequency equal to the maximum running speed. This corresponds to a sinusoidal flux variation of 0.6 Tesla amplitude at a frequency of 1200 Hz. The total magnet iron mass associated with the radial bearing pole structures is 0.232 kG, while the combined mass of the two journal laminations is 0.073 kG. Given these figures and the manufacturer's data on iron losses for the lamination material, the expected losses are 5.1 watts in the pole pieces and 40.2 watts in the journals.

The total power losses for these bearings are thus conservatively estimated at under 200 watts. This is a reduction of at least 90% from the 2 kW reported with fluid film bearings.

Effect on Shaft Dynamics

One of the most troublesome aspects of applying magnetic bearings to small flexible rotors is that significant rotating mass must be added to the rotor in the form of laminated journals and the thrust disk. This tends to reduce the frequencies of the higher critical speeds and may bring them into the operating range. Figure 4 shows the critical speed map of this rotor prior to addition of the magnetic bearing components. The operating speed range is neatly tucked in between the third and fourth critical speeds, providing an acceptable speed margin at both of the extremes. After adding the magnetic components, Figure 5 shows that the fourth critical speed has been dropped into the upper extreme of the operating range. If the bearing is capable of providing sufficient damping at this speed, then this condition will be acceptable although not desirable. For such a design, rigorous analysis of the unbalance response near the fourth critical is mandated.

Conclusions

The magnetic component design for magnetic bearings in a typical high speed rotor has been presented. Three primary design factors were considered: load capacity, thermal performance, and size. The relations presented for analyzing the design are approximate, but are useful in the preliminary design stage. Any bearing design should ultimately be examined using finite element methods.

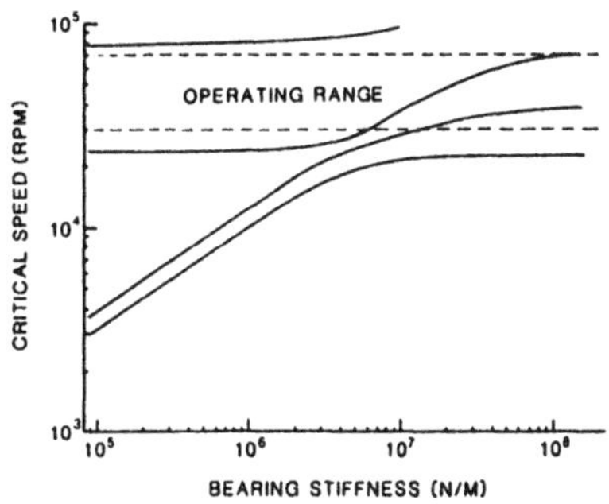

Figure 4 - Critical Speed Map of Original Rotor

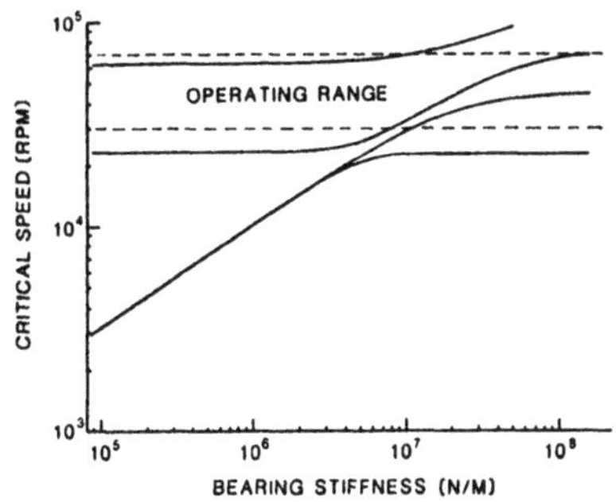

Figure 5 - Critical Speed Map with Magnetic Bearing Components

To date, very few magnetic bearings have been employed in high speed rotors. Much research and design work needs to be done to make the design procedure as routine as it is for more conventional bearings. The development of computer programs using the type of analysis presented in this paper will greatly assist in this process. Advanced finite element packages will also be required and are beginning to emerge. In particular, there is a need for a method of estimating eddy current and hysteretic losses which is more convenient than the detailed analyses currently in the literature, but more accurate than the very crude method desribed in this paper.

References

1. Foster, E. G., V. Kulle, and R. A. Peterson. "The Application of Active Magnetic Bearings to a Natural Gas Pipeline Compressor". ASME paper 86-6T-61.

2. Allaire, P.E., A. Mikula, B. Banerjee, D.W. Lewis, and J. Imlach. "Design and Testing of a Magnetic Thrust Bearing". Proceedings of the NASA Workshop on Magnetic Bearings, Langley, VA, March 1988.

3. Sudo, H. and A. Takahashi. "Rotary Anode X-Ray Tube". Patent No. 4468801, issued Aug. 28,1984. assignee: Toshiba Corp.

4. Maslen, E., P. Hermann, M. Scott, and R. Humphris. "Practical Limits to the Performance of Magnetic Bearings: Peak Force, Slew Rate, and Displacement Sensitivity". Proceedings of the NASA Workshop on Magnetic Bearings, Langley, VA, March 1988.

5. Hebbale, K.V. "A Theoretical Model for the Study of Nonlinear Dynamics of Magnetic Bearings." Doctoral Dissertation, Cornell University, Ithaca, N.Y., USA, January, 1985.

6. Schweitzer, G. and R. Lange. "Characteristics of a Magnetic Rotor Bearing for Active Vibration Control", Technishe Universitat Munchen, Institut fur Mechanik, IMECHE 1976.

7. Smith, S. *Magnetic Components Design and Applications.* Van Nostrand Reinhold, New York, p.172.

8. Kreith, F. *Principles of Heat Transfer.* Harper and Row, New York, 1973. p.487.

Vibration Control of Flexible Rotors

Vibration Reduction in a Multimass Flexible Rotor Using a Midspan Magnetic Damper

P. E. Allaire, Professor
M. E. F. Kasarda, Research Associate
R. R. Humphris, Research Professor
D. W. Lewis, Professor

Department of Mechanical and Aerospace Engineering
University of Virginia
Charlottesville, VA 22901 USA

Summary

The purpose of this study is to investigate the effects of a magnetic damper on a multimass rotor. A small three mass rotor which would simulate the dynamics of a centrally mounted steam turbine or compressor was built and a magnetic damper placed near the rotor midspan. The magnetic damper was constructed of four separate radially arranged silicon iron magnets with coil windings. A proportional and derivative control circuit was employed to control the damper. The rotor was run at speeds from 0 to 10,000 rpm, through three natural rotor-bearing-support system resonances.

Introduction

Rotating machines with long thin shafts often have vibration problems. Modern steam turbines, compressor, and gas turbines operate above the first critical speed in many cases. Large clearances must be maintained in the seals to avoid contact during running through the critical speed. Some high speed machines even operate above two or more critical speeds. The trend is to higher speeds and thinner shafts to improve operating efficiencies so that vibration problems will continue to be present.

A damper placed at the center of such machines could control the high levels of vibration. Until recent years, no damper could operate in the working fluid of the rotating machine because the temperature was too high (steam turbine) or the gas would become contaminated with the oil used in the damper (compressor) unless complicated seals were used.

---

*Partial support for this work was obtained from the U.S. Army Research Office and the Center for Innovative Technology of the Commonwealth of Virginia.

Magnetic dampers can now be used in industrial machines for vibration reduction. Magnetic bearings are being employed in compressors such as natural gas pipeline compressors. Many other papers are available in the literature on applications or research related to magnetic bearings used for support of rotating machines. Due to length restrictions, these will not be discussed in this paper except when they relate directly to the damper employed in this study.

Magnetic dampers have been reported in the literature in only a few cases. Nikolajsen et al. [1] described a magnetic damper used to control vibrations in a long transmission shaft. This pioneering work discussed both the design of the damper and its optimum location along the shaft. Another paper by Gondhalekar et al. [2] gave further developments in this project.

Burrows and Sahinkaya [3,4] applied an electromagnetic damper to a laboratory rotor supported in fluid film bearings. They considered the effects of employing a limited number of measurement sites on controller performance. Vibration reduction of 90% was obtained for the first mode. Wai and Morel [5] used a large midspan damper to control a single mass test rig.

Allaire, et al. [6] presented a paper on the use of a centrally mounted magnetic damper. An experimental test rig was set up with a long thin shaft and several masses to represent a flexible shaft machine. A dramatic reduction in vibration amplitude was obtained at the rotor center with the damper on. Kasarda, et al. [7] reported further work with that damper in controlling vibrations through the rotor first mode. The magnetic damper was placed in three locations: near the midspan, near one end disk, and close to the bearing. With typical control parameter settings, the midspan location reduced the first mode vibration 82%, the disk location reduced it 75% and the bearing location attained a 74% reduction. Magnetic damper stiffness and damping values used to obtain these reductions were only a few percent of the bearing stiffness and damping values. A theoretical model of both the rotor and the damper was developed and compared to the measured results with good agreement.

Some aspects of the control algorithms used in the works [6,7] were discussed by Humphris et al. [8]. The same magnetic bearing/damper configuration was employed in other papers [9,10,11]. All of the above work was carried out with an analog controller as described in [8]. Recently a digital controller has also been developed for the magnetic bearing/damper [12] and successfully rotor test rig runs carried out.

The purpose of this paper is to report on some new magnetic damper research. The damper magnetic component remained the same as [6,7] but a new electronic controller was employed. The initial controller had some amplifiers which would saturate under high speed and amplitude of vibration conditions. Thus it did not operate well above a speed of about 4,000 rpm. The transfer function for the new controller is given in Appendix A. It allowed for damper test rig operation up to 10,000 rpm as well as more independent variation of the damper proportional and derivative gains.

<u>Magnetic Damper</u>

The magnetic damper employed in this study consisted of four electromagnets located radially around the rotor. Figure 1 illustrates the geometry as employed in this study [6,7,8]. It is an active damper, as compared to a passive damper, because the electromagnets have currents continually adjusted by an automatic feedback control circuit. The feedback signal is provided by induction probes which sense the displacement of the shaft [8]. A damper differs from a bearing in that it does not support a steady state load in any direction. Rather, it provides damping, and possibly stiffness, in response to shaft lateral vibrations.

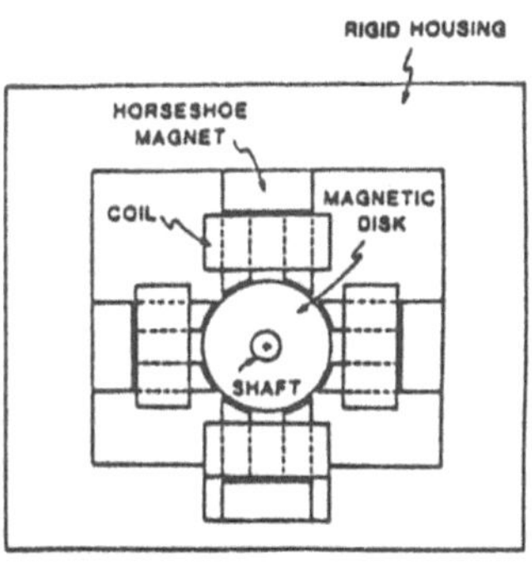

Fig. 1 Magnetic Damper Model

Fig. 2 Pump Bearing

Figure 2 shows a magnetic bearing/damper suitable for use in a rotating machine. It is employed in a canned motor pump at the University of Virginia. The prototype damper shown in Figure 1 used in this study is an older version which was available for this study.

The four electromagnets were located radially around a laminated ferromagnetic disk attached to the rotor shaft. Each magnet core consisted of a silicon-iron horseshoe with pole faces cut to a diameter of 60.5 mm (2.38 inches). The radial clearance was 1.0 mm (0.040 inches). Each leg of the coil was wound with 920 turns of wire. The disk provided a continuous flux path between the magnets and the rotor with a minimum of eddy current losses. All steady state bias currents in the four magnets were the same.

Two position sensing eddy current probes of standard commercial design were placed in the horizontal and vertical directions. They were located axially near the damper for feedback control. Various tests were conducted to insure that there was no coupling between the probes and the magnetic fields from the magnets.

The dynamic properties of the damper are very important to the level of vibration control of the rotor. A discussion of these properties is given in [7]. The two basic control gains in the damper were the proportional gain, denoted as K_g, and the derivative gain, denoted as K_r. It is shown in [7] that the proportional gain is directly related to the effective stiffness coefficient of the damper while the derivative gain is related to the effective damping coefficient. Other studies with this bearing/damper have verified that the stiffness is linear with K_g and that the damping is linear with K_r [8].

The three mass laboratory rotor, shown in Figure 3, was assembled with a 19.0 mm (0.75 inch) shaft supported in conventional sleeve bearings with a bearing span of 660.4 mm (26.0 inches). The total rotor length and weight, including the laminated magnetic disk, was 762 mm (30.0 inches) and 107.2 N (24.1 lb), respectively. Three steel disks were placed four inches apart at the center of the shaft leaving 190.5 mm (7.5 inches) between each outboard disk and the closest support bearing. Each of these three disks was 139.7 mm (5.5 inches) in diameter, 25.4 mm (1.0 inch) in thickness, and weighed 28.48 N (6.4 lbs).

Experimental and Analytical Results vs. Damping

The rotor was operated through a speed range of 0 to 10,000 rpm as shown in Figure 4. Only the vertical probe measurements on the rotor at the

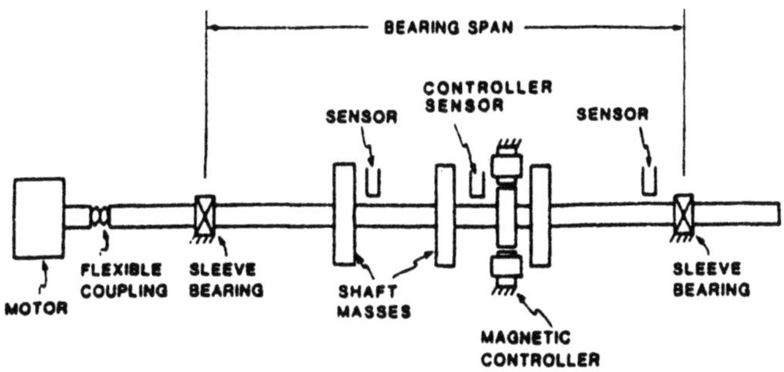

Fig. 3 Schematic of 3-Mass Rotor with Magnetic Damper at Midspan (Not to Scale)

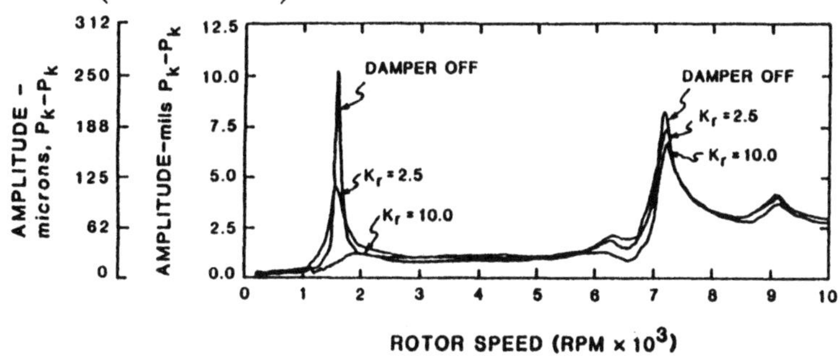

Fig. 4 Experimental Response of 3-Mass Rotor when the Damper is at the Midspan (Effect of Increasing Rate Gain (K_r)) - Response of Vertical Disk One Probe (K_g = 0.0)

left end disk are shown for simplicity. With the damper off, three critical speeds were observed in the rotor. These were at approximately 1600, 7200, and 9000 rpm. An undamped critical speed analysis predicted critical speeds at 1641, 6176, 8465, 8879, and 16004 rpm. The first two matched the observed first and second critical fairly well. The third critical calculated at 8465 rpm was a well damped cantilever mode of the motor which did not appear in the measurements. The undamped fourth critical speed at 8879 rpm corresponded to a pedestal mode measured at 9000 rpm. It was modeled with a support flexibility in the analytical model for the critical speeds. The 16004 rpm mode was above the running speed range.

Bearing properties for this test rig were estimated by comparison with analysis. The bearing stiffness was approximately 2,100,000 N/m (12,000 lb/in) and the bearing damping was estimated at 1100 N-s/m (6.4 lb-s/in).

Damper properties were obtained by the method described in Appendix B. In the speed range of the first critical speed, approximately 1600 rpm, the damper had an estimated stiffness of -9,000 N/m (-50 lb/in) for a proportional gain K_g = 0.0 and a damping of 31 N-s/m (0.18 lb-s/in) for a rate gain K_r = 2.5. An increase in the rate gain setting to K_r = 10 corresponded to a damping of 66 N-s/m (0.38 lb-s/in).

At 1600 rpm, the damper had the effect of reducing the amplitude at the probe from 260 microns (10 mils) pk-pk to 120 microns (4.5 mils) pk-pk with K_r = 2.5. A slight decrease in the critical speed was due to the negative stiffness of the damper. At K_r = 10, the amplitude was reduced further to approximately 32 microns (1.3 mils) pk-pk. This was an overall reduction of 88% over the case with no damper.

For the second rotor critical speed at 7200 rpm, the damper had much less effect. The damper stiffness was estimated at 28,000 N/m (160 lb/in). The estimated damping was 30 N-s/m (0.17 lb-s/in) at K_r = 2.5 and 5.3 N-s/m (0.03 lb-s/in) at K_r = 10. With no damper, the amplitude was 210 microns (8.5 mils) pk-pk. With the damper at K_r = 10, the amplitude was 180 microns (7.4 mils) pk-pk. This was only a 13% overall decrease.

The pedestal mode at 9,000 rpm showed a small reduction as well. At K_r = 10, the overall reduction was about 18%.

An analytical unbalance reponse computer run was carried out for each of the speed ranges near critical speeds. Figure 5 shows the results obtained near the first critical. The overall reduction in amplitude is 81% for K_r = 10 as compared with the measured reduction of 88%. Results

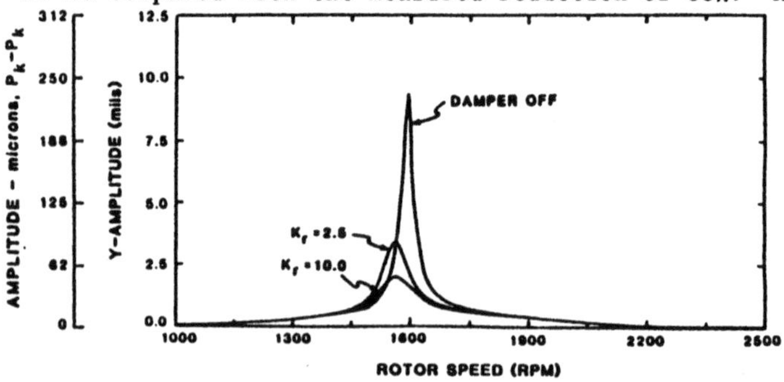

Fig. 5 Forced response predictions when Damper is at the Midspan (Effect of Increasing Rate Gain (K_r)) - First Mode Response at Vertical Disk One Probe (K_g = 0.0)

for other speed ranges were also obtained but the plots are not shown here because of space limitations. Generally the agreement was good when the estimated damper stiffness and damping coefficients were employed.

Experimental Results Vs. Stiffness

The effect of proportional gain K_g for the damper at the midspan is shown in Figure 6. The estimated damper stiffness with K_g = 2.5 was 140,000 N/m (800 lb/in). No estimate of the stiffness was made for K_g = 10. The rate gain was held at K_r = 10.

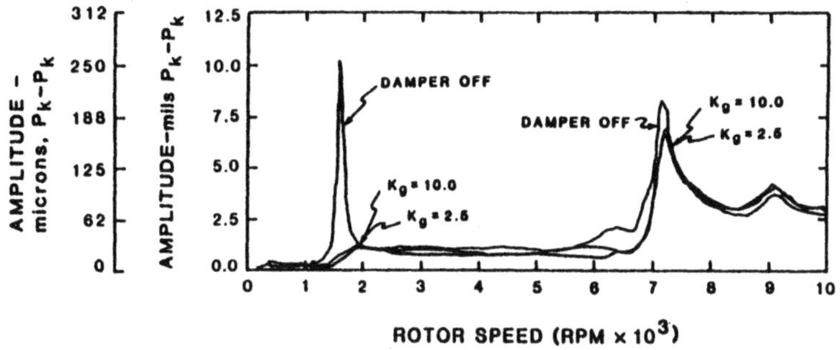

Fig. 6 Experimental Response of 3-Mass Rotor when the Damper is at the Midspan (Effect of Increasing Proportional Gain (K_g)) Response at Vertical Disk One Probe (K_r = 10.0)

The measured result for both K_g cases is overdamped so the exact frequency is difficult to determine. There appears to be approximately a 50 rpm increase for both runs. There is little effect on either the second critical speed or the pedestal mode.

Conclusions

Experimental and analytic studies have been carried out on the forced response characteristics of a three mass rotor with a midspan magnetic damper. Vibration reductions were observed near the left end disk in all cases to be consistent. With an increase in damping as discussed earlier in the paper, the amplitude of vibration was reduced by 88%, 13%, and 8% for the three critical speeds of the rotor. Small shifts in frequency were found with changes in damper stiffness, at the highest damping value.

In all cases presented in this paper, the damper was located near the midspan of the rotor. The first mode of vibration was the largest at the midspan so it was reasonable to expect the largest reduction in amplitude for that mode. The second mode of vibration had a node point near the midspan. Thus it was expected that the vibration reduction would not be as good. Similarly, in the pedestal mode, the shaft and rotor base are moving together. Thus the damper, fixed to the base, did not have much effect on that mode. Other damper locations will be discussed in future papers.

References

1. Nikolajsen, J., Holmes, R., and Gondhalekar, V., "Investigation of an Electromagnetic Damper for Vibration Control of a Transmission Shaft," IMechE, 1979, pp. 331-336.

2. Gondhalekar, V. and Holmes, R., "Design of an Electromagnetic Bearing for the Vibration Control of a Flexible Transmission Shaft," Rotor Dynamics Instability Problems in High Performance Turbomachinery, Texas A&M University, May 1984.

3. Sahinkaya, M. N., and Burrow, C. R., "Control of Stability and the Synchronous Vibration of a Flexible Rotor Supported on Oil-Film Bearings," ASME Journal of Dynamic Systems, Measurement and Controls, Vol. 107, 1985, pp. 139-144.

4. Burrows, C. R., Sahinkaya, M. N., and Clements, S., "Electromagnetic Control of Oil-Film Supported Rotors Using Sparse Measurements," Rotating Machinery Dynamics, ASME Publication, September 27-30, 1987, pp. 127-132.

5. Wai, C. C. H., and Morel, J., "Application of an Active Magnetic Bearing for Vibration Control Near Critical Speeds in Large Rotating Machines," Rotating Machinery Dynamics, ASME Publication, September 27-30, 1987, pp. 151-158.

6. Allaire, P. E., Humphris, R. R., Kasarda, M. E. F., and Koolman, M. I., "Magnetic Bearing/Damper Effects on Unbalance Response of Flexible Rotors," Proc. AIAA Conference, Philadelphia, PA, August 10-14, 1987.

7. Kasarda, M. E. F., Allaire, P. E., Humphris, R. R., and Barrett, L. E., "A Magnetic Damper For First Mode Vibration Reduction in Multimass Flexible Rotors," Workshop on Rotordynamic Problems in High-Performance Turbomachinery, Texas A&M University, May 1988.

8. Humphris, R. R., Kelm, R. D., Lewis, D. W., and Allaire, P. E., "The Effect of Control Algorithms on Magnetic Journal Bearing Properties," Journal of Engineering for Gas Turbines and Power, Trans. ASME, Vol. 108, No. 4, October 1986, pp. 624-632.

9. Allaire, P. E., Lewis, D. W., and Knight, J. D., "Active Vibration Control of a Single Mass Rotor on Flexible Supports," Journal of the Franklin Institute, Vol. 315, 1983, pp. 211-222.

10. Allaire, P. E., Humphris, R. R., and Barrett, L. E., "Critical Speeds and Unbalance Response of a Flexible Rotor in Magnetic Bearings," Proc. of European Turbomachinery Symposium, October 27-28, 1986.

11. Allaire, P. E., Humphris, R. R., and Imlach, J., "Vibration Control of Flexible Rotors with Magnetic Bearing Supports," AFOSR/ARO Conference, Nonlinear Vibrations, Stability, and Dynamics of Structures and Mechanisms, Virginia Tech, Blacksburg, Virginia, March 23-25, 1987.

12. Keith, F. J., Williams, R. D., Allaire, P. E., and Schafer, R. M., "Digital Control of Magnetic Bearings Supporting a Multimass Flexible Rotor," Presented at NASA Conference on Magnetic Suspension Technology, NASA Langley, February 2-4, 1988.

Appendix A - Controller For Active Magnetic Damper

A feedback control diagram of the controller is given in Figure 7. The controller transfer function is

$$G(s) = - K_p (0.1 K_t) \left[\frac{sR_1CK_r}{(1+RCs)(1+R_1C_1s)} \right] \left(\frac{1}{1+\tau s}\right)$$

The nomenclature is the same as that in earlier papers.

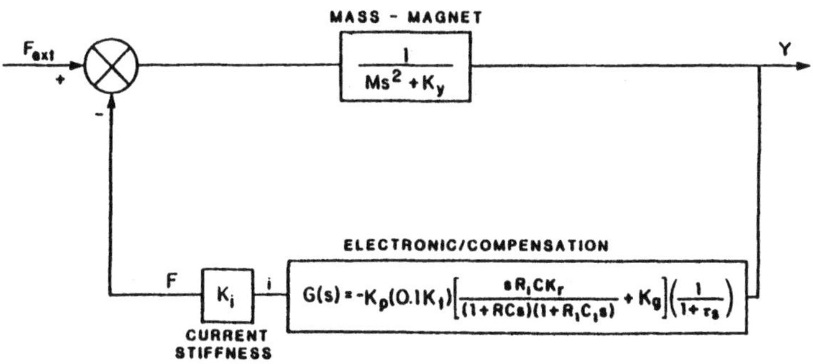

Fig. 7 Diagram of Magnetic Damper System

Appendix B - Determination of Damper Dynamic Properties

Damping and stiffness characteristics of the magnetic damper for various controller settings of proportional gain (K_g) and rate gain (K_r) were determined experimentally by examining the vibration characteristics of a single mass rotor as shown in [7]. The single mass rotor was intentionally unbalanced and the effect of the electromagnetic damper (with various control settings) was compared to an analtyical model of the rotor-bearing system.

The effective stiffness of the damper is determined by the proportional gain (K_g) setting only. The effective damping of the magnetic damper is determined by the rate gain (K_r) setting.

A rotor model of the single mass rotor was established including support bearing characteristics, shaft bow, and unbalance in the rotor. The support bearings had a vertical stiffness of 4,030,000 N/m (23,000 lb/in) and a vertical damping of 10,000 N-sec/m (57 lb-sec/in). This model was used in a forced response program and the predicted response was compared to experimentally measured rotor response. Various values of stiffness and damping were added to the rotor model at the electromagnetic damper location until predicted response (i.e. frequency and amplitude of the first mode response) matched the experimental rotor response to within approximately 5%.

Design and Application of a Magnetic Bearing for Vibration Control and Stabilization of a Flexible Rotor

C.R. BURROWS*, N. SAHINKAYA*, A. TRAXLER+, G. SCHWEITZER+

* School of Mechanical Engineeering, University of Bath, U.K.
+ ETH Zurich, Switzerland

Abstract

This paper describes the design of a magnetic actuator designed to control the synchronous vibration of a 2.3m long 100mm diameter rotor supported on oil-film bearings. An algorithm is described which allows the control force to be computed to minimise the synchronous rotor vibration without prior knowledge of oil-film or rotor characteristics nor of the inherent out-of-balance distribution along the rotor. In this approach the magnetic actuator is first used to apply sequential test signals along two orthogonal axes whilst the shaft is rotating. The corresponding displacements are measured and the algorithm used to determine the control forces to be applied by the actuator. The process is applied sequentially as the rotor is run-up from rest to the required operating speed. Some experimental results are presented to demonstrate the ability of this technique to significantly reduce the synchronous vibration. It is also discussed how the electromagnetic actuator can be used to avoid the instability associated with oil-whirl.

1. Actuator design

Electromagnetic actuators can be used to apply forces to a rotor without contact. These forces may be controlled by an independent external source, or by the rotor vibration. Thus the actuators can be used as a contactless excitation device e.g. for parameter estimation studies, as an active damper, or as an active bearing.

Magnetic actuators can be designed for quite different applications[1]. The objectives of this design are small overall dimensions of the actuator, high force output over a wide frequency range, and inherent capability for the dynamic requirements of vibration control.

Fig. 1 shows the electromagnetic actuator. It consists mainly of a magnet unit (a), power amplifiers with some additional electronic network (b), a rotor sleeve (c) and sensors (d). The inner diameter of the actuator is 190 mm, the gap between rotor sleeve and actuator is 1.0 mm, the length including a gap limiting roller bearing is 150 mm and the maximum force is 1200 N. The displacement sensors are of the eddy-current type. They have a measurement range of two millimetres and measure signals up to 10 kHz. Due to the direct measurement of displacements, the controllability of the displacements in the subsequent vibration control will be optimal.

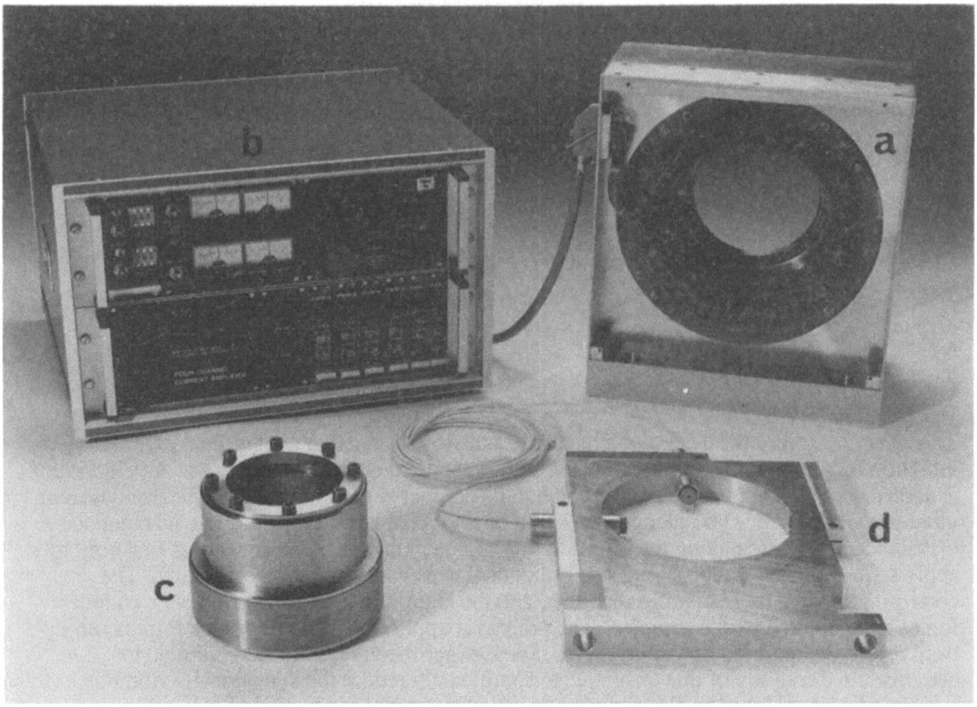

Fig. 1: Magnetic Actuator ETH-190/50/1
a) Magnet unit b) Power amplifier and signal conditioning c) Rotor sleeve d) Sensor unit.

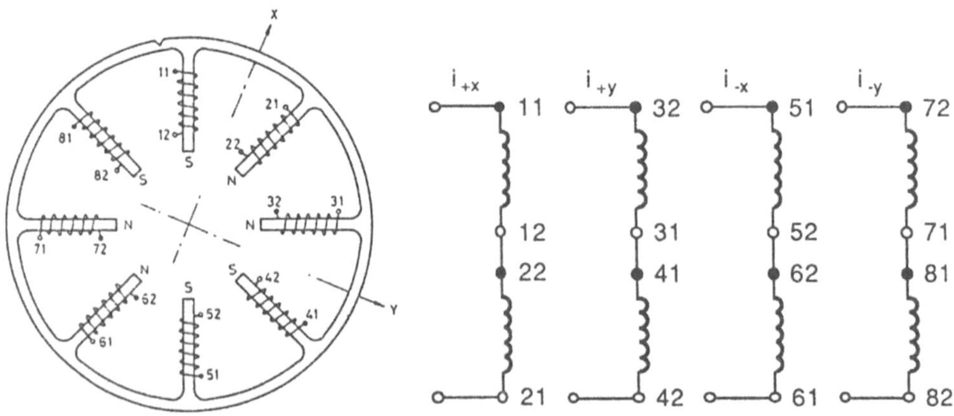

Fig. 2: a) Magnet configuration of the actuator b) Connection-diagram of the coils

The magnet configuration of the actuator is shown in Fig. 2a. The magnet has eight poles, carrying coils. The configuration corresponds to four U-shaped magnets and therefore the coils of each pole-pair are connected together as shown in Fig. 2b. Each pole-pair is energized by one of four power amplifier-channels. The number of coils and amplifiers could be minimized down to three[2], but the magnet configuration with eight poles needs smaller coil-heads and thus less space. And the eight-pole design with its inherent decoupling of the x and the y-axis simplifies the control, and makes it faster.

One of the objectives of the design of an electromagnetic actuator is to make optimum use of the available space to maximise the attainable control force. This is achieved by judicious alocation of the space between that needed by the iron core and that for the windings. The optimisation process has been done with the help of a CAD program.

2. Actuator control

The actuator is designed to give linear input/output characteristics, that is the generated forces are proportional to the input signals to the power amplifiers. This simplifies design of the control system for the vibration control at the expense of slightly more complex hardware. This can be justified on the basis that there is a large body of knowledge to aid with the design of linear control systems but the design of nonlinear systems is still less well-defined. The problem is magnified because of the complexity of the equations needed to model a flexible rotor, especially if it is carried on oil-film bearings which introduce co-ordinate coupling[3].

The nonlinear characteristics of the magnets are linearized, using bias currents and differential currents in corresponding pairs of magnets [1]. An electronic circuit calculates for the two corresponding power amplifiers of the x and the y direction the output currents i+ and i- from the force control signal U_f.

$$i_{+x} = k\,(U_o + U_{fx}); \quad i_{+y} = k\,(U_o + U_{fy}) \tag{1}$$

$$i_{-x} = k\,(U_o - U_{fx}); \quad i_{-y} = k\,(U_o - U_{fy}) \tag{2}$$

Usually, the bias signal U_o is equal to the maximum of the force control signal U_f. Neglecting saturation effects and leakage flux, the generated forces F_x and F_y are proportional to the force control signal U_{fx} and U_{fy}. They depend significantly on the position of the rotor within the magnet unit. Furthermore this linearization method leads to the desired linear force-displacement characteristics of the actuator[1]. The linearized input-output relation of the actuator can be expressed by

$$F_x = k_i\,U_{fx} + k_s\,x \tag{3}$$

$$F_y = k_i\,U_{fy} + k_s\,y \tag{4}$$

where k_i and k_s are hardware dependend constants and x and y are the displacements of the rotor from its zero-position within the magnet. Fig. 3 shows the measured force characteristics of the actuator.

For vibration control the actuator forces should be controllable in a large frequency range.

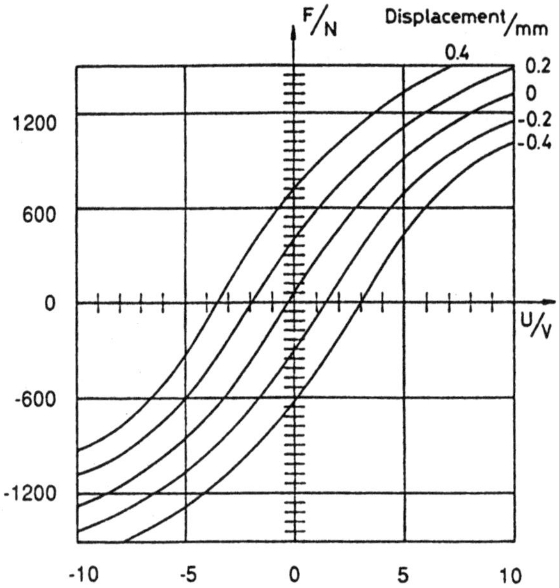

Fig. 3: Measured force characteristics of the actuator.

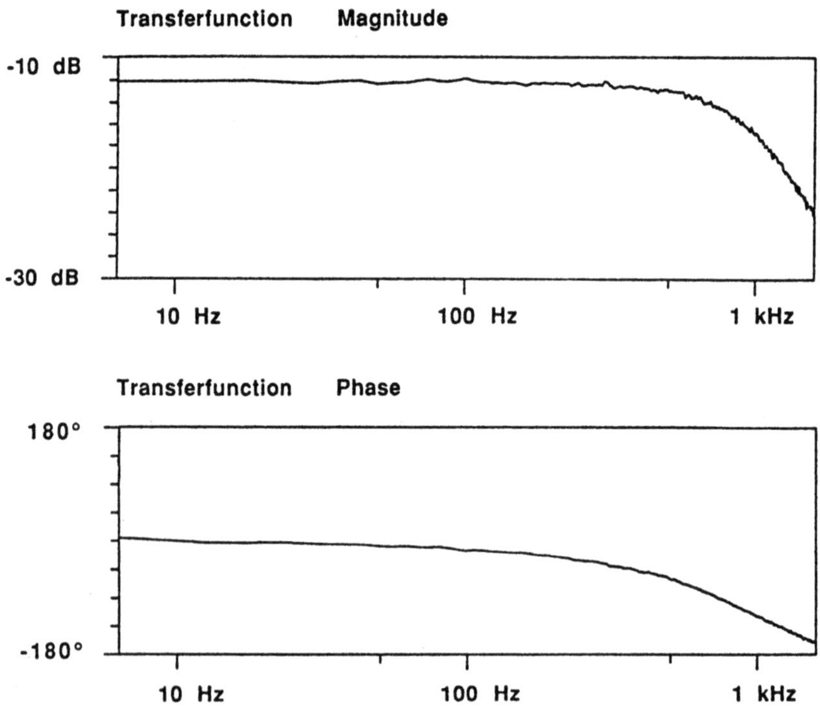

Fig. 4: Measured transfer function: Force output of the actuator / Input signal.

The amplitude of the forces will be limited by the current and the voltage that can be supplied by the power amplifiers. In simple terms: the higher the maximum current output, the higher the attainable force amplitude, and the higher the maximum voltage, the higher the maximum frequency at which this maximum force amplitude can be achieved.

To keep the losses low, the power amplifiers are of a chopper type. They work with pulse-width-modulation at a pulse frequency of 50 kHz. The applied voltage is 310 V and the maximum current of each channel is 2.8 A. With this power, the maximum force amplitude of 1200 N can be applied up to 24 Hz. At frequencies higher than 24 Hz, the maximum force amplitude reduces 20 dB per decade, i.e. the maximum force amplitude at 240 Hz is 120N.

The small signal performance of the actuator is given by the linear transfer function of the electromagnet-power amplifier combination[2]. The measured transfer function is shown in Fig. 4.

3. Rig.

The actuator described above was applied to a 2.3m long 100m diameter steel rotor carried on two oil-film film bearings. These bearings cause coupling between axes [3] as well as introducing the possibility of instability due to oil-whirl. It has been shown elsewhere that the actuator can stabilise the system in addition to reducing the synchronous vibration [4]. The arrangement is shown in Fig. 5 from which it is clear that the rotor carried two large overhanging discs of mass 93kg. each.

The rotor was driven by a 25kW d.c. variable speed motor over a speed range of 0-5000 rev/min.

The journal bearings had a 10 mm circumferential central groove and 35 mm land along either side. The measured diametral clearance was between 0.305 mm and 0.279 mm. The static load per land was 828 N. Shell Tellus oil T15 (average viscosity 0.013 Nsm^{-2} at the average operating temperature) was gravity fed to the grooves from a 180 litre reservoir. No end seals were employed.

Shaft vibration was monitored by eight eddy current proximity probes. The signals were conditioned before being fed to an ADC. The computed control forces were fed back to the actuator through a DAC. These convertors were synchronised by two signals (one pulse/rev and 36 pulse/rev) generated by optical switches actuated by a slotted disc mounted on the drive shaft.

4. Vibration Control Algorithm

The rotor-bearing system can be modelled as a multimass linear system. If $\underline{q}$ represents the complex displacement vector for the measured stations, then the synchronous response to a change $\delta\underline{u}(jw)$ in the control vector $\underline{u}(jw) = (u_x\ u_y)^T$ can be formulated as follows:

$$\underline{q}_- = \underline{q}_o - (\underline{R}_x, \underline{R}_y)\ \delta\underline{u} = \underline{q} - \underline{R}\ \delta\underline{u} \qquad (5)$$

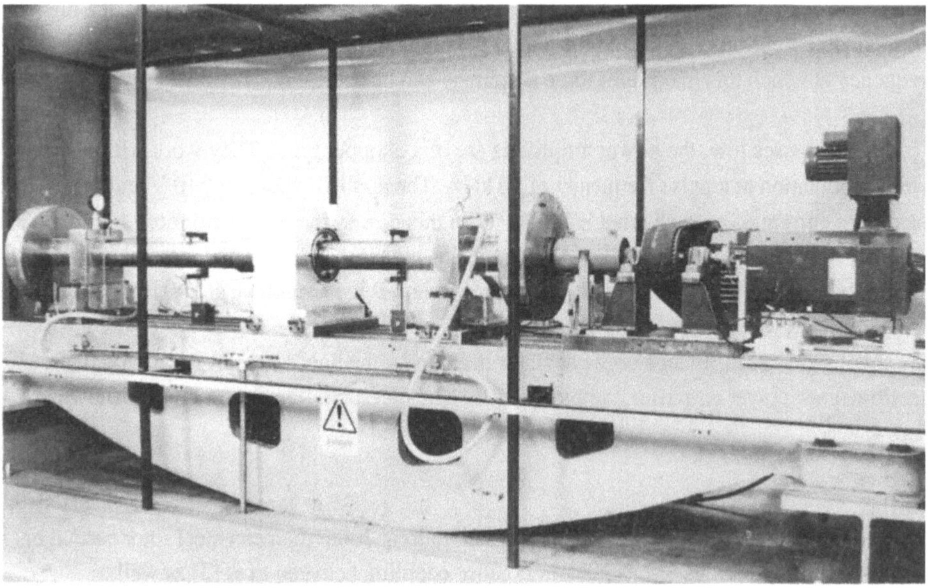

Fig. 5: Test Rig

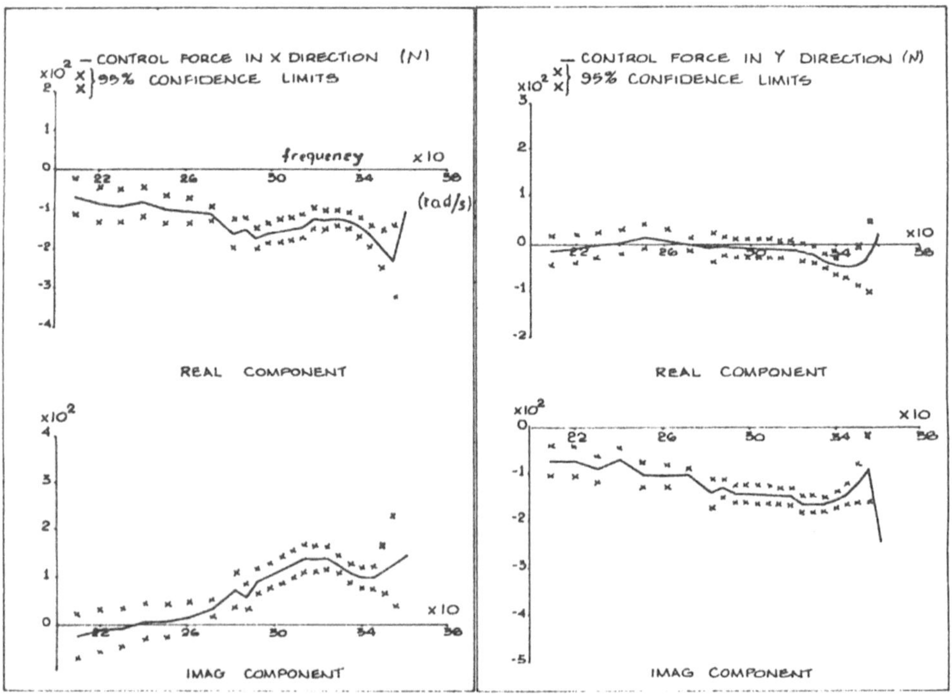

Fig. 6: Control forces calculated on-line

The complex scalars u_x, u_y are the control forces along x and y axes.The change $\delta \underline{u}$ is not constrained to be a small *variation* providing conditions for linearity are not violated. The corresponding complex vectors of influence coefficients are $\underline{R}_x$, $\underline{R}_y$. The displacement vector is $\underline{q}_o$.

Equation (5) can be rewritten as follows

$$\underline{q}_o = (\underline{R}_x, \underline{R}_y)\, \delta \underline{u} + \underline{q} \tag{6}$$

Then, the Least Square estimation of $\delta \underline{u}$ to minimize the performance index $(\bar{\underline{q}}^T \underline{q})$ is as follows

$$\delta \hat{\underline{u}} = (\bar{\underline{R}}^T \underline{R})^{-1}\, \bar{\underline{R}}^T \underline{q}_o \tag{7}$$

where $(\bar{\underline{R}})$ represents the complex conjugate, and T denotes the transpose. The above optimum force expression depends on the measured response and the corresponding influence coefficients of the system. The difficulty of assigning reliable theoretical values to the elements of $\underline{R}(jw)$ associated with the oil-film bearing coefficients is well known. This problem is overcome by estimating the $\underline{R}$ matrix on-line.

The overall control is achieved by the help of a digital computer. Pulses are generated once and 36 times per revolution to synchronise the input(displacement) and output(force) to and from the computer. The overall control algorithm can be summarised as follows

(i) sample displacements and perform FFT to construct $\underline{q}_o\,(jw)$

(ii) set $\delta \underline{u} = \begin{pmatrix} 0.1\, u_x \\ 0 \end{pmatrix}$, sample displacements and perform an FFT to construct $\underline{q}_{o,1}\,(jw)$ and calculate $\underline{R}_x$ as follows

$$\underline{R}_x = (\underline{q}_{o,1} - \underline{q}_o)/(0.1\, u_x)$$

(iii) set $\delta \underline{u} = \begin{pmatrix} 0 \\ 0.1\, u_y \end{pmatrix}$, sample displacements and perform an FFT to construct $\underline{q}_{o,2}(jw)$ and estimate $\underline{R}_y$ as follows

$$\underline{R}_y = (\underline{q}_{o,2} - \underline{q}_o)/(0.1\, u_y)$$

(iv) estimate optimum control force adjustments from equation (7)

(v) check the statistical data on $\delta \hat{\underline{u}}$ (goodness of fit and standard deviation) to assess its validity.

If not suitable set $\delta \underline{u} = \begin{pmatrix} 0 \\ 0 \end{pmatrix}$ and go to (i)

If suitable set $\delta \underline{u} = \delta \hat{\underline{u}}$

(vi) monitor speed and displacements. Whenever updating is required go to (i)

It is important to note that a background program ensures that the synchronous control forces are continuously fed to the magnetic actuator independent of the speed of the above control algorithm. (i.e. the previous value of $\underline{u}$ is retained until a change is required). Initially, typically at low speed, $\delta \underline{u}$ in (ii) and (iii) is set at an arbitrary value. This is updated after the first implementation of the above algorithm. Since the force is applied in open loop there is no danger of causing instability. The statistical data obtained from the Least Square estimator is discussed in detail in Reference [3].

5. Results

The actuator is able to apply forces along the horizontal and vertical axes. The control algorithm determines the correct amplitude and phase of each force to minimise the performance index. It is stressed that this is done without knowledge of the bearing and rotor parameters nor of the unbalance distribution. Some typical optimum control forces obtained experimentally are shown in Fig. 6. This figure also includes confidence bounds generated by the algorithm. The corresponding controlled synchronous response is compared with the uncontrolled response in Figure 7. This shows the amplitude response at two different sections along the rotor. The significant reduction in vibration due to the magnetic actuator is apparent. The results are obtained by use of horizontal and vertical displacement measurements at three rotor stations. The consumed power by the magnetic actuator to provide the reduction in amplitude was typically 250W. It is significant to note that unlike many results presented in the literature, the controlled response is compared with the damped response of the rotor and not the undamped vibration. There is a ninety per cent reduction in amplitude at the critical speed. It has been shown also [5] that the reduction in synchronous response could not be achieved using *local* closed-loop control because that would require negative stiffness and damping values which would lead to instability.

Although oil-whirl did not occur it can be shown that feedback can be used to increase the stable operating range(before minimising the synchronous vibration). This leads to the idea of combined open loop and closed loop control using one actuator to both improve stability and reduce vibration.

6. Conclusions

The paper describes the design of a magnetic actuator for use with a 100 mm diameter and 2.3 m long rotor suported on oil-film bearings. A control algorithm has been implemented for in-situ tuning of the magnetic actuator to reduce synchronous vibration. The control has been achieved without prior knowledge of any system parameters or unbalance distribution. This is particularly significant because oil-film bearing parameters cannot be accurately predicted from theoretical modelling.

The approach described here allows automatic on-line tuning of the actuator to the system. Monitoring the actuator signals provides a facility for condition monitoring to detect changes in system characteristics associated with the rotor or bearings.

Acknowledgements

The results presented here were obtained by Dr. S Clements, as part of his Doctoral Studies in the University of Strathclyde, under supervision of C R Burrows and M N Sahinkaya. This work was supported by SERC.

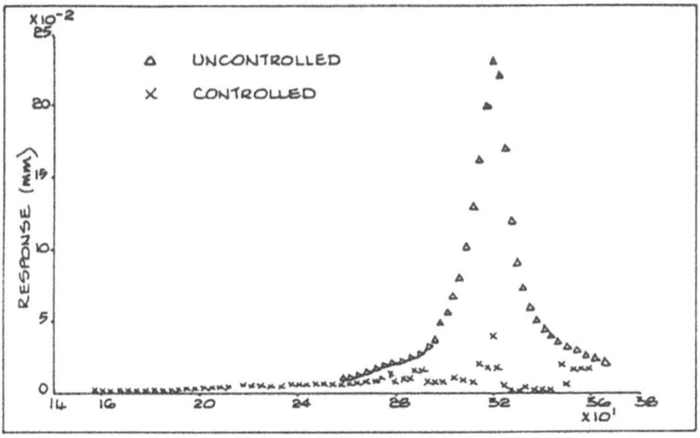

Fig. 7: a) Response at the rotor station (a)

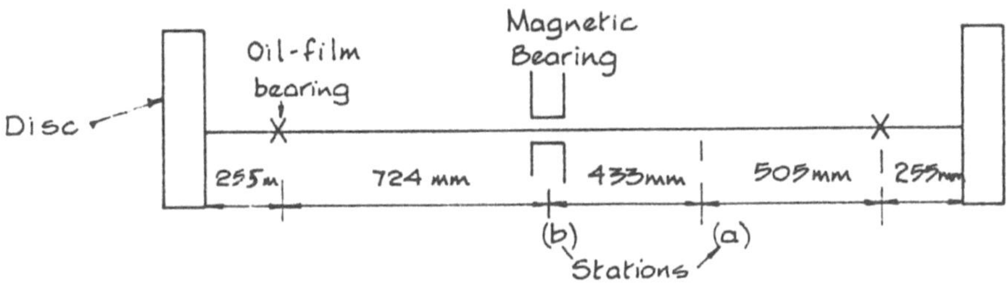

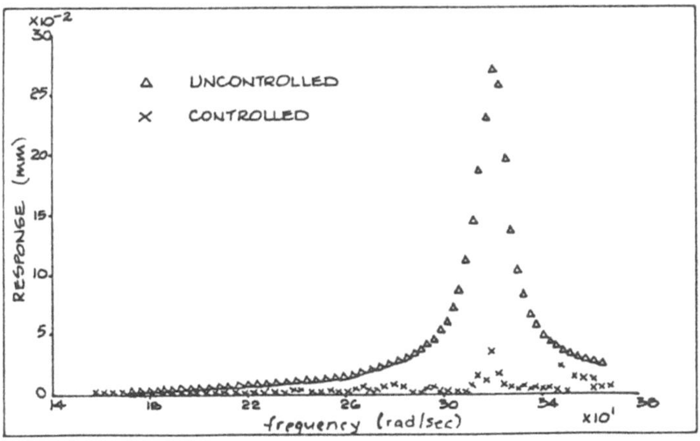

Fig. 7: b) Response at the rotor station (b)

References

1. SCHWEITZER, G & TRAXLER, A.
Design of Magnetic Bearings,
International Symposium on Design and Synthesis, The Japanese Society of Precision Engineers, Tokyo, July 1984.

2. BRADFIELD, C D, ROBERTS J B & KARUNENDIRAN, P.
Performance of an electromagnetic bearing for the vibration control of a supercritical shaft. Proc. Instn. Mech. Engrs., Vol. 201, NoC3, I MechE, 1987.

3. BURROWS, C R & SAHINKAYA, M N.
Vibration control of multi-mode rotor-bearing systems, Proc. Royal Soc., Lon. A386, 1983, p77-94.

4. SAHINKAYA, M N & BURROWS, C R.
Control of stability and the synchronous vibration of a flexible rotor supported on oil-film bearings, Trans. ASME, JDSMC, Vol. 107, 1985, p139-144.

5. BURROWS, C R & SAHINKAYA, M N.
Control strategies for use with magnetic bearings, 4th International Conference on Vibrations in Rotating Machinery, I. Mech. E, September 1988, Edinburgh, Scotland.

Vibration and Control in Outer Rotor Type Magnetic Bearings

Ichiju SATOH*1, Jinzo NAGAHIRO*2
Chikara MURAKAMI*3 and Atsushi NAKAJIMA*4

*1 Graduate Student of Tokyo Denki University
*2 Tokyo Denki University
*3 Tokyo Metropolitan Institute of Technology
*4 National Aerospace Laboratory

Abstract

The magnetic bearings described here has been developed for high speed rotating machinery. The manufactured experimental model is 4-axis-active type with simple structure. Since the model is so called outer-rotor type, resonance frequency of the rotor could set higher than rotating frequency, and only first resonance frequency of the stator shaft is in the rotating region. In this paper, we describe the analysis and experimental results to obtain stable high speed rotation. Firstly, stability of the gyroscopic motion of rigid mode is examined at low speed rotation. Subsequently, influence of flexibility of the stator shaft on stability is discussed. Then we indicate that nonlinearity of the power amplifier plays an important role in stability of the system. Finally, a suppression method using notch-filters and inter-axis cross coupling feedback at relatively high speed rotation is discussed.

1. Introduction

The 4-axis-active control magnetic bearings is composed of two radially-controlled magnetic bearings which utilized permanent magnets. As the results, the rotor is passively stabilized by the restoring force of the magnets in the thrust axis direction. Since no active thrust bearing is used, this model has a simple structure and easy to assemble compared to five-axis control type. The permanent magnets enable virtually zero power(VZP) method, which can levitate rotors with zero DC power and also can compensate the sensor drift.

Thus, the model has several advantages, so we have been researching the model in order to realize the practical use. The most difficult problem is control in high-speed rotating region. In this paper, we describe analysis and experimental results which are performed to investigate and to suppress the whirling motions of the model. Mainly, we deal with the gyroscopic and flexible stator shaft problems which must be resolved to obtain stable high speed rotation.

2. 4-Axis Active Control Magnetic Bearings

2.1 Structure and Static Characteristics of the Model

The model, shown in Fig.1, is composed of two radially-active type magnetic bearings which are located on both ends of the high speed motor. This model is so called outer rotor type and has an oblong shape (moment of inertia ratio = 0.504). Axial stiffness of the bearings keeps the rotor 0.3mm drop from geometrically symmetric point due to the gravity force when the rotating axis is vertical. These static characteristics of the model are shown in table.1.

Table.1 Static characteristics

Rotor Mass	$M = 5.7$	[kg]
Rotor Inertia		
Spin Axis	$I_p = 15.9\times10^{-3}$	[kg・m²]
Radial Axis	$I_d = 31.5\times10^{-3}$	[kg・m²]
Ratio	$\gamma = I_p / I_d = 0.504$	
Axial Stiffness	$K_z = 2.46\times10^{5}$	[N/m]
Radial Unbalance Stiffness	$K_u = 12.7\times10^{5}$	[N/m]

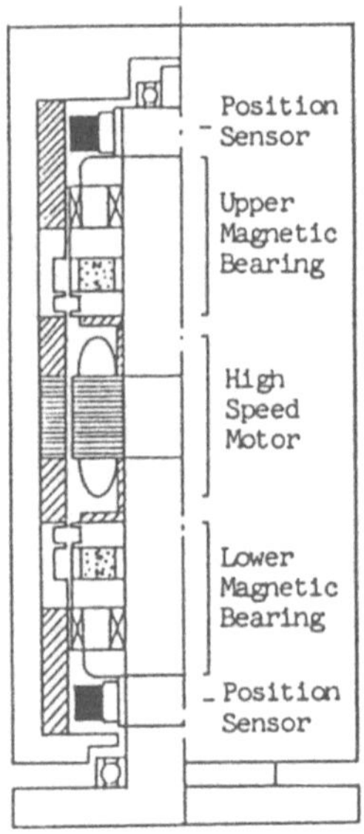

Fig.1 Structure of the model.

2.2 Block diagram of the system

The motion of the rigid rotor which rotate at ω is separated into translational and rotational motion. A block diagram of the system including gyroscopic motion is shown in Fig.2, where, l_1 and l_2 are the distance from the top and the bottom sensors and L_1 and L_2 are the ones from the top and bottom actuators, to the center of the gravity, respectively. g_{x1}, g_{x2}, g_{y1} and g_{y2} are distances of the rotor from neutral position of the gap for each axis, respectively. F_{cx1}, F_{cx2}, F_{cy1} and F_{cy2} are the control force of actuators for each axis, respectively. H(s) represents a transfer function of each control circuit.

2.3 Control circuit

The independent control method, which does not separate the motion into translational and rotational one, was adopted for simplicity in structure. Every axis has the same compensator, which consists of a lead-lag element and a first order lag power amplifier positively fedback by an integrator for VZP.

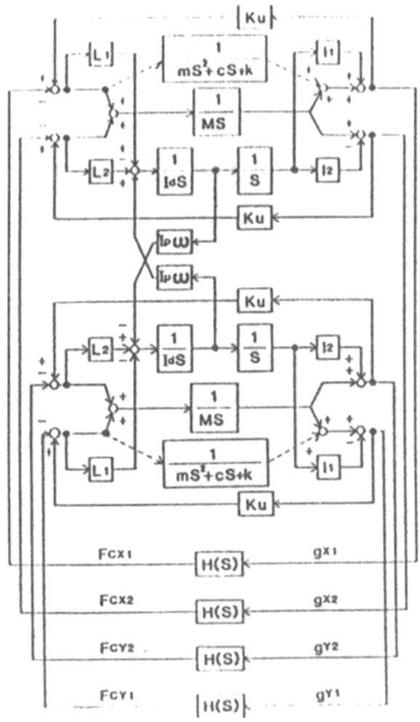

Fig.2 Block diagram of the system.

3. Rigid mode

3.1 Stability of conical motion

Though the system has unstable poles caused by control force of the permanent magnets and VZP, adjustment of loop gain of every axis stabilize these roots which start from the unstable poles. When rotation of the rotor is taken into consideration, the roots move due to the gyroscopic effect. Root-loci of the conical motion for variations of rotational speed are shown in Fig.3. The characteristic roots of precession mode(long period mode) move into the unstable region at approx. 20,000 rpm. This instability is caused by the phase lag at low frequency of the control circuit on account of VZP.

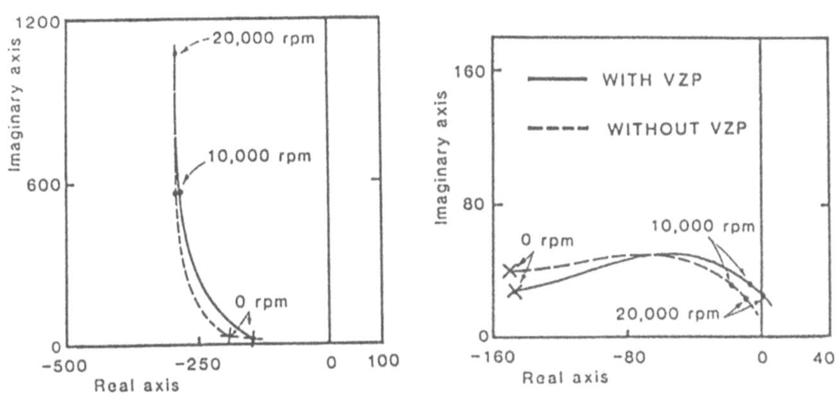

(a) Nutation (b) Precession

Fig.3 Root-loci of the conical motion (influence of VZP).

3.2 Modification of control circuit

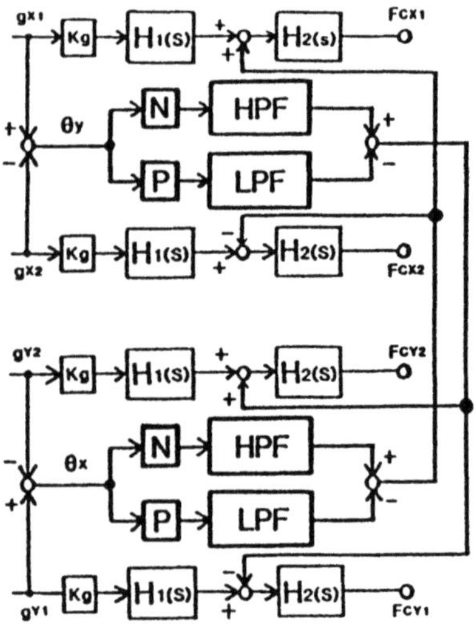

$$H_1(s) = \frac{1+T_1 s}{1+T_2 s}$$

$$H_2(s) = \frac{K_c T_i s}{T_c T_i s^2 + T_i s - K_c K_i}$$

Fig.4 Control circuit including cross-feedback.

In rigid mode with restoring control force, there are two conical or whirling motions i.e. precession(long period, backward) and nutation(short period, forward) motion [1], which should be suppressed by adequate control circuits. If we can get rate information of the rotor, the control circuit becomes rather simple, but we have no rate sensor in the model. Instead of the rate signals, we can use inter-axis cross coupling feedback(cross-feedback) method for that, provided that rotational directions of the coning motion are known. Fortunately, as each mode has different frequency at high rotation speed, we can separate each mode by simple filters. A low-pass-filter(LPF) is used for precession control(P-circuit) and a high-pass-filter(HPF) is used for nutation control(N-circuit). The separated signals can be mixed in opposite sign and cross-fedback. A control circuit including the cross-feedback is shown in Fig.4. As analytical results show unstable precession at high rotation speed, P-circuit is applied to suppress it. The root-loci of

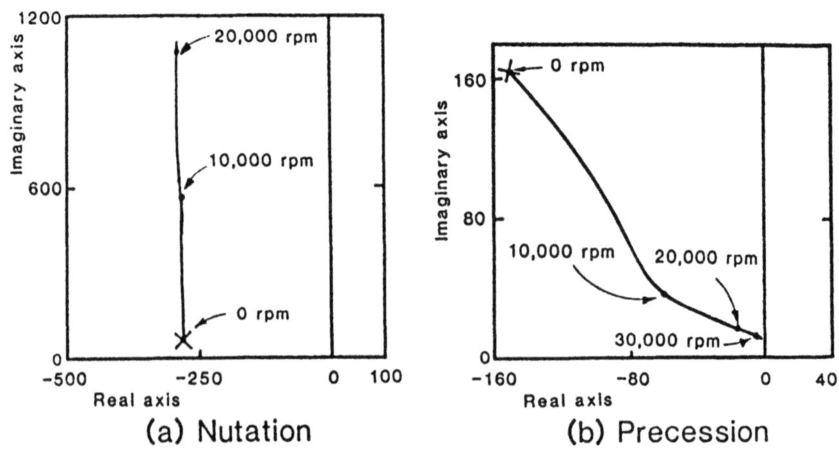

Fig.5 Root-loci of the conical motion (effect of P-circuit).

conical motion in case of cross-feedback of P-circuit are shown in Fig.5, where the precession mode is clearly improved.

4. Experiments

In experiments without P-circuit, amplitude increase of the precession at about 10,000 rpm was observed. By insertion of the cross-feedback of P-circuit, problem of the precession is cleared. However, when the rotation frequency approaches to a resonance frequency of the stator shaft(about 530 Hz), the system became unstable.

5. Flexible mode

As the model is an outer-rotor type, resonance frequency of the rotor could be set at rather high value. But the first bending mode frequency of the stator shaft is in the rotational region.

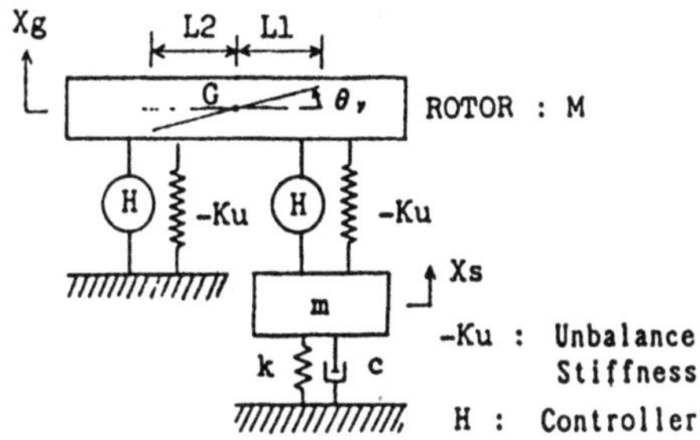

Fig.6 Structural model.

5.1 Structural model

As the stator shaft has a cantilever shape, the top part swings large, whereas the bottom little. Therefore, we adopt a structural model shown in Fig.6, where the flexibility is considered only at the upper side magnetic bearings. In block diagram of Fig.2, the flexibility is shown by dotted line.

5.2 Analysis of characteristic roots

The root-loci of the system is shown in Fig.7, where the feedback gain K_g of all compensator is varied at a time. Since the damping ratio of the flexible mode characteristic roots at operating point(△ mark in Fig.7) is not so small, we don't think the flexible mode alone is a cause of

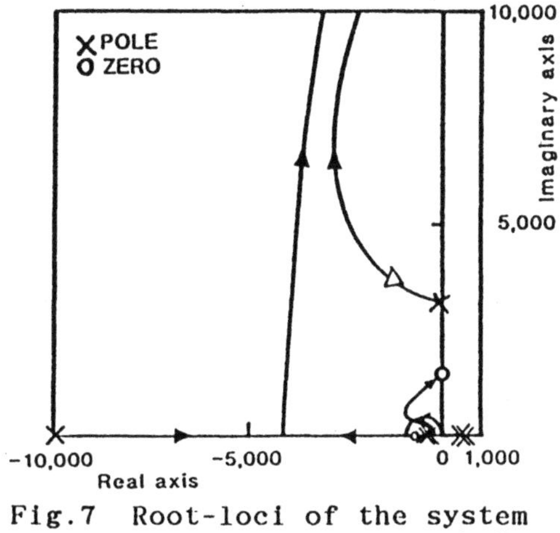

Fig.7 Root-loci of the system with flexible stator shaft.

the instability.

5.3 Nonlinear characteristics of the power amplifier

Transfer functions of power amplifiers have been expressed by linear lag elements so far. However, the results of detailed experiments about the system clarified the nonlinear characteristics of the power amplifier. The circuit of the power amplifier used in the system is shown in Fig. 8.

It has such nonlinearity as its phase characteristic is getting worse in the case where high frequency with large amplitude signals are applied. Fig.9 shows the experimental results of frequency response of the power amplifier for variations of input amplitudes. As is clear from the Fig.9, the larger the amplitude of input signals become, the worse the phase characteristics of the power amplifier become.

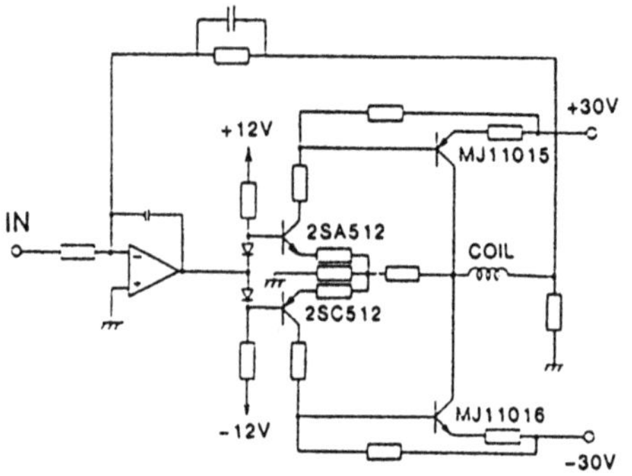

Fig.8 Circuit diagram of the power amplifier.

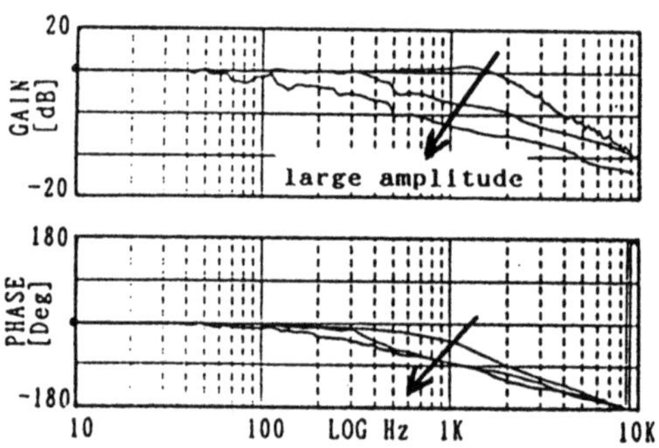

Fig.9 Frequency response of the power amplifier (experiments).

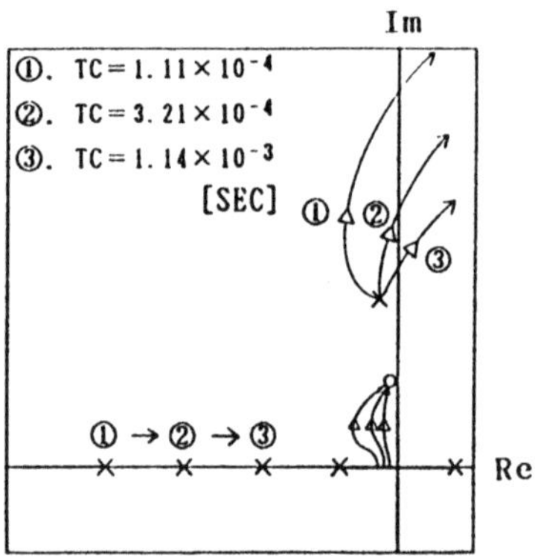

Fig.10 Variation of root-loci to TC.

In the case, where the frequency characteristics of the power amplifiers are approximated to a first order lag element, the variations of root-loci to three step changes of time constant(TC) of the element are shown in Fig.10. When TC is small(①), the flexible mode has wide stable region. But when TC becomes large(③), the roots of flexible mode move quickly to unstable region.

5.4 Stabilization of stator shaft vibration

The system is stable unless the resonance signals of the shaft are applied to the power amplifiers. The band-elimination-filters(BEFs or notch filters) are adopted as a compensator. According to the mode shape of the stator shaft, the BEFs are inserted in the detection circuits of upper bearings only. Fig.11 shows the root-loci of the system when the flexible mode is compensated by BEFs. The power amplifiers keep small time constants because of the BEFs, and the poles of flexible mode construct dipoles. The new roots, starting from the poles of the BEFs, have rather good damping ratio.

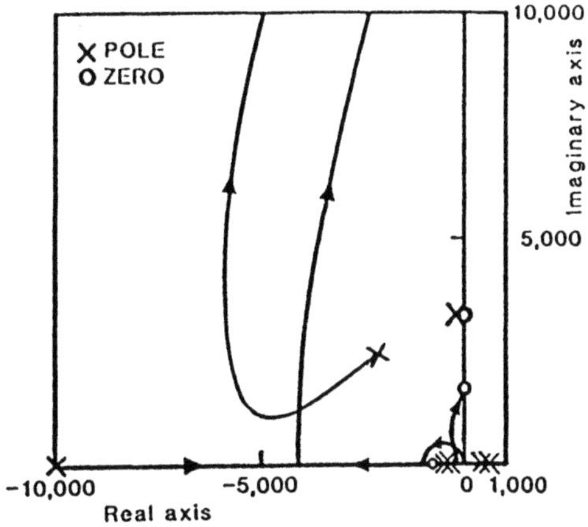

Fig.11 Root-loci of the system with BEF.

5.5 Influence of the BEFs on the conical motion

The root-loci of the conical motion of the system including P-circuit with and without BEFs for variations of rotation speed are shown in Fig.12. Influence of BEFs on precession mode is only little but the one on the damping ratio of the nutation mode is rather large.

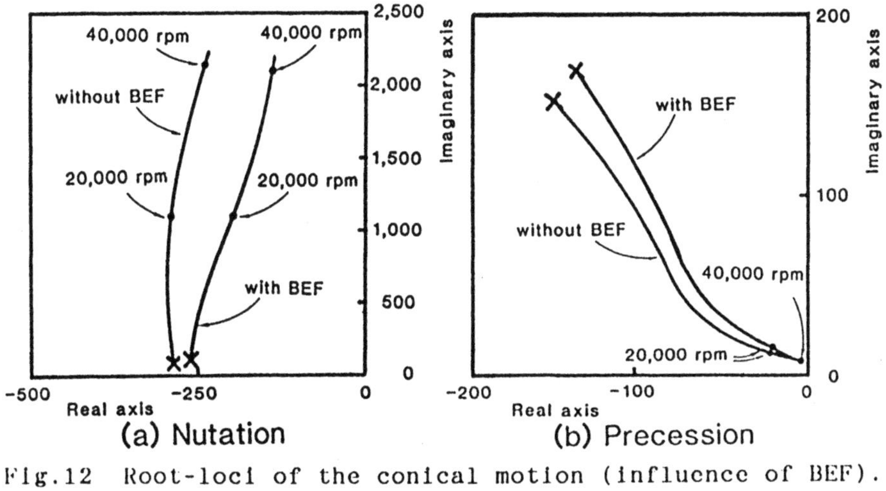

Fig.12 Root-loci of the conical motion (influence of BEF).

5.6 Suppression of the nutation mode

In addition to the P-circuit, N-circuit is applied to compensate the characteristic root of the nutation mode, and the second order HPF were adopted to the N-circuit. Fig.13 is simulation results of upper side whirling motion, and we can see good convergence of the nutation mode but small flexible mode remains. The flexible mode is decaying by its own damping.

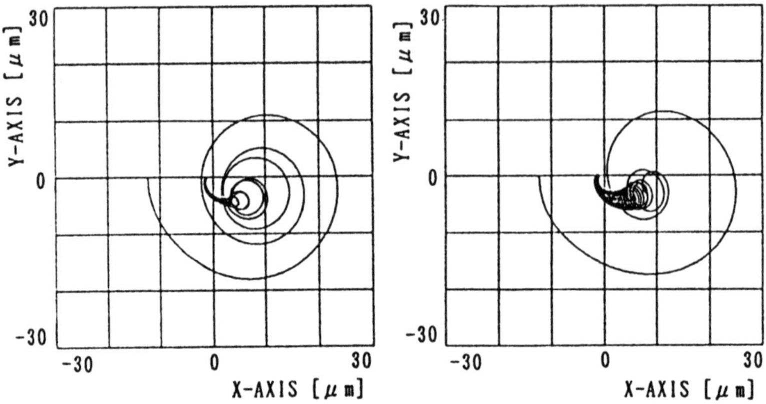

Fig.13 Transient response of relative motion of rotor and flexible stator shaft.

6. Conclusions

The results of this research are summarized as follows.

(1). In rigid mode, virtually zero power method cause unstable precession mode, which could be stabilized by cross-feedback of low-pass-filtered signals (P-circuit).

(2). The system falls into unstable due to interaction between flexible mode of the stator shaft and nonlinearity of the power amplifiers.

(3). Though notch filter(BEF) can eliminate nonlinearity of the power amplifiers, but may give bad influence on nutation mode.

(4). Cross-feedback of high-pass-filtered signals(N-circuit) can increase damping of nutation mode. Small flexible mode remains, but decays by its own damping.

Reference

[1].C.Murakami and I.Satoh : "A Suppression Method of Conical Motion of an Axi-symmetrical Spinning Rigid Rotor Suspended by Magnetic Bearings" , First International Symposium on Magnetic Bearings, June 6-8, 1988, ZURICH.Switzerland.

Vibration and Control of Flexible Rotor Supported by Magnetic Bearings

National Aeronautics and Space Administration
Lewis Research Center
Cleveland Ohio 44135

1. INTRODUCTION

Many papers have been written on magnetic bearings. These papers are divided into two categories: One is the use of magnetic bearings taking the place of contact bearings, and the other is the use of noncontact actuators for active vibration control. These categories are combined in applications. The research on using magnetic bearings in place of contact bearings has been going on for years. As early as 1842 it was known that magnetic bearings should have at least one direction actively controlled [1]. For the last century and a half there has been a great deal of research and development. Magnetic bearings have been used mainly in experiments and in aircraft and missile inertial guidance control. However, these bearings are now coming into use in the industrial world [2] for turbo-molecular pumps [3], machine tool spindles, etc. Although these magnetic bearings are for rigid rotors [4 to 6], they recently have been used with flexible rotors [7 to 9]. In such applications, the magnetic bearings suppress unbalance vibration. Many papers have been published on this subject [10 to 16].

This paper is concerned with active vibration control of flexible rotors supported by magnetic bearings. Using a finite-element method for the mathematical model of a flexible rotor, the author has formulated the eigenvalue problem taking into account the interaction between the mechanical system of a flexible rotor and the electrical system of magnetic bearings and a controller. However, for the sake of simplicity, gyroscopic effects are disregarded herein. It is possible to adapt this formulation to a general flexible rotor - magnetic bearing system. Controllability with and without collocation sensors and actuators located

*National Research Council - NASA Research Associate; on leave from Chiba University, 1-33 Yayoi-cho, Chiba 260 Japan.

at the same distance along the rotor axis is discussed for the higher order flexible modes of the test rig. In conclusion, the author has proposed that it is necessary to add new active control loops for the higher flexible modes even in the case of collocation. It is then possible to stabilize for the case of uncollocation by means of this method.

2. Analysis of Rotor - Magnetic Bearings System for Flexible Rotor [17]

The mathematical model for this flexible rotor is given by using a finite element method:

$$[M]\{\ddot{q}\} + [G]\{\dot{q}\} + [K]\{q\} = \{0\} \tag{1}$$

where [M], [G], and [K] are the mass, gyroscopic, and stiffness matrices, respectively. We assume [G] = 0 for simplicity. Therefore, the X and Y directions are uncoupled. This flexible rotor is now divided into six elements as shown in Fig. 1, and the state values are set as x_i and θ_i at each station. [M], and [K] become 14x14 matrices and {X} vector (instead of {q} vector) of length 14 in the X direction. The flexible rotor is supported by magnetic bearings at stations 2 and 6 and sensors are located at stations 1 and 7. From these results, the formulations in the horizontal direction of the left magnetic bearing shown in Fig. 1 are defined by

$$\left.\begin{aligned}
u_{x\ell} &= \frac{4P_\ell}{H_\ell} x_2 - \frac{4P_\ell}{I_\ell} i_{x\ell} \\
E_{1x\ell} - V_{x\ell} &= L\frac{di_{x\ell}}{dt} + Ri_{x\ell} \\
\dot{V}_{x\ell} &= \frac{E_{1x\ell}}{CR_1} - \frac{R_1 + R_2}{CR_1R_2} V_{x\ell} \\
E_{1x\ell} &= -f_{1\ell}x_1 - f_{2\ell}\dot{x}_1 - f_{3\ell}\alpha_\ell
\end{aligned}\right\} \tag{2}^*$$

where the force $u_{x\ell}$ acts on station 2, and the subscript ℓ means the left magnetic bearing. The steady-state values p_0, h_0, and i_0 are replaced by P_ℓ, H_ℓ, and I_ℓ. The expressions are derived about the right magnetic bearing in the same manner. Combining Eqs. (1) and (2), we can obtain the 34th order state equation as shown in Eq. (3):

$$
\frac{d}{dt}\begin{bmatrix} \{X\} \\ \{\dot{X}\} \\ \hline i_{x\ell} \\ \alpha_{x\ell} \\ v_{x\ell} \\ \hline i_{xr} \\ \alpha_{xr} \\ v_{xr} \end{bmatrix}
=
\left[\begin{array}{cc:ccc:ccc}
[0] & [I] & 0 & 0 & 0 & 0 & 0 & 0 \\
-[M]^{-1}[K] + \overset{*1}{\boxed{\frac{4P_\ell}{H_\ell}}} + \overset{*6}{\boxed{\frac{4P_r}{H_r}}} & [0] & -\overset{*2}{\boxed{\frac{4P_\ell}{I_\ell}}} & 0 & 0 & -\overset{*7}{\boxed{\frac{4P_r}{I_r}}} & 0 & 0 \\
\hdashline
-\overset{*4}{\boxed{\frac{f_{1\ell}}{L}}} - \overset{*15}{\boxed{\frac{f'_{1\ell}}{L}}} & -\overset{*5}{\boxed{\frac{f_{2\ell}}{L}}} - \overset{*16}{\boxed{\frac{f'_{2\ell}}{L}}} & -\frac{R}{L} & -\frac{f_{3\ell}}{L} & -\frac{1}{L} & & & \\
\overset{*3}{\boxed{1}} & [0] & 0 & 0 & 0 & & [0] & \\
-\overset{*11}{\boxed{\frac{f_{1\ell}}{CR_1}}} - \overset{*17}{\boxed{\frac{f'_{1\ell}}{CR_1}}} & -\overset{*12}{\boxed{\frac{f_{2\ell}}{CR_1}}} - \overset{*18}{\boxed{\frac{f'_{2\ell}}{CR_1}}} & 0 & -\frac{f_{3\ell}}{CR_1} & -\frac{R_1+R_2}{CR_1R_2} & & & \\
\hdashline
-\overset{*9}{\boxed{\frac{f_{1r}}{L}}} - \overset{*19}{\boxed{\frac{f'_{1r}}{L}}} & -\overset{*10}{\boxed{\frac{f_{2r}}{L}}} - \overset{*20}{\boxed{\frac{f'_{2r}}{L}}} & & & & -\frac{R}{L} & -\frac{f_{3r}}{L} & -\frac{1}{L} \\
\overset{*8}{\boxed{1}} & [0] & & [0] & & 0 & 0 & 0 \\
-\overset{*13}{\boxed{\frac{f_{1r}}{CR_1}}} - \overset{*21}{\boxed{\frac{f'_{1r}}{CR_1}}} & -\overset{*14}{\boxed{\frac{f_{2r}}{CR_1}}} - \overset{*22}{\boxed{\frac{f'_{2r}}{CR_1}}} & & & & 0 & -\frac{f_{3r}}{CR_1} & -\frac{R_1+R_2}{CR_1R_2}
\end{array}\right]
\begin{bmatrix} \{X\} \\ \{\dot{X}\} \\ \hline i_{x\ell} \\ \alpha_{x\ell} \\ v_{x\ell} \\ \hline i_{xr} \\ \alpha_{xr} \\ v_{xr} \end{bmatrix}
\qquad (3)
$$

(Column blocks: 28, 3, 3; row blocks: 28, 3, 3.)

*1: element (17,3)	*6: element (25,11)	*11: element (31,1)	*16: element (29,19)	*21: element (34,9)
*2: element (17,29)	*7: element (25,32)	*12: element (31,15)	*17: element (31,5)	*22: element (34,23)
*3: element (30,1)	*8: element (33,13)	*13: element (34,13)	*18: element (31,19)	
*4: element (29,1)	*9: element (32,13)	*14: element (34,27)	*19: element (32,9)	
*5: element (29,15)	*10: element (32,27)	*15: element (29,5)	*20: element (32,23)	

where the subscript r means the right magnetic bearing. In addition, $f'_{1\ell}$, $f'_{2\ell}$, f'_{1r} and f'_{2r} show feedback gains added for active vibration control loops for higher modes. Similarly, we get almost the same equation for the vertical direction. Equation (3) is systematically formulated for a general flexible rotor with magnetic bearings. We can design the control system by solving the eigenvalue problem of Eq.(3).

3. Experiments

3.1 Test Rig

The laboratory test rotor is shown schematically in Fig. 1. The flexible shaft, which is a 10-mm-diameter drill rod that is 1000 mm long, has five disks with masses as indicated in Fig. 1. The total mass of the flexible rotor is about 3.3 kg. Two of the disks are used for magnetic bearings and two others are used for sensors - i.e, the bearings and sensors are not collocated. The rotor is supported by two magnetic bearings with a

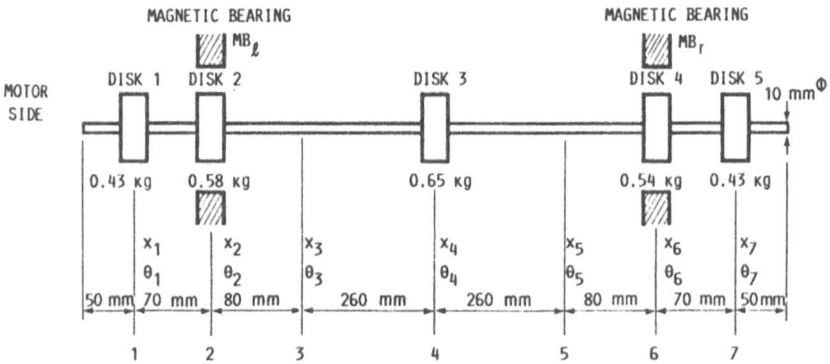

FIG. 1. - SCHEMATIC OF TEST ROTOR.

bearing span length of 680 mm. The shaft is connected to the driving motor by a flexible coupling. This is a radial magnetic bearing system with four degrees of freedom only because the radial directions are non-contacting. The gap between the electromagnets and the disk is about 2.0 mm. The backup bearings protect in an emergency when the amplitudes exceed 1.0 mm. The shaft displacements, which are measured by gap sensors of the eddy current type, are measured as close as possible to the magnetic bearings. These signals are the inputs for the analog controller. The analog controller consists of proportional, differential, and integral circuits, the phase lead circuit, and some filters. The control signals from the analog controllers are supplied to the electromagnets through servoamplifiers.

3.2 Experiments for Flexible Rotor

3.2.1 Critical Speed Map

Figure 2 shows the critical speed map of the rotor model shown in Fig. 1. In the case shown, both stations 2 and 6 are supported by the equivalent spring K. For zero bearing stiffness, there are two rigid modes at 0 Hz, the first bending mode at 20 Hz, the second at 86 Hz, and the third at 130 Hz. The computed free-free data have good agreement with the measured free-free test data shown in Fig. 3. From this, it seems that the mathematical model is correct. The actual critical speeds are estimated as the dot-dash line shown in Fig. 2 using the measured spring constant.

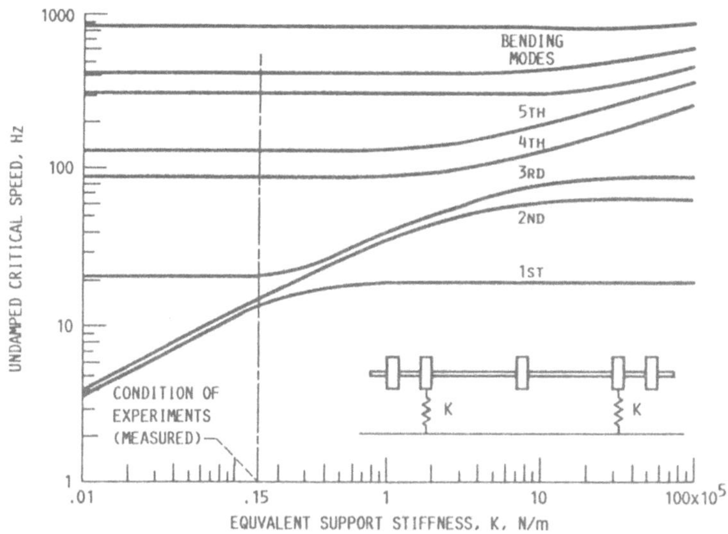

FIG. 2. - CRITICAL SPEED MAP OF THE ROTOR MODEL SHOWN IN FIG. 1.

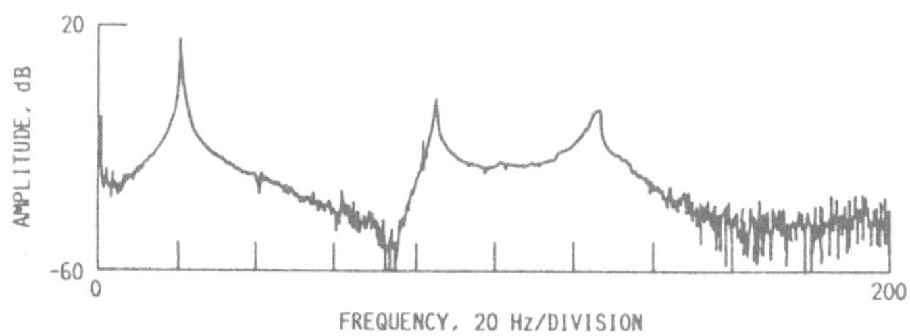

FIG. 3. - MEASURED FREE-FREE TEST DATA.

3.2.2 Frequency Response of Flexible Rotor - Magnetic Bearing System

Figure 4 shows the measured frequency responses of the FFT analyzer based on an impulse response using the same conditions as in Fig. 2. Figure 5 shows measured mode shapes. The first mode at 10.40 Hz is omitted because

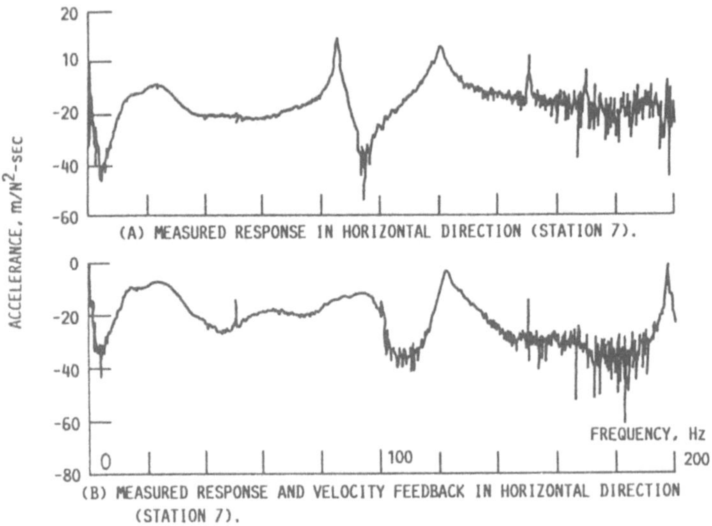

FIG. 4. - MEASURED FREQUENCY, RESPONSE.

it is similar to the second mode. The modal damping ratio at the first mode was ζ = 15.04 percent. The measured critical speeds up to the fifth in Figs. 4 and 5 coincide with the computed critical speeds shown in Fig. 2. The resonance peaks from the first through the third are not clearly recognized because of their high modal damping ratios as shown in Fig. 5.

3.2.3 Eigenvalue Analysis

In this section, we determine the eigenvalues by using Eq. (3). Table I lists the eigenvalues with collocation in the ideal case and Table II shows the eigenvalues without collocation in the case of this experiment. Only complex eigenvalues are listed in the tables and the real roots are omitted. The parameters and feedback gains used for the eigenvalue analysis are the same as those in Section 3.2.2. The results in Table II almost coincide with the results of the critical speeds shown in Figs. 2 and 5. The fourth mode is unstable in Table II. The measured modal damping ratio at 123.13 Hz (as shown in Fig. 5) is too small. Moreover, we cannot pass the fourth critical speed (as shown in Fig. 6) because the amplitudes at locations near the magnetic bearings are growing at speeds greater than 5500 rpm. Therefore, the results of the eigenvalue analysis of Eq. (3) are reasonable. It is possible to design the control system of a flexible rotor - magnetic bearing system by using the eigenvalue analysis of Eq. (3). It is very clear that every mode is stable with collocation. However, the roots of the higher order modes approach the imaginary axis as shown in Table I. Table III shows the improvement of the higher modes

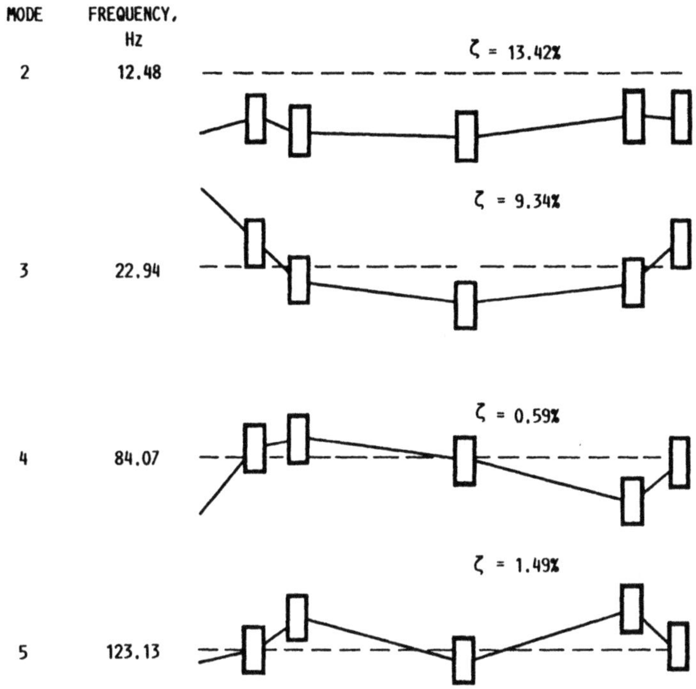

FIG. 5. - MEASURED MODE SHAPES AND MODAL DAMPING RATIOS.

TABLE I. - EIGENVALUES WITH COLLOCATION

Real	Imaginary	Frequency, Hz
-0.22273D-2	0.36128D+5	5750.0
-.22359D-1	.15725D+5	2502.7
-.70242D-1	.12784D+5	2034.6
-.57519D+0	.59381D+4	945.1
-.89774D+1	.29159D+4	464.1
-.22630D+2	.20626D+4	328.3
-.64734D+2	.94692D+3	150.7
-.47460D+2	.66731D+3	106.2
-.82129D+2	.15805D+3	25.2
-.94146D+2	.15287D+3	24.3
-.49705D+1	.12235D+3	19.5

TABLE II. - EIGENVALUES WITHOUT COLLOCATION

Real	Imaginary	Frequency, Hz
0.16181D+2	0.20101D+4	319.9
.74842D+2	.86925D+3	138.3
.71604D+2	.62756D+3	99.9
-.11111D+3	.17393D+3	27.7
-.12722D+3	.15032D+3	23.9
-.65155D+1	.10677D+3	17.0

TABLE III. - EIGENVALUES WITH COLLOCATION IN ADDITION TO VELOCITY FEEDBACK

Real	Imaginary	Frequency, Hz
-0.46911D+3	0.11555D+4	183.9
-.59602D+3	.91103D+3	145.0
-.14625D+3	.36865D+3	58.7
-.28629D+3	.34515D+3	54.9
-.46764D+0	.13729D+3	21.9

with collocations in addition to new velocity feedback loops. The difficulty of passing the higher order critical speeds with collocation is a very important result. Namely, it points out that the usual design method for a rigid rotor has a natural limitation for control of flexible modes. New active control loops are needed for the higher flexible modes even with collocation.

3.2.4 Unbalance Response

Figure 6 shows the unbalance responses for this experiment for the conditions of Section 3.2.2. From these results, both the horizontal and the vertical responses have similar characteristics. There are only low

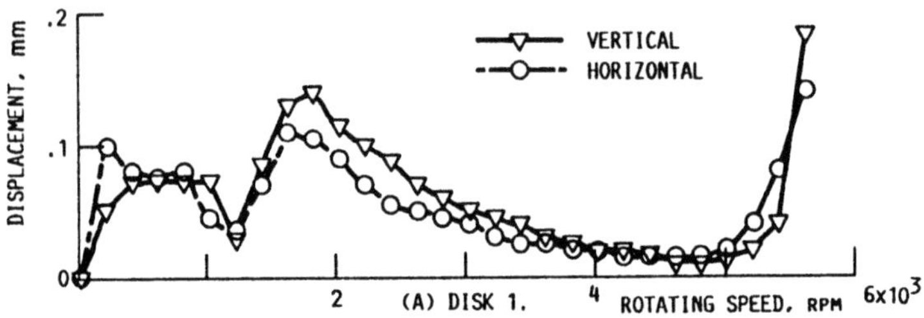

FIG. 6. - UNBALANCE RESPONSES.

amplitude resonances up to 5500 rpm. The rotor should have passed the fourth critical speed by 5500 rpm since the fourth critical speed exists at 84 Hz, 5040 rpm as shown in Fig. 5. Some results indicate that this is caused by gyroscopic effects.

3.2.5 Active Vibration Control

Improving the damping characteristics at the fourth mode is needed in addition to considering the velocity feedbacks $f_{2\ell}'$ and f_{2r}'. Namely, the vibration velocities measured at locations 3 and 5 are used as the control inputs at locations 2 and 6. The results of the eigenvalue analysis for this are shown in Fig. 7. The measured frequency responses by an FFT analyzer are shown in Figs. 4(b). The analysis and experiments show that the characteristics of the fourth mode are improved by active velocity feedback loops. It is possible to pass the fourth critical speed if the dynamic characteristics are as shown in Figs. 4(b). However, it is impossible to pass the fourth critical speed in the actual test rig because of gyroscopic effects. We must design the active vibration control system taking into account the gyroscopic effects of matrix [G] in Eq. (1). The higher order critical speeds can be passed if the flexible rotor - magnetic bearing control system is designed using the stable eigenvalues of Eq. (3) that take into account gyroscopic effects.

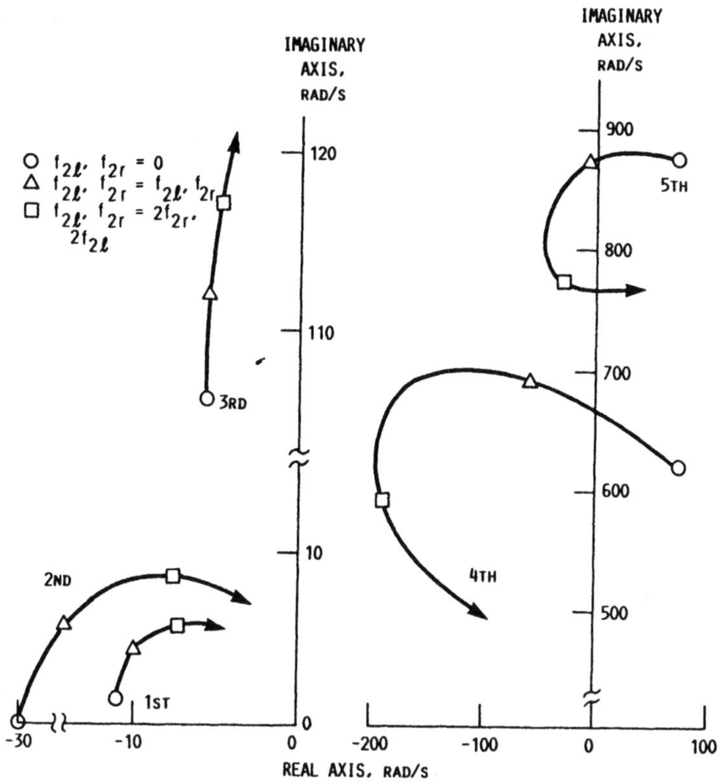

(A) MODES 1 TO 3. (B) MODES 4 AND 5.

FIG. 7. - ROOT LOCI WITHOUT COLLOCATION IN ADDITION TO VELOCITY FEEDBACK.

4. Conclusions

This paper deals with vibration problems and control methods without gyroscopic effects for a flexible rotor supported by radial magnetic bearings with a four-degree-of-freedom system. The conclusions obtained are as follows:

(1) The eigenvalue problem, taking into account the interaction between the mechanical system of a flexible rotor and the electrical system of magnetic bearings and the controller can be formulated as shown herein. The control system of a flexible rotor - magnetic bearing system was designed by an eigenvalue analysis.

(2) If the control system is designed for a rigid rotor, the flexible modes do not become unstable with collocation.

(3) New active control loops must be added for the higher flexible modes even with collocation because the damping characteristics degenerate for the higher order modes. It is possible to stabilize for the case of no collocation by means of this method.

(4) It is not possible to pass the critical speed in the actual test rig even if frequency responses by a FFT analyzer are satisfactory mainly because of gyroscopic effects with rotation.

References

1. Habermann, H.; and Brunet, M.: The Active Magnetic Bearing Enables Optimum Damping of Flexible Rotor. ASME Paper 84-GT-117, 1984.
2. Habermann, H.; and Liard, G.: An Active Magnetic Bearing System. Tribology Int. 13 (1980) 85-89.
3. Foster, E.G.; Kulle, V.; and Peterson, R.A.: The Application of Active Magnetic Bearings to a Natural Gas Pipeline Compressor. ASME Paper 86-GT-61, 1986.
4. Matsumura, F.; Kaobayashi, H.; and Akiyama, Y.: Fundamental Equation for Horizontal Shaft Magnetic Bearing and Its Control System Design. Electr. Eng. Jpn., 101 (3) (1981) 123-130.
5. Matsumura, F.: System and Control, 26 (1982), 209 (in Japanese).
6. Higuchi, T.; and Mizuno, T.: Control System Design for Totally Active DC-Type Magnetic Bearings - Structures of the Optimal Regulator for Systems with Gyroscopic Coupling. Trans. Soc. Instrum. Control Eng., 18 (1982), 507-513 (in Japanese).
7. Allaire, P.E.; Humphris, R.R.; and Kelm, R.D.: Dynamics of a Flexible Rotor in Magnetic Bearings. Rotordynamic Instability Problems in High-Performance Turbomachinery, NASA CP-2443, Washington, D.C.: NASA, 1986, pp. 419-430.
8. Matsushita, O., et al.: Proceedings International Conference on Rotordynamics (1986), 421.
9. Salm, J.R.: Active Electromagnetic Suspension of an Elastic Rotor: Modeling Control and Experimental Results. Rotating Machinery Dynamics. Vol.1, A. Muszynska and J.C. Simonis (eds.) New York: ASME, 1987, pp. 141-149.
10. Schweitzer, G.: Dynamics of Rotors. F.I. Niordson (ed.) New York: Springer-Verlag, 1975, pp. 472.
11. Nikolajsen, J.L.; Holmes, R.; and Gondhalekar, V.: Investigation of an Electromagnetic Damper for Vibration Control of a Transmission Shaft. Proc. Inst. Mech. Eng., 193 (1979), 331-336.
12. Ulbrich, H.; and Anton, E.: Inst. Mech. Eng. Conf. C299/84 (1984) 543.
13. Anton, E.; and Ulbrich, H.: Active Control of Vibrations in the Case of Asymmetrical High-Speed Rotors by Using Magnetic Bearings. J. Vibration Acoustics Stress and Reliability in Design, 107 (1985) 410-415.
14. Salm, J.R.; and Schweitzer, G.: Inst. Mech. Eng. Conf. C277/84 (1984) 553
15. Okada.,Y., et al.: Trans. Jpn. Soc. Mech. Eng., Vol. 52, No. 467 (1985) 1760 (in Japanese).
16. Matsushita, O., et al.: Trans. Jpn. Soc. Mech. Eng., Vol. 53, No. 496 (1987), 2453 (in Japanese).
17. Nonami, K.: Vibration and Control of Flexible Rotor Supported by Magnetic Bearings. NASA TM-100888, 1988, pp.1-21.

*notes

p_0 : the steady-state attractive force
i_0 : the steady-state current
h_0 : the steady-state gap length
L : the coil inductance
I : the coil current
E_1 : the input voltage to the phase lead circuit
R : the coil resistance
V : the phase lead voltage
f_1, f_2, f_3 : feedback gains

Industrial Applications

The Electromagnetic Damper – Towards a First Large-Scale Industrial Application

CHAN HEW WAI C. ; MOREL J.

Electricité de France
Direction des Etudes et Recherches
1 av. du Général de Gaulle
92141 Clamart. FRANCE

Summary :
Studies concerning the magnetic bearing were begun at the Research and Development Department when the first industrial applications on large machines appeared. Our work concerned, in particular, the modelling of this device and the construction of models used in performing validation tests. Furthermore these studies involved the search for new applications. Thus a large-sized electromagnetic damper was constructed and tested on a second test bench to demonstrate that this solution may be applied to a large 900 MWe turbo-generator in operation.

Since advances in power amplifiers indicated the possibility of supporting large sized rotors, the Research and Development Department at Electricité de France became actively interested in the magnetic bearing. From 1979 studies have enabled the potential of this new technology to be evaluated. In consideration of all that is at stake and the serious constraints concerning the availability and reliability of machines, considerable efforts are being made in the areas of modelling, behaviour prediction and reliability. Validation tests on large-scale models complete this work.

The project included three phases :

1) The development of software enabling the dynamic behaviour of the active magnetic bearing to be predicted,
2) The construction of test benches designed to verify the computing programmes and to measure bearing characteristics.
3) Application tests.

The requirements with respect to machine operating conditions and reliability are such at E.D.F. that it is impossible to put a new machine into operation unless its operating characteristics, under all possible abnormal conditions, are known, including exceptional transient currents (broken turbine blades, earthquakes, etc...). It is therefore indispensable to be able to simulate the behaviour of the machines under these conditions. Computing programmes which the E.D.F. has had in its possesion for many years now make this analysis possible. They were adapted with regard to the particular type of support provided by the magnetic bearing. The models used are capable of handling the following types of problems :

. linear (small displacements) at steady state,

. linear transient currents,

. non-linear transient currents,

. stability.

The above concern machines with several types of bearing.

FIRST CONCLUSIVE TEST

Along with theoretical studies and simulation calculations, tests involving real equipment were conducted, either to evaluate the characteristics of the active magnetic bearings (AMB), or with the intention of direct applications to machines in operation.
To this end, the Electricité de France constructed a test bench around a 4 t supporting capacity bearing. Its purpose was :

. to verify the computing programmes

. to measure bearing characteristics

. to enable reliability testing

. to test the emergency devices

The test bench consists of a 4 m long 2 t rotor, which is supported by two oil-film bearings (end bearings). The magnetic bearing, with a load capacity of 4 t, manufactured by the French company S2M (Société de Mécanique Magnétique) is located in the middle. This arrangement is best adapted to the first two points of the study. The first tests conducted in 1984 are very encouraging since there is an accurate correspondance between the measured and computed characteristics (see table). The dynamic coefficients of the magnetic bearing were determined by finding the first two natural frequencies of the rotor. For the second natural frequency, the magnetic bearing is located on a vibration node . This enables the dynamic coefficients of fluid bearings to be evaluated. This, along with the knowledge of the first natural frequency, enables the dynamic coefficients of the magnetic bearing to be estimated.

	Measurements		Computations	
Control loop gain	Magnetic bearing stiffness K (N/m)	Magnetic bearing damping C (N.s/m)	Magnetic bearing stiffness K (N/m)	Magnetic bearing damping C (N.s/m)
0,25	$2{,}25 \cdot 10^7$	$9{,}09 \cdot 10^4$	$2{,}15 \cdot 10^7$	$8{,}39 \cdot 10^4$
0,5	$4{,}59 \cdot 10^7$	$1{,}62 \cdot 10^5$	$4{,}50 \cdot 10^7$	$1{,}64 \cdot 10^5$
0,7	$6{,}23 \cdot 10^7$	$2{,}22 \cdot 10^5$	$6{,}51 \cdot 10^7$	$2{,}29 \cdot 10^5$
1,0	$9{,}17 \cdot 10^7$	$3{,}20 \cdot 10^5$	$9{,}81 \cdot 10^7$	$3{,}24 \cdot 10^5$

Table. Determination of the dynamic coefficients of the magnetic bearing as a function of the control loop gain. This table shows a good concordance between the computed values and measurements on the test bench with respect to magnetic bearing stiffness and damping.

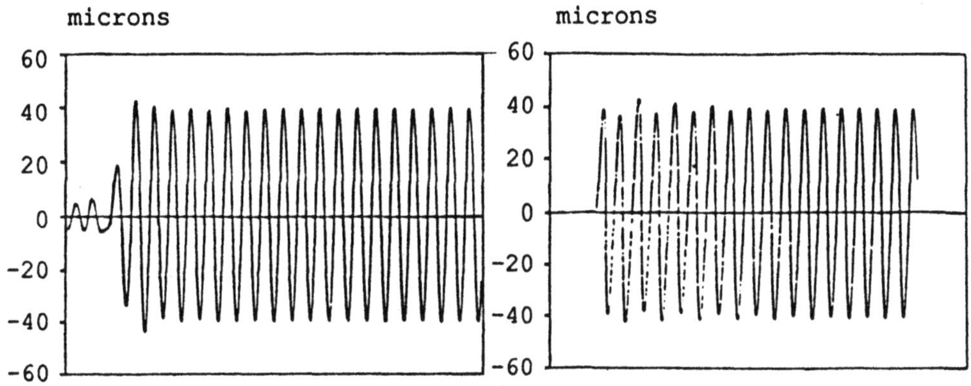

a) Measurement b) Simulation

Fig.1. Comparison between simulation and measurement on the test bench during the detachment of a rotating explosive bolt (transient unbalanced mass). Overvoltage levels are very similar in both cases. However, numeric computations lead to greater amplitudes, especially in the actual transient phase.

Furthermore, the response to a disturbance, (the detachment of an explosive bolt during rotation) has proven to be in conformance with predictions (see figure 1). These results, obtained for small displacements (linear) still need to be validated with respect to large displacements, and especially in the vicinity of saturation areas (magnetic fields greater than 1.3 Tesla).

SUPPORTING OR CORRECTING BEARING? STEPS TOWARDS A FIRST LARGE-SCALE APPLICATION

Another application of the AMB consists in using it not as a support, but as a corrector or damper. Indeed, traditional bearings generally offer little damping and exceeding of critical speeds may lead to vibratory problems. In these cases, one solution would be to insert an AMB in the shaft line which serves as a damper facilitating the crossing of these critical speeds. In this instance the bearing does not have to support the rotor. Its only function is to develop a force in constant opposition to that created by eventual unbalances. This type of application is already being considered by Electricité de France on high-power turbogenerators in operation. The idea is not new. Several papers have described the use of electromagnetic forces to dampen vibrations, but only for small-size machines.

Nikolajsen and Holmes (1) have demonstrated the effectiveness of a controlled electromagnet in reducing vibration in a transmission shaft.

Allaire and Humphris (2) have developed magnetic bearings which were tested on a flexible rotor as both supporting bearings and dampers. Our approach is original for three main reasons :

1) Our purpose is to show that this method may be applied to a 900 Mwe turbogenerator in operation.
2) The electromagnetic damper only uses the upper 270° of the rotor to facilitate mounting and dismounting at the site.
3) The rotor is massive and evaluation of rotor heating is an important point of our study.

In order to demonstrate the feasability of such an undertaking, a test bench has been developed. This test bench takes into account the dimensional restrictions of the actual shaft line with respect to the planned installation point of the electromagnetic damper (coupling plates).

DIMENSIONING OF A LARGE SIZE ELECTROMAGNETIC DAMPER

The first phase of the study consisted in determining (by digital simulation on shaft line models) possible installation locations and the performance of the electromagnetic dampers so as to provide the required impedance to the shaft lines.

These preliminary computations have shown that on the machine under consideration, a device which only provides damping offers the optimal solution with respect to vibration reduction at all critical speeds. Furthermore, they reveal that a rotating force of approximately 10 tons would be sufficient to significantly reduce vibration when passing the most difficult critical speeds. Figure 2 shows the peak vibrations obtained by digital simulation in the horizontal direction at coupling 2 and in the vertical direction with respect to coupling 3 during a deceleration. With magnetic dampers, the horizontal amplitude is divided by 7 and the vertical amplitudes by 3.5 during the crossing of critical speeds. In this simulation, the damping of each damper is 8.10^6 N.s/m.

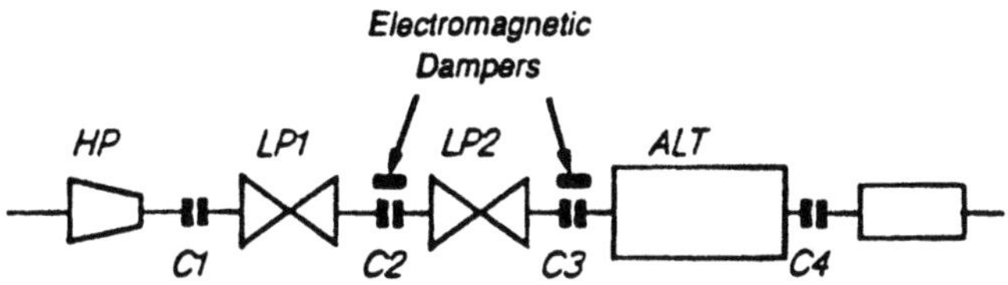

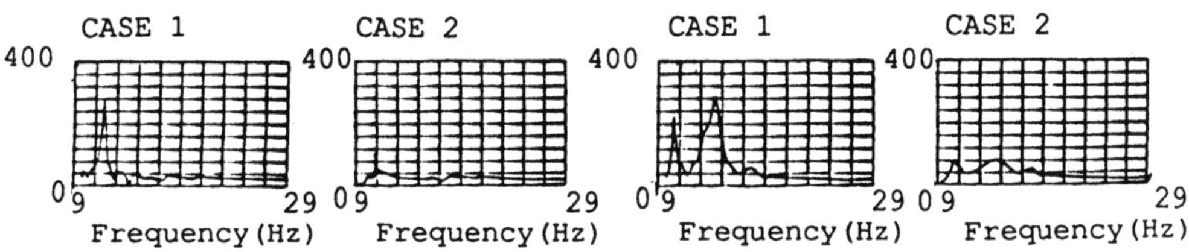

a) Horizontal vibration at coupling 2 b) Vertical vibration at coupling 3

Fig.2. Digital simulation (finite element method) of the effect of 2 electromagnetic dampers installed on couplings 2 and 3 during deceleration.
Case 1 : Initial
Case 2 : With 2 magnetic dampers each producing 8.10^6 N.s/m.

The second phase therefore consisted in developing and constructing a large size damper for controlling the vibrations of an existing shaft line.

The damper which was proposed, following all of the above considerations, is described in figure 3.

It includes 6 poles with 4 magnetic axes. It only covers the upper 270° of the rotor to facilitate installation on the actual machine. This particular characteristic means that it must operate in the vicinity of an average static force so as to create the necessary rotating force to dampen vibrations. Each 45° pole contains 3 independant electromagnets. Each of the 18 power amplifiers which drive each electromagnet operates on 160 V dc with a maximum current of 50 A. In our application, flux circulation is axial in order to reduce losses in the rotor.

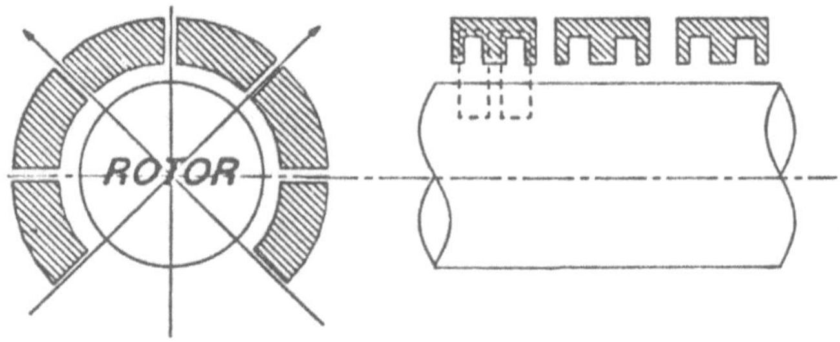

6 poles of 45° each   Axial flux circulation

Fig.3. Electromagnetic damper working principle.

The rotating force is optimised via a compensation method which consists in supplementing to the detector signals of the various poles with part of the signals that the missing poles would have detected if they existed. Use of the latter improves the modulus of the rotating force approximately 43 % (3). Thus, each pole must provide 7,000 daN in order to develop 10,000 daN of rotating force.

The stability and dynamic characteristics of an active magnetic damper (or bearing) are determined by the electronic control system. In our application, the various filters used provide a phase of approximately 90° in the 10 to 20 Hz frequency range. The damping function is assured in this way. To obtain a high damping coefficient without reducing stability, a synchronous peak gain filter is used. This type of filter (synchronised with the rotation speed) significantly increases the loop gain only at the rotation speed without modifying the phase. The overall control system transfer function which represents damper impedance and the resulting damping curves are represented as a function of frequency for two different gains (see fig.4). To minimize rotor heating due to eddy currents, the static force is automatically adjusted by the control system according to the amplitude of the vibrations.

TEST BENCH

The rotor is 5 meters long and its total weight is approximately 7 tons. The middle of the rotor represents the coupling of a LP

rotor in operation and therefore has the same dimensions. The rotor is supported by two circular hydrodynamic bearings. It is driven by a dc motor and may attain a maximum speed of 1,200 RPM. The frame consists of stiff metal beams. Its total weight is approximately 70 tons. It stands upon isolating devices so as to avoid transmitting its vibrations to the building.
Since the rotor is massive, the knowledge of rotor temperature is essential in order to establish the operating limits of the damper. These temperatures are measured via 12 sensors distributed along the surface and at different depths (see fig. 5). These signals are transmitted via a remote sensor and plotted in real time on a paper recorder.

MODELLING

It is not always possible to determine all of a machine's parameters through experiments. Modelling is one way of achieving this. Thus the rotor and frame were modeled so as to :

1) Verify that damping of the electromagnetic damper conforms to the theoretical value predicted.
2) Predict test bench behaviour for different values of the exciter energy (unbalanced mass) and the damping supplied by the electromagnetic damper.

The rotor was modeled using 100 continous beam elements under plane deflection. The rotor model was validated using the first two natural free-free frequencies experimentally obtained by modal analysis as shown in the table below :

Rotor free-free Frequency	Computed	Measured	Difference
1st natural frequency	43,43 Hz	43,55 Hz	0,3 %
2nd natural frequency	99,57 Hz	99,61 Hz	0,04 %

The model of the frame consists of continuous beams which are interconnected and which have the same main dimensions and mass as the real frame.

MAIN RESULTS

The main results are shown in figures 6 to 8. They make possible the comparison between the values measured on the test bench and the values computed with the model.

Figure 6 indicates rotor vibration with a residual unbalanced mass of approximately 0.07 m.kg during deceleration for different damping values provided by the electromagnetic damper.

Firstly, we observe that the rotor's critical speed, which is located at approximately 17 Hz, does not shift when the damper is operating, and secondly, that vibrations when this speed is exceeded are well damped. Therefore the electromagnetic damper well fulfills its damper function (stiffness is practically zero). The curves computed make it possible to estimate the damping values and to readjust rotating models. The distance between the fluid bearing axes was 4,5 m.

The electromagnetic damper was designed to provide a rotating force of 10 tons at 15 Hz and maximum damping of 8.10^6 N.s/m.

To verify these figures, the distance between the fluid bearing axes was reduced to 3 m, and tests were conducted on the test bench by applying larger and larger unbalanced masses up to 11 m.kg which represent a force of approximately 10 tons at 15 Hz. Figure 7 shows that the results obtained on the test bench correspond well with those first calculated.

The various measurements made on the test bench indicate that the increase in temperature at the hottest point of the rotor (across from the unbalanced mass) is approximately 2° C per ton and per minute at 15 Hz during the first moments.

Figure 8 represents the results of a long duration test. It indicates that the slope of the graph decreases as rotor thermal equilibrium is approached. An estimation of the power loss was made using a relatively simplified calculation (bidimensional calculation - localised source of heat...).

CONCLUSIONS

The results obtained from the various series of tests show that the main objectives aimed for were attained.

The device does behave as a genuine damper since the rotor's critical speed is indeed damped and not shifted.

We have observed a good concordance between measurements and calculations. Differences arise primarily from measurement uncertainties (sensitivity of the sensors to type of material and to temperature, attenuation in the filters...) and from the simplified modelling of the test bench frame. It should be noted that differences are especially significant below 10 Hz and that above this level theoretical and measured values are very close. Thus we can say that the theoretical damping curves of the device are completely in accordance with real operation, especially within the range of 10 hz to 20 Hz which is the most important.Tests with large unbalanced masses have shown that the damper is well capable of providing a force of 10 tons at 15 Hz. Temperature sensors in the rotor also made it possible to have a good idea of the change in temperature at different points of the rotor according to the unbalanced mass applied and according to time. However, the total power loss in the rotor could not be calculated with accuracy, since the existing thermal calculation codes do not enable exact simulation of the actual phenomenon (due to its three-dimensional and transient nature and non-sinusoidal source). The simplified calculation made still provides a good indication of the order of magnitude of rotor power loss.

In conclusion, the tests conducted show that the electromagnetic damper in question meets the requirements of design specifications even if certain improvements may still be made. In particular, the installation of a control system gain limiter would enable saturation of the amplifiers to be avoided. The stability of the unit may be improved so as to increase damper effectiveness.

Power loss in the rotor is not neglegible, even if it is lower than the value predicted by the manufacturer. It could be significantly lowered by the development of a full damper.

BIBLIOGRAPHY

(1) NIKOLAJSEN J., HOLMES R., GONDHALEKAR V. "Investigation of an electromagnetic damper for vibration control of a transmission shaft".University of Sussex, March 1979.

(2) ALLAIRE P.E., HUMPHRIS R., KELM R.D. "Dynamics of a flexible rotor in magnetic bearings" presented at the 4th Workshop on Rotor Dynamics Instability Problems in High Speed Turbomachinery, Texas A and M University, June 1986.

(3) CHAN HEW WAI C., MOREL J., NEAU P. Presentation of the compensation method used to optimise the rotating force of a damping magnetic bearing only covering 270° of the rotor. EDF report P34-85/359

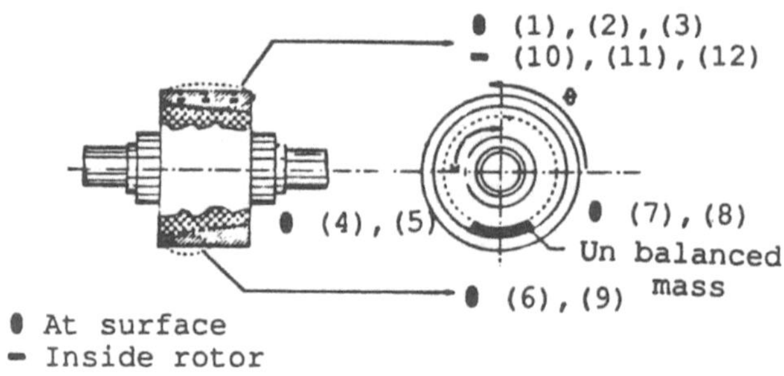

Fig.5. Placement of temperature sensors.

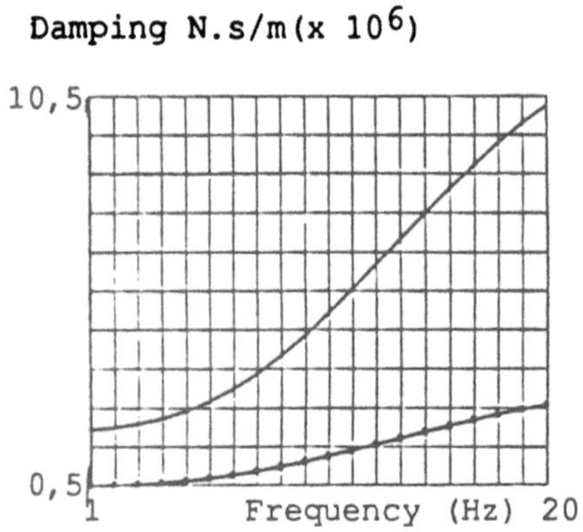

——— Gain = 4 -x-x-x-x- Gain = 1

Fig.4. Damper damping curves as a function of frequency for 2 different gains.

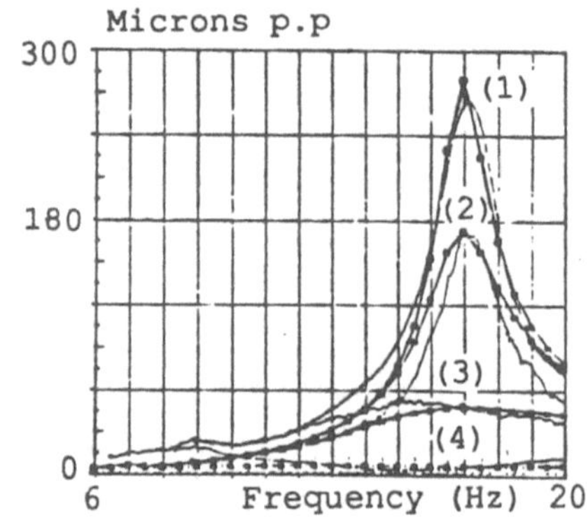

-x-x-x-x- Calculated ——— Measured

1) Cd=0 2) Cd=3 10^4 Ns/m 3) Cd=2 10^5 Ns/m 4) Cd=2 10^6 Ns/

Fig.6. Vertical rotor vibrations with a residual unbalanced mass of 0,07 m.kg for differe damping values Cd.

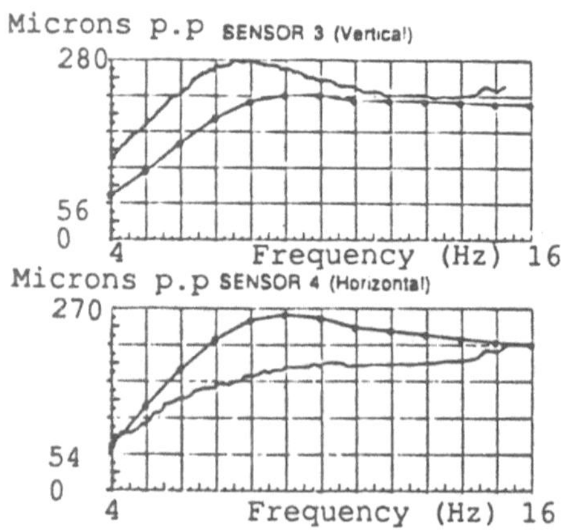

——— Measured -o-o-o-o- Calculated

Fig.7. Rotor vibration with an unbalanced mass of 11,22 m.kg and damping of 7,6 10^6 Ns/m. Distance between bearing axes is 3 m.

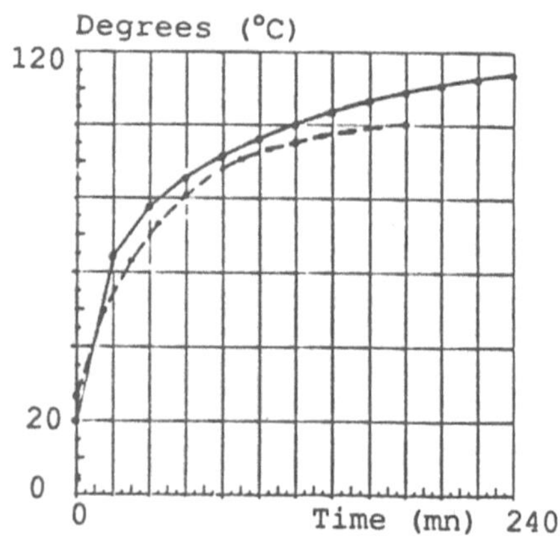

-x-x-x-x- Measured -o-o-o-o- Calculate

Fig.8. Change in rotor temperature as a function of time for an unbalanced mass of 1,8 m Power loss is estimated at 5 kW at 15 Hz.

Implementation of Active Magnetic Bearings on Advanced Rocket Engine Turbopumps

J.P. GIRAULT

Société Européenne de Propulsion
Division Propulsion à Liquides et Espace
Forêt de Vernon
BP 802 - 27207 VERNON, FRANCE

Summary

Preliminary studies are conducted at SEP to clarify the advantages and drawbacks of using S2M ACTIVE MAGNETIC BEARINGS (AMB) to suspend rotors of rocket engine turbopumps.
After having pointed out their specific interest for rocket propulsion, technological keys for development will be presented. These latter are found among the following subjects: AMB materials properties (mechanical and chemical), miniaturization of the electronic controller, rotordynamic prediction, fluid and electrical interfaces.

1. Introduction

The constant care of every engineer in propulsion field is to increase the machine efficiency and to minimize the weight and overall dimension of the engine and of the feeder turbopumps.

This care most often results in a search for higher rotor speed, beyond the limitations of the components currently used, especially for the bearings. This basic rule is worsened by the requirements adherent to modern engines : increased life, high safety, extended reusability, low development and recurrent costs. Thus is layed the necessity to use alternates found among the fluid bearings (hybrid/hydrostatic/ hydrodynamic) or active magnetic bearings (AMB).

The first option is explored since twenty years, particularly in the United States where advanced turbopumps have been tested with hybrid/hydrostatic bearings [1] [2].

However, the use of such bearings has drawbacks due to their complexity and recycled flow rates.

Therefore, beside trade-off studies on fluid bearings, detailed investigations are conducted at SEP since 1986 to determine specific interest of AMB in space propulsion. These works are based on conceptual and technological studies, both being tightly connected. Some of them are linked to cooperation works with american companies.

This paper is intended for a brief description of the works trends and the main results obtained during this preliminary phase.

2. General studies

2.1 Realm of interest

Most of the features of the AMB are attractive for rocket engine components as turbopump bearings. The high level of reliability obtained in previous space applications [3] [4], the potential high life, an increased rotational speed, the shaft dynamic control, are among their most interesting characteristics [5].

But they have also some counterparts, as they need power supply, on-board control cabinet, dedicated transfer function and safe compatibility with the propellants used.

Company funding were then committed to conduct applicability studies precising the requested efforts to implement AMB on rocket engine turbopumps. These investigations start from the system specifications to obtain the AMB parameters in terms of mass, power consumption, cost, life, reliability, load capacity (fig. 1)

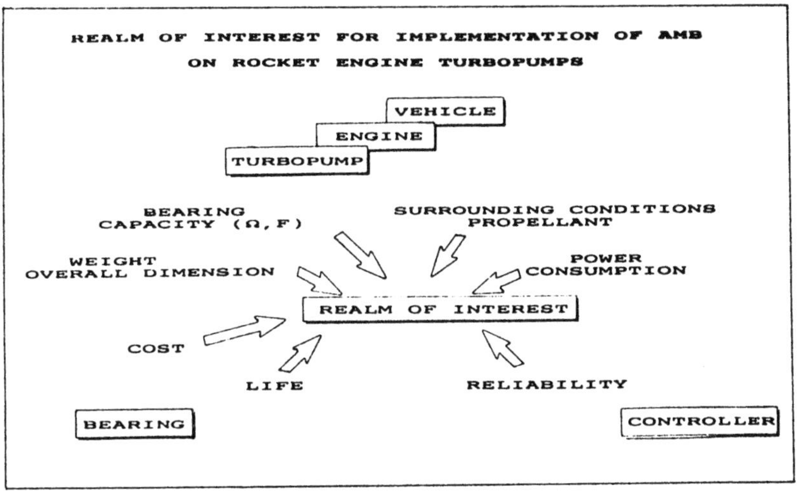

Fig. 1 AMB's applicability on rocket engine turbopomps

Those specifications are related to each particular engine, but generally speaking, we have scanned an engine thrust from 30 kN to 3000 kN, life requirement up to 500 cycles/20 hours with no fail occuring before 2 years, and cryogenic/storable propellants.

The main results show that :

(a) weight and power dissipations increase largely with the AMB load capacity, i.e. the engine thrust (fig 2,3)
(b) cryogenic propellant allow smaller sizes (increasing the current density by a factor of 6 compared to ambient conditions)
(c) higher rotor speeds can be obtained with an increase of about 50 % compared to cryogenic ball bearings. Expected improvements in mechanical strength of the rotor sheets should increase this value.
(d) AMB expected life seems consistent with the system requirements. On-going works on fatigue behaviour will complete this topic.

(e) axial thrust capacity appears too low regarding the loads encountered in start/shutdown transients. Improved designs for axial AMB has to be compared to mechanical/hydraulic devices currently used on turbopumps.

(f) recurrent costs remain relatively high today, due to a low number of parts intended to be produced. This point turns AMB application towards high life, reusable advanced turbomachineries.

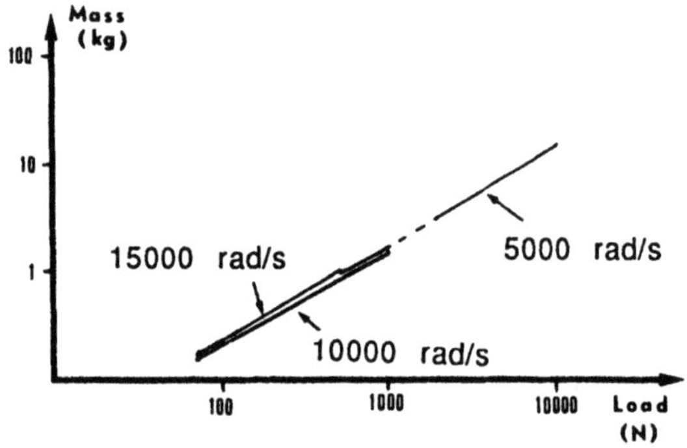

Fig. 2 AMB's mass vs radial load

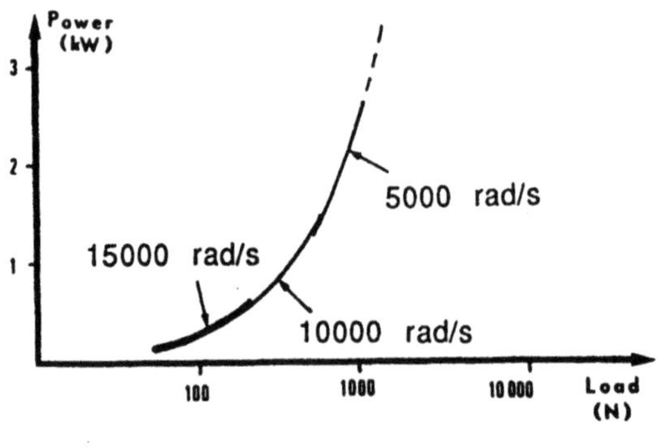

Fig. 3 AMB's power dissipation vs radial load

2.2 Turbopump Designs

These designs were undertaken to complete the applicability study, to point out the technological assessments and to illustrate the possible concepts.

Two exemples are proposed here, one imaging a high thrust engine, liquid oxygene turbopump (HTOTP Fig 4 (a) (b)), the other picturing a low thrust engine, hydrogen turbopump (LTHTP Fig 5 (a) (b)). Their main features are reported Table 6 and the AMB characteristics are found Table 7.

The location of the forward bearing on HTOTP Fig. 4(a), placed on the inducer shroud, shortens the axial length and enables better dynamic control. All bearings are cooled with a small propellant flow rate through them.

The main technological keys are detailed in the following chapter.

Table 6 . Advanced Turbopumps Features

CHARACTERISTICS	HTOTP	LTHTP
Pressure rise (MPa)	28	24
Shaft Power (kw)	17,000	670
Rotor Speed (rpm)	30,000	190,000
Overall length (m)	1.0	0.25
TP weight (kg)	260	4.8

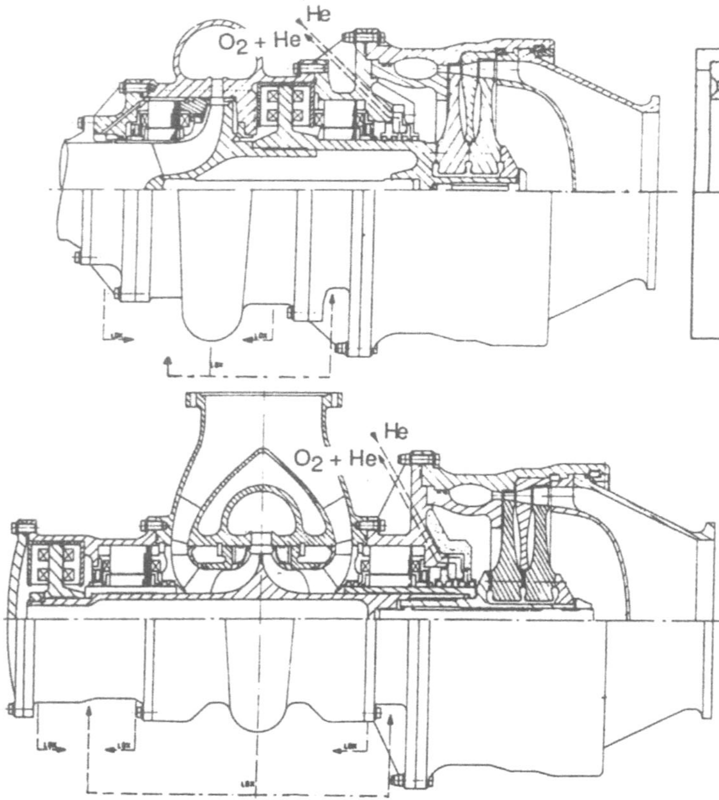

Fig. 4 (a) (b) HTOTP Concepts

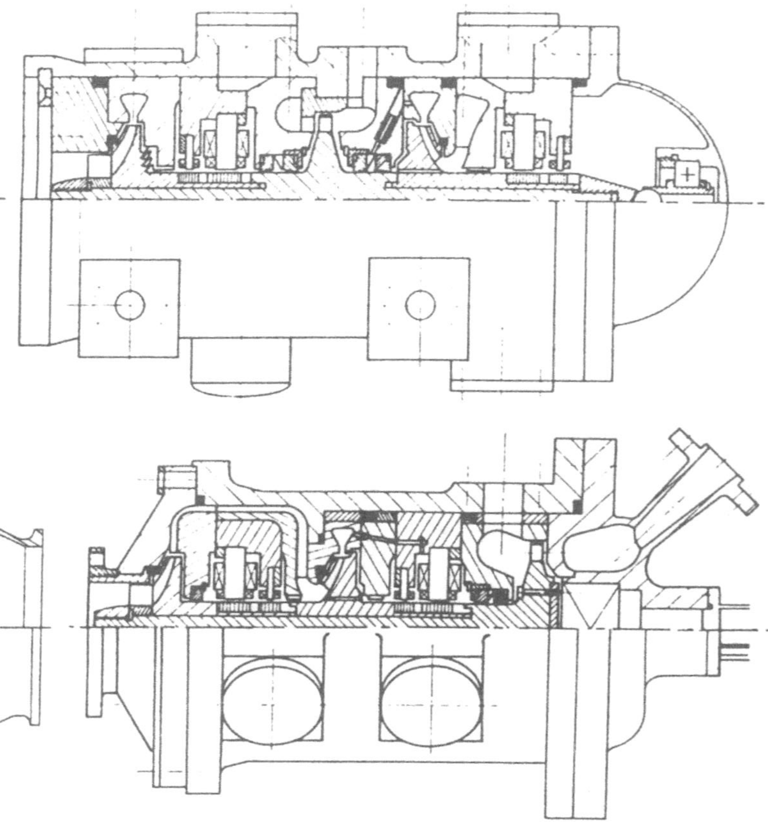

Fig. 5 (a) (b) LTHTP Concepts

Table 7 . Radial AMB's Characteristics

CHARACTERISTICS	HTOTP	LTHTP
Airgap (mm)	0.4	0.2
Materials (rotor) (stator)	Fe-Si3 Fe-Co50	Fe-Si3 Fe-Co50
Stator O.D. (m)	0.21	0.06
Load (N)	4,600	65
Weight (kg)	30	0.6
Power consumption (W)	5,000	500

3. Technological works

3.1 Size reduction of the control cabinet

This is obviously a major point, allowing or not the use of AMB on flight systems. The challenge is to miniaturize the circuitry and the power amplifiers while keeping the high level of reliability obtained with industrial cabinets.
The expected size reduction is about 5 to 7 times, implying the use of digital processing, deep hybridation and optimisation of the control box design. Those are trends of our works.
This effort will also give us ability to use complicated transfer functions and real-time health diagnostic capability.

3.2 Rotordynamic

As far as turbopump design are concerned, an accurate mean of predicting the rotor responses is needed. Significant effort has been done at SEP to obtain such an efficient tool taking into account :

- any transfer function
- non-isotropic behaviour of the bearing
- all the hydraulic connections between stator and rotor parts: pump and turbine stages, seals, AMB airgaps
- any location for the bearings and the displacement probes.

This computer code, developped with assistance of the University of Liege (LTAS), is a finite element program with formulation allowing non-linearities. This control system can be adequately expressed in the standard quadratic form : $M\ddot{x} + C\dot{x} + Kx = F(t)$, by mean of extra degrees of freedom [6].

An operating code is currently used for quasi-steady calculations, with shafts controlled plane by plane or axis by axis. Improvements are on-going yet to allow for non-linearity of the control loop and start/shutdown transients.

The shaft behaviour is analyzed with :

(a) Campbell diagrams (Critical speeds and damping coefficients, fig 8 (a) (b))

(b) the mode shapes showing appropriate deflections at the bearing/detector locations (fig 9)

(c) the synchronous response (unbalance response) allowing to check gaps vs displacements.

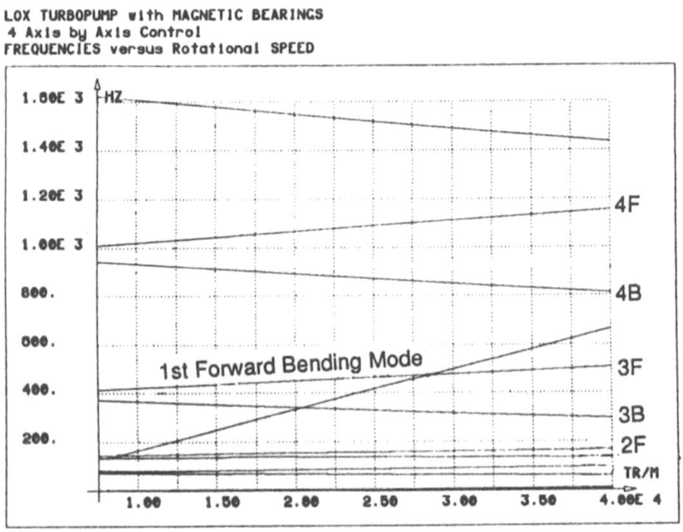

Fig. 8(a) HTOTP Critical speeds

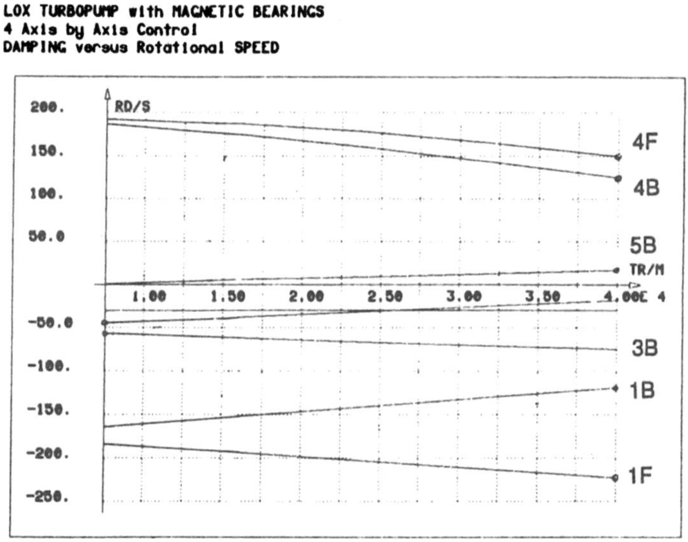

Fig. 8 (b) HTOTP Damping coefficients

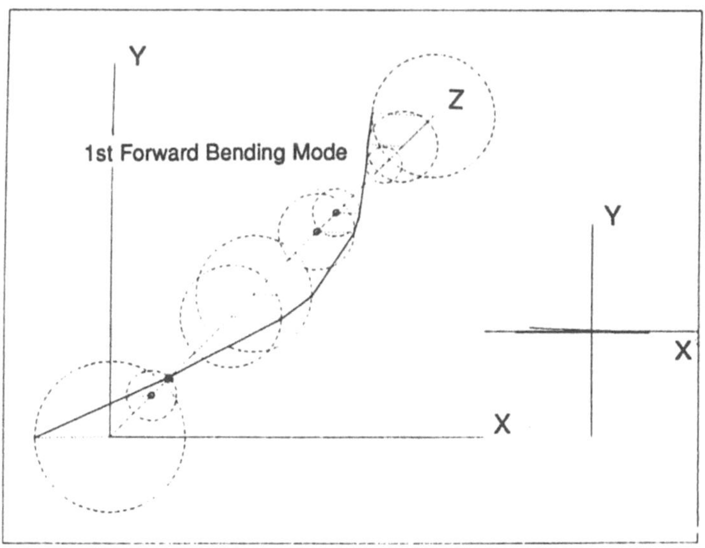

Fig. 9 HTOTP 3rd Mode shape

Those given exemples illustrate potential problems encountered on these machines : broad frequency range control and tight relationship between the turbopump design and the stability of the machine.

3.3 Material compatibility

Following extensive literature review and flight profiles required in future, priority propellants has been retained, i.e. liquid oxygen (LOX) and liquid hydrogen (LH2). Preliminary tests then have been performed to check :

(a) cryogenic behaviour of the Fe-Si3 sheets employed

(b) compatibility with GOX and LOX

(c) hydrogen brittleness for the Fe-Si3 alloy

Those tests were conducted with the assistance of the National Test Laboratory (LNE) and Air Liquide company.

(a) Tensile tests carried out with notched and unnotched Fe-Si3 samples confirm the usual results [7], with rupture stress higher than 500 MPa and no yield for cryogenic temperature ($\theta<77K$). These alloys has a light sensitivity to the notch. High-cycle fatigue tests done also at cryogenic temperatures show rupture stresses higher than 350 MPa for a number of cycles equal or higher than 10^7.

(b) 16 organic materials amoung insulations, coatings, various protections were tested by normalized check-out in GOX. Following these first choices, 4 materials have been retained and used to make small components similar to magnetic bearings (fig 10).

They were tested successfully during adiabatic compression tests, opening the way to a full compatibility with liquid oxygen.

(c) The foreseen implementation on hydrogen turbopump compel to measure GH2 brittleness, especially for the highly stressed rotor sheets. This so-called "external brittleness" study was conducted with disk ruptures in GHe and GH2 [8].

The results show classical curves with a peak around 200K and its disappearance at cryogenic temperatures.

Fig. 10 Subscaled Components for LOX compatibility tests

4. Conclusion

A preliminary program was started by SEP in 1986 to verify a possible implementation of AMB on advanced turbopumps.

These studies has confirmed their potential interests :

(a) high reliability, with an expected MTBF higher than 40 000 h. This will allow high life in orbit and high cycles reusability.
(b) possible use into cryogenic propellants (LOX and LH2), related to attractive turbopump design.
(c) active control of the rotor dynamic, with effective engine thrust throttling, realtime diagnostic and easier development.

These advantages should balance for numerous applications the two major drawbacks of these bearings : a mass penalty and a relatively high recurrent cost.
These both disadvantages come mainly from the control cabinet, whose required spatialisation is possible for mean term.

According to these results, the design of a high speed test device was initiated beginning of 1988. Extensive tests planned for 1990 should confirm and precise this balance.

Those global results will be compared to similar investigations conducted on fluid bearings leading SEP to proper choices on future turbopumps

References

1. REDDECLIFF, J.M. ; VOHR, J.H. : Hydrostatic bearings for cryogenic Rocket Engine Turbopumps. Jal of Lubrication Technology, July 1969, pp 557-575

2. NIELSON, C.E. : Hybrid Hydrostatic/ball Bearings in High-Speed Turbomachinery, NASA-CR-168124, Jan 1983

3. GAUTHIER, M. ; ROUYER, C. : Semi Active Magnetic Bearings for Space Applications. ESA 2ème Symposium Europeen sur les Mecanismes Spatiaux - RFA Oct. 1985.

4. STUDER, P.A. : "Magnetic Bearings for Instruments in the Space Environment" NASA-NT-78-15492.

5. HABERMANN, H. ; BRUNET, M. : The Active Magnetic Bearing enables Optimum Damping of flexible rotor, Trans. ASME, 84-GT-117.

6. BRUNE, C. : Rotordynamics and Active Magnetic Bearings. Paper presented at 8th annual ROMAC Industrial Research Conference, June 20/24, 1988.

7. Rutherford and Saclay Laboratories Results , private communications.

8. BARTHELEMY, H. ; BRYSELBOUT, J. : BARBE, C. : Testing methods to select steels for GH2 storage and transport vessels. World Hydrogen Energy Conf. Pasadena, June 1982.

Developments and Tests on Retainer Bearings for Large Active Magnetic Bearings

H. Dell[1], J. Engel[2], R. Faber[1], D. Glass[2]

1) Hochtemperatur-Reaktorbau GmbH, Dortmund
2) ASEA Brown Boveri Aktiengesellschaft, Mannheim

1. Reactor Coolant Gas Circulators with Magnetic Bearings and the Retainer Bearing Test Facility

Radial and axial active magnetic bearings due to their specific features and operating characteristics can meet technical requirements beyond the scope of conventional bearings. With these bearings rotating machines can be arranged encapsulated in the process medium without any risk of interaction with lubrication oil or stationary component parts during normal operation. They are thus predestined for use in gas-cooled high temperature reactors. This is true in particular because the demand of freedom from maintenance is met here in a way not reached in mechanical engineering before.

Fig. 1 is a cross section through a coolant gas circulator prototype being developed at present. It includes the same proven assemblies used in the circulators for gas-cooled reactors designed and built by ABB to date /1/. A basically new feature is the vertical circulator axis with rotor weights up to several tons and, naturally, the non-contacting active magnetic bearings /2/.

The circulator is designed as a withdrawable unit for a reactor service life of forty years, with an overhung impeller arranged at the top. Drive motor and cooling device are integrated. The complete unit is arranged in the helium loop of the reactor and is under operating pressure.

An essential prerequisite for the use of active magnetic bearings are retainer bearings, needed for preventing undesired contacts between the rotor and fixed parts of the machine when the magnets are de-energized. In view of the large dimensions and the shaft weights to be kept under control design and construction of the retainer bearing are the subject of a research and development project backed by experiments.

Fig. 2 depicts the functioning mode of the retainer bearing test facility with active magnetic bearings designed by ABB and set up at HRB in Jülich. Its configuration and shaft dimensions have been chosen such that they correspond to the conditions of the prototype circulator as far as possible. The locations shown in red and dark blue illustrate the principle: if the active magnetic bearings fail the shaft will drop onto conical parts which themselves run in rolling-contact bearings.

Within a few hundred ms the retainer bearing is accelerated to the shaft speed when the shaft is dropped. The shaft may then safely coast to a halt according to the power consumed in the impeller wheel or with additional electric braking by the motor.

The advantage of the active magnetic bearing in the retainer bearing test facility is not only that shaft drops take place under realistic boundary conditions. It is also possible in addition to make rotor loads associated with larger rotor dimensions act upon the retainer bearings via magnetic forces. And finally it is possible to re-energize the magnetic bearing during an experiment in the event of an impending failure of a retainer bearing. The rotor will then return to the levitation position and further damage of conical parts and retainer bearing components is avoided.

Fig. 3 is a view of the retainer bearing test facility. The switch cabinets house the retainer electronic devices for the magnetic bearing and the converter as well as the instrumentation. The actual rotating system is inside the container coated with light paint; this container also allows experiments in a helium atmosphere. Nearly 100 rotor drops have been carried out to date.

2. Design Principles for Retainer Bearings

Fig. 4 illustrates the design features. In the axial-radial retainer bearing the shaft forces are directed via friction cones into the retainer bearing bushing which in this case runs in three rolling-contact bearings. Depending on the failure mode the shaft can be supported by the pairs of cones RKI or RKII.

In a radial retainer bearing radial shaft motions are guided by the cylindrical friction surfaces RSIV. For centering of the shaft the pair of cones RKIII may be used.

The retainer bearing units thus basically consist of rolling-contact bearings and friction surfaces. By means of the friction coefficient μ a major influence on the angular acceleration of the retainer bearing bushing is exerted right at the start of the drop. Fairly smooth transitions from standstill to rated rotational speed are to be aimed at in order to keep the ball and ball retainer ring accelerations occurring inside the rolling-contact bearings within acceptable limits.

The friction surfaces furthermore have the additional basic task of acting as an emergency bearing and of bridging a potential rolling-contact bearing failure.

The friction cones may be made of various materials, such as e.g. sintered materials based on bronze, metallic and ceramic coatings, and metallic materials of high surface hardness. For the technology of helium-cooled reactors the considerably differing friction behaviour - compared with normal atmospheric conditions - is the main factor for the choice of materials. Generally the values for the friction coefficient μ in helium under reactor conditions are much higher than in air.

Fig. 5 shows an axial-radial retainer bearing unit in assembled condition with mounted telemetering instrumentation for the non-contacting transmission of temperatures measured in the rotating part. A part of the friction cone with a radial slot arranged for collecting abraded material is clearly visible. The balls, running grooves and retainer borings shown represent the parts of a dry-lubrication bearing after exposure to loads. The matt luster of the balls and the traces of tarnishing are clear indications that ball bearing wear has already started.

As the presentation of the test results to date will also show, great importance is attached to the tribology of rolling-contact bearings. For the application of retainer bearings in machines with active magnetic bearings used for nuclear reactors dry lubrication has a wide application potential. In view of a reactor service life of 40 years the maintenance requirements and disadvantages involved in the limited stability time of lubricating grease justify almost any effort for developing serviceable dry-lubrication retainer bearings with reproducible behaviour.

In our opinion the application of the active magnetic bearings technology for large machines in nuclear engineering depends decisively on the number of safely guaranteed drops into the retainer bearings. For conventional mechanical engineering this restrictions does not apply to the same extent, since reactor-related and conservative availability considerations for active magnetic bearings and their components need not be made. The difficulties involved in the handling of irradiated machine components do not exist, and the bearings are always accessible. For conventional engineering the operational and economic benefits of active magnetic bearings are therefore much more distinct.

3. Phenomenology of the Test Parameters

Fig. 6 shows the typical curves of a number of test parameters for a drop from an initial speed of 6100 r.p.m.

From the curve trace "ARF bush speed" it can be seen that the run-up of the axial-radial retainer bearing (ARF) occurs approximately 1.1 s. For the radial retainer bearing (RF) an analog run-up with a time delay of approximately 0.5 s takes place according to the curve trace "RF bush speed".

The curve of the axial force taken up by the ARF is most interesting. At a rotor weight of 1320 kg in the test facility an additional magnetic load was introduced for this test. As the curve trace "axial force" shows an axial load of approx. 25 kN was present after the dynamic processes involved in the drop had died down.

The shock load occurring at the first moment of the drop amounting to 104 kN was four times as large as the quasi-static load occurring subsequently.

The curve trace "Rotor Displacement" for the axial shaft position clearly reflects the dynamic dropping process. The elasticity of the design chosen by us permits an overshooting of more than 0.2 mm beyond the axial "drop path" of the rotor amounting to approx. 0.7 mm in the equilibrium condition.

In Fig. 7 the experimental findings for the run-up time of the ARF cone with retainer bearing bushing are depicted as a function of the drop speed of the shaft. The scattering of the measured values as shown reflects, apart from the effect of the damage to the rolling contact bearing which increases with the number of drops the assumption that the friction processes during the drop are much more complex than might be expected according to the classical friction law.

With Fig. 8 the important influencing quantity temperature increase in the friction cone comes in. It also rises considerably with rising drop speed. If the plotted data are adapted by means of a power function one obtains nearly the theoretically expected speed exponent 2 for the approximation of the ΔT figures ($\ln \Delta T = - 12.6 + 1.93 \ln n$).

In Fig. 9 the evolution of the temperatures in the 5 rolling-contact bearings L1 to L5 of the entire retainer bearing arrangement for a drop from approximately 6000 r.p.m. is shown. The shaft coastdown time amounts to slightly more than 15 minutes. Distinct maximum temperature differentials of up to 12 K appear between internal and external ring of the ball bearings with a lubricating grease film.

For dry-lubrication bearings the respective ΔT values are higher by a factor 3 to 5. The corresponding material expansions are a major input variable for the detail design of circulators, because they directly affect the tolerances and clearances between rotating and fixed machine parts.

Since additional reserves for temperature gradients in the circulator resulting from operational influences must always be taken into consideration, it is quite obvious that this is one of the major problem areas for the design of retainer bearings for large machines.

This brief survey makes clear that even with the use of commercial rolling-contact bearings with a grease film no satisfactory results could be observed at first during the drops. For a successful application considerable modifications and adaptations of the design parameters are necessary. This concerns in particular:

- Bearing play, pressure angle and osculation
- Ball bearing spacer guidance, design and material
- Surface configuration of the elements involved in the rolling movement
- Type, quantity and method of application of lubricants
- Damping of the bearings.

In case of dry bearings special attention must be paid to some of the above-mentioned parameters acc. to Fig. 10, since the temperature developments are much more pronounced during a drop, and the impact of the tribological properties is much more serious. Therefore experimental parameter optimizations are necessary which are being carried out by us at present.

The interactions of the various material gliding upon each other in the rolling-contact bearing are a key problem, with the proportion of the masses to be accelerated in the starting phase exerting a great influence. Dry lubrication films can help to avoid the rapid destruction of the surface structure of balls, rings and retainers in a helium atmosphere with metallic con–tacts and cold welding.

From a phenomenological point of view the bearing load is best determined by a comparison of two surface peak-to-valley measurements. Fig. 11 shows two extremes: rolling-contact bearing surfaces in new condition and after failure due to overloading. The shown orders of magnitude for the surface roughness illustrate the experimental tolerance, with the failure value for the surface roughness of a rolling-contact bearing being dependent on the dry lubricant used. It is our objective to discover recipes and procedures by means of which, in a similar way as with a film of grease, the metallic contact is reproducibly prevented for a large number of drops. A prerequisite for this is the development of suitable non-destructive inspection procedures for in-process manufacturing control.

4. Metrology Assisting the Development of Retainer Bearings

Irrespective of the influences of dry lubrication discussed so far there is a large improvement potential in the configuration of the heat removal when the retainer bearings are loaded. There are structural possibilities available as well as the possibility to exert influence through forced cooling.

For solving the difficult optimization tasks we are assisted in the retainer bearing test facility by the sophisticated instrumentation which has been installed. In Fig. 12 the significant measured variables and the means for recording them are illustrated. The temperature measuring devices based on complex telemetric equipment shall be specially mentioned, since thermocouples can be placed in the immediate vicinity of the rotating heat sources, and there are no disturbing factors when the thermoelectric voltage is picked up.

In Fig. 13 the design data of the retainer bearing test facility are compiled. The large variation scope of the individual parameters arose from the requirements involved in the different circulator variants of future HTR lines, from the heating reactor up to the HTR 500.

We thus have a universally useful test equipment at our disposal allowing us to investigate even special features of active magnetic bearings.

The success already achieved with the detail design of retainer bearings for coolant gas circulators is an indication of the fact that the retainer bearing test facility may in the medium term also be used for optimizing retainer bearings for other applications. Under this aspect the R&D funds invested will have a "spin off" effect beyond the current objective. We wish to thank the Federal Ministry of Research and Technology for their sponsorship of the experimental work.

References

/1/ IAEA Specialists Meeting on Gas-Cooled Reactor Coolant Circulator and Blower Technology, San Diego, Nov. 30 - Dec 2, 1987

D. STOELZL, "25 Years of Brown Boveri Experience in Development, Design, and Fabrication of Circulators for HTGR"

E.L. ZIERMANN, J. ENGEL, "Two Decades of Excellent AGR Circulator Operating Performance at AVR Jülich"

E. GLAHE, D. STOELZL, "Circulator Performance during the THTR Trial and Post-Commissiong Operation"

/2/ J. ENGEL, D. GLASS, VGB Conference 'Research in Power Plant Technology 1988 , Essen, 16/17 March 1988;
"Circulators with Active Magnetic Bearings for High Temperature Reactors"

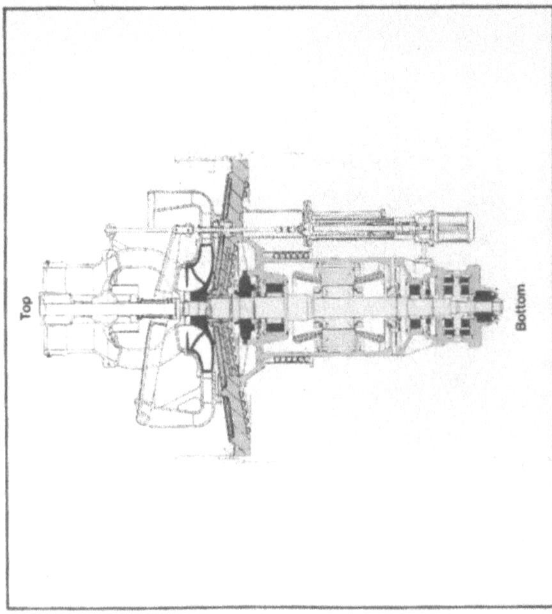

BBC BROWN BOVERI | Cross Section of the NWA-Circulator for the HTR 500 | Fig. 1

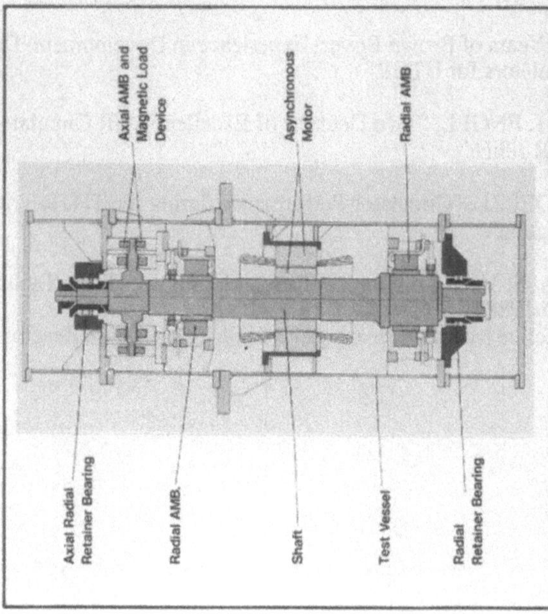

BBC BROWN BOVERI | Arrangement for the Retainer Bearing Test Facility, Schematic | Fig. 2

BBC | Retainer bearing test facility with active magnetic bearings | Fig. 3

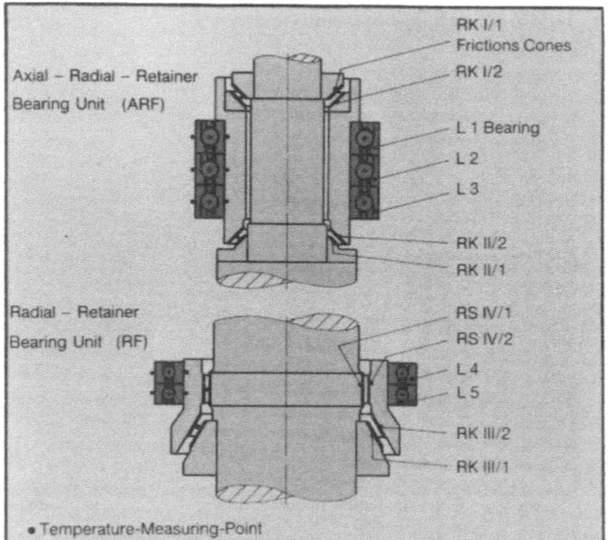

HRB | Schematic of the Retainer Bearing Elem.– Ball Bearings and Friction Cones | Fig. 4 | 88 2-59

HRB | Complete Axial – Radial Retainer Bearing Unit and Ball Bearing Details | Fig. 5 88.2–62

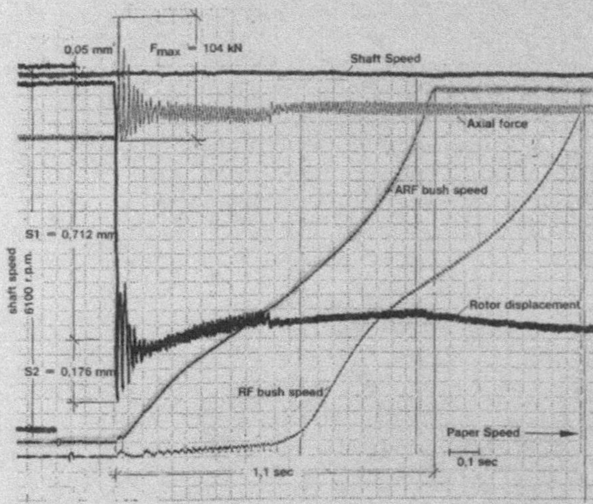

BBC BROWN BOVERI | Test Parameters at the Beginning of a Shaft Drop | Fig. 6

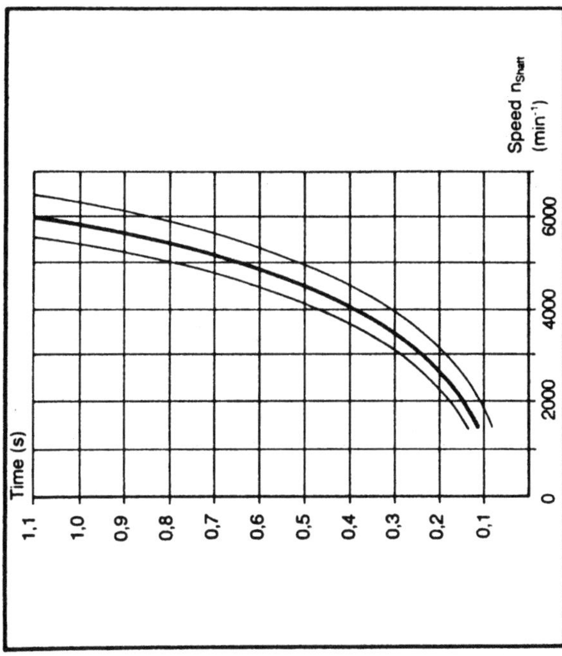

HRB Cone Run-up Time as a Function of Initial Shaft Drop Speed — Fig. 7 — 88.2–61

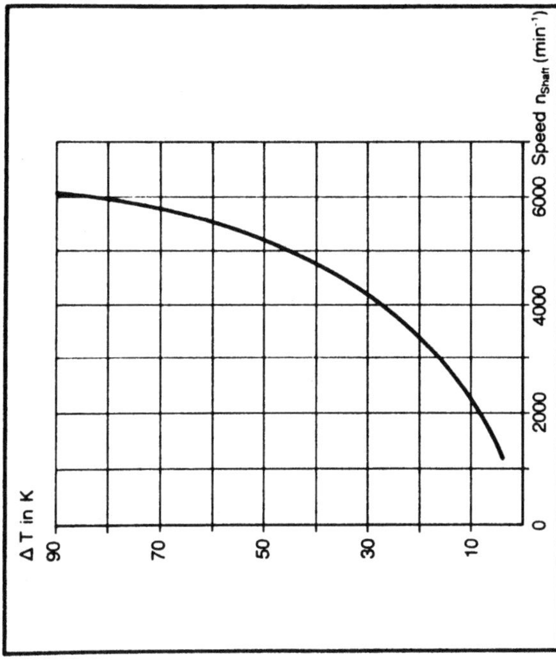

HRB Temperature Rise in the Friction Cones as a Function of Initial Shaft Drop Speed — Fig. 8 — 88.2–65

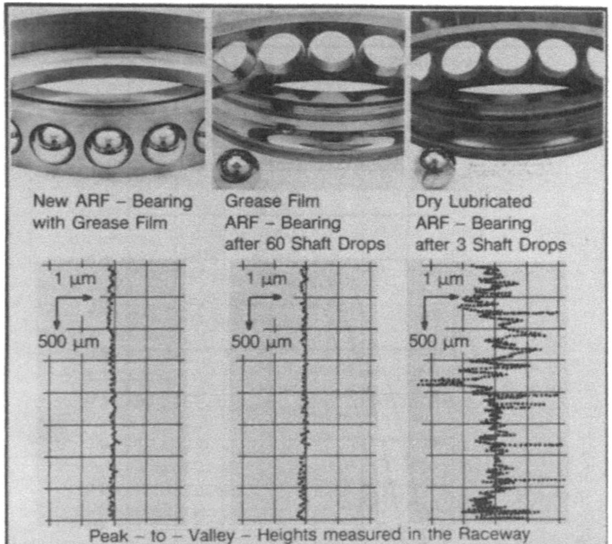

HRB | Peak-To-Valley-Heights after Various Ball Bearing Loads | Fig. 11 | 88.2-60

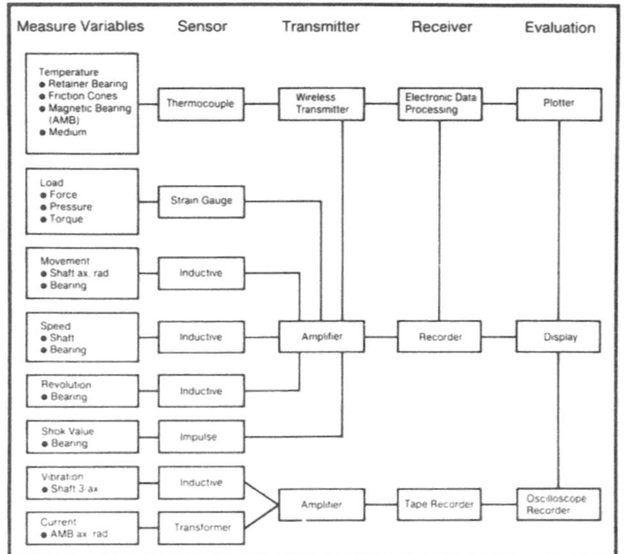

HRB | Retainer Bearings Test Bed Measuring Techniques | Fig. 12 | 88.2-67

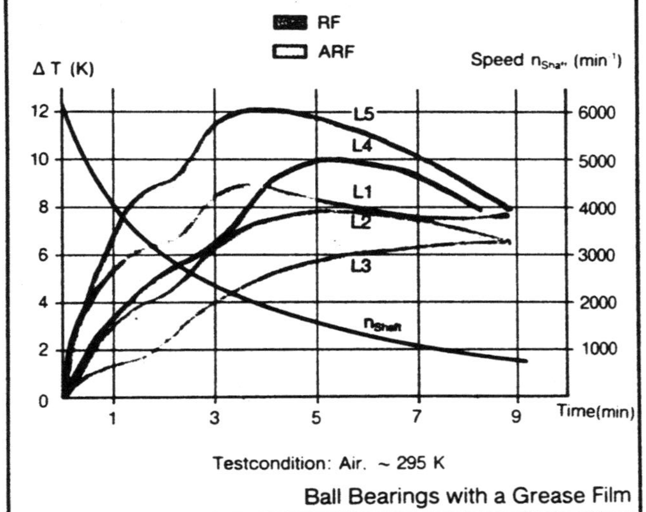

HRB | Typical Temperature Rise Curves ΔT During Coastdown from ~6000 r.p.m | Fig.9 | 88.2-64

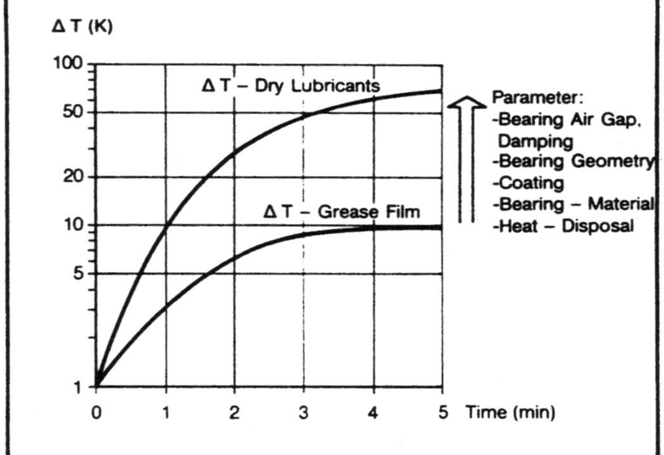

HRB | Parameters Affecting the Temperature Differences ΔT in Ball Bearings | Fig.10 | 88.2-66

Test Bearing -Dimensions (max.)		di ~ 300 mm
Speed		≤ 7.200 min^{-1}
Loads	-Pressure	300 kN
	-Tensile	100 kN
	-Moment of Torsion	4000 Nm
Kind of Loads		axial + radial
Pressure		$0 < p < 3$ bar
Temperature		333K (393K)
Medium		He, N_2, Air, Vakuum
Engine Power		Pmax = 300 kW
Shaft Mass		2000 kg
Testbox – Dimensions		Ø 1 m Height 3 m

HRB — Significant Desgin Data of the Retainer Bearings Test Bed — Fig. 13 — 88.2–63

Practical Applications of the Active Magnetic Bearings to the Industrial World

Maurice BRUNET, Technical Manager

Société de Mécanique Magnétique (S2M)
Vernon/Saint-Marcel France

Summary
Active Magnetic Bearings (AMB) are developed, manufactured and marketed by Société de Mécanique Magnétique (S2M) since 1976. This company has made major improvements to the AMB fundamental technology in order to meet industrial requirements. Each component of the AMB system has been improved through a simple, robust, and patented solution. Among others, inductive sensors, flux feedback control, control loops (automatic balancing system), auxiliary bearings,... Many industrial applications in various and severe environmental conditions have been set into operation. They have been into operation for years and have shown their reliability.

Introduction

This paper does not deal much with theory of Active Magnetic Bearings (AMB). Many other papers have previously described the subject; for an introduction to AMB one should refer to them.

We shall focus here mainly on:

- some of the technological improvements brought by S2M to the AMB,

- industrial applications.

The Société de Mécanique Magnétique (S2M) was founded in 1976 by the Société Européenne de Propulsion (SEP) and the SKF group. Its aim was to continue development and to commercialize active magnetic bearings which had been the subject of research programs starting in the 1960s.

S2M is located in the Vernon/Saint-Marcel industrial area in Normandy, France. The S2M plant has been built on 2 hectares (5 acres) and occupies 3,000 m^2 (27,000 sp.ft.). S2M employs 75 persons (January 1st, 1988 figures).

Fig. 1. S2M plant in Vernon/Saint-Marcel

More than 35 patents registered since 1970 (in France, in Europe, in the United States and in Japan) protect SEP and S2M inventions in this field.

For over 15 years the S2M shareholders have invested more than 50 MFF directly in the development of active magnetic bearings, in the form of capital or services. S2M invests approximately 10 % of its turnover in research and development.

S2M is equipped with computerized design and management equipment and has production capabilities both in the electromechanical and electronic fields. S2M also ensures the implementation and after sales service of its active magnetic bearings throughout the world.

S2M's wealth is in its team of men and women, whose work and experience have built an unequalled technological advance in the design and production of active magnetic bearings.

S2M has created two subsidiaries to implement the process in America and in the Pacific and South-East zones:

- MBI (MAGNETIC BEARINGS, INC.) in the United States, with the KOLLMORGEN group,
- JMB (JAPAN MAGNETIC BEARINGS CO., LTD.) in Japan, with the SEIKO group.

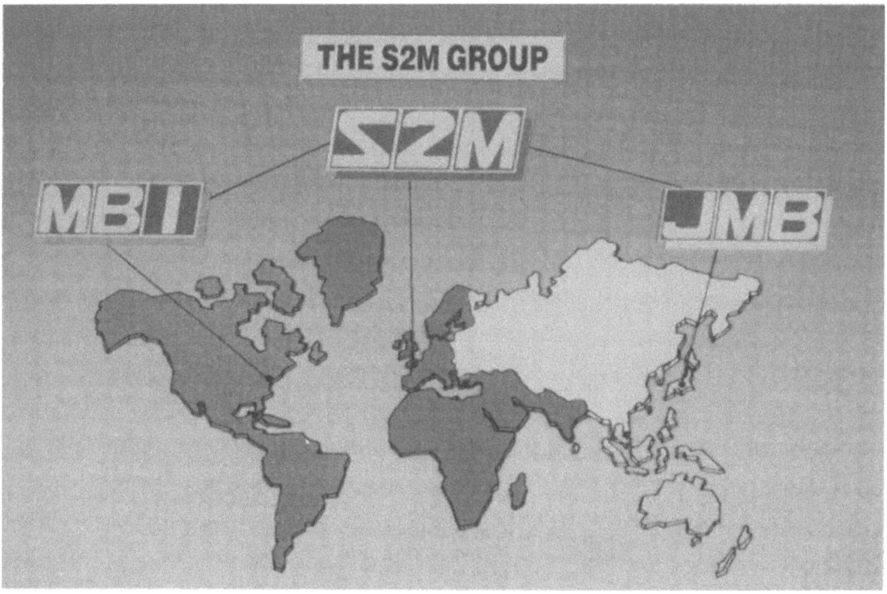

Fig. 2. S2M group in the world

These two subsidiaries make up a worldwide organization based on three poles:

- S2M which is especially orientated towards the European, African and Middle East markets. S2M is, of course, the leader both in research and development and promotion of new applications,
- JMB licensed for the Pacific area and handling the heavy industry market while the sub-licensee SEIKO SEIKI handles the light industry and machine tool markets.

The S2M group, the growth rate of which is about 40% per year, is the only group in the world able to design, manufacture and commercialize active magnetic bearings on an industrial level.

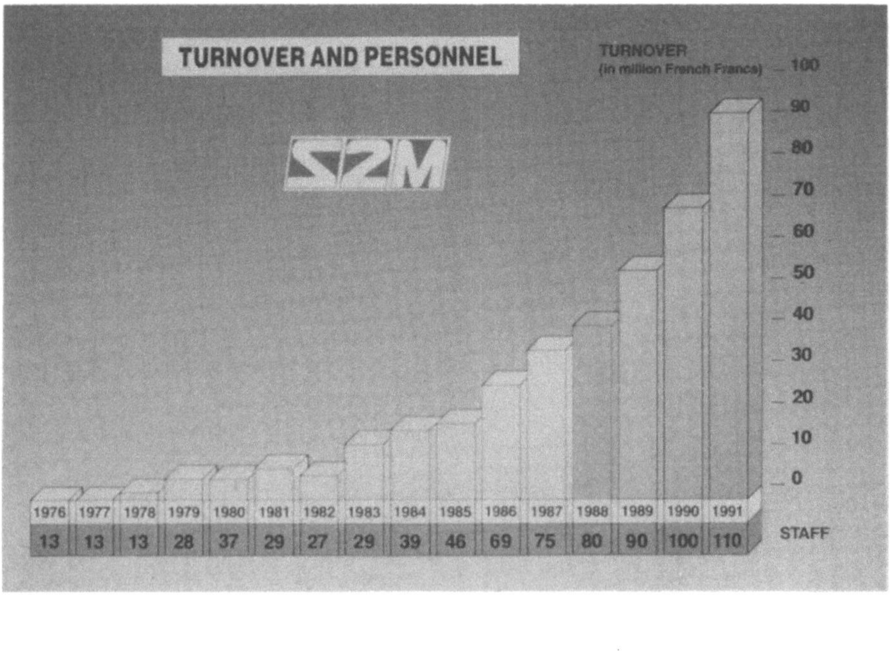

Fig. 3. S2M growth, historic and perspectives

MAJOR IMPROVEMENTS OF AMB'S FUNDAMENTAL TECHNOLOGY BY S2M

The basic AMB technology using electromagnets in attraction controlled by a servo-loop is well-known. However, various major improvements to the AMB fundamental technology are necessary to meet industrial requirements.

Each component of the AMB system has to be improved through a simple and robust solution, to avoid weak points.

Figure 4. shows some of the S2M patented improvements. Thanks to them really operational AMB are manufactured.

MAJOR IMPROVEMENTS OF AMB'S FUNDAMENTAL TECHNOLOGY BY S2M

INDUCTIVE SENSORS TECHNOLOGY

ELECTROMAGNETS WITH FLUX FEEDBACK CONTROL

CONTROL LOOPS:
AUTOMATIC BALANCING SYSTEM

AUXILIARY BEARINGS

Fig. 4. Major improvements of AMB's fundamental technology

Inductive sensors technology

The technology selected by S2M for the transducer of the servo-loop is the inductive sensors in a Wheatstone bridge. This principle is known as especially accurate (principle of measuring equipment) and simple. It provides a high level signal which does not required to be preamplified before line transmission to the control electronics. It is a noise-free signal, from which rotation harmonics are filtered through a simple, static mechanical device, also patented. Getting rid of any preamplifier on the machine itself increases the reliability of the AMB system. All sensitive parts are under protection in the control room, remote from the machine (up to 50 meters or even more).

The most impressive application in the S2M sensors know-how is the ultra-precision spindle for optical lenses machining. The spindle drives the lense (lathe principle) at 2,000 rpm, with a power of 1 kW. On this spindle an accuracy of 0.05 micron on the machined part is achieved. This value is not dependent of the material.

Fig. 5. Example of ultra-precision achieved with S2M inductive sensors technology: spindle for optical lenses machining. Machining accuracy: 0.05 micrometer.

The electromagnets with flux feedback control

The well-known drawback of the electromagnets is the non-linearity of the electromagnetic force in a variable airgap. The S2M solution (patented) is the flux feedback control by means of a simple flux measurement coil. This technology avoids any fragile sensor on the machine.

Most of S2M applications are equipped with this device. The figure hereafter shows one example. This is a turbo expander unit used for air liquefaction. On the turbine wheel side the gas is expanded and cooled to -190°C, the shaft transmits the power (300 kW, 38,000 rpm) to the compressor wheel on the other side. The thrust reaches 7,500 N, i.e. 25 times the rotor weight. This high load factor would not be possible without the flux control.

Fig. 6. Example of high load factor: turbo expander with a 7,500 N thrust load on a 30 kg rotor weight.

Control loops: Automatic Balancing System

The classic PID (Proportional, Integral, Derivative) network has been improved by various filters and cross-coupling devices.

One of the S2M patents is the Automatic Balancing System (ABS). Target of this system is the elimination of stator vibrations at variable speeds, even in case of high and variable unbalance by a rotation of the rotor on its inertial axis. The ABS system uses rotating coordinates driven by the rotation speed. The "frequency filter" of the bearing is thus directly indexed on the rotation speed.

The ABS system has been used many times, it is integrated in the majority of the S2M applications, especially the turbo-machines.

For example, this turbogenerator equipped with a 8,000 kg rotor rotating at 3,000 rpm in a very dusty environment of hot blast furnace exhaust gas (energy recovery expander). The process dust by sticking on or abrasing the expansion wheel changes permanently the unbalance. Thanks to the ABS, the rotor rotates around its inertial axis and no vibration is transmitted to the stator. The unbalance allowed is 250 g.m.

Fig. 7. Example of high unbalance: turbogenerator with 250 g.m. acceptable unbalance.

Auxiliary bearings technology

S2M has now a great practical experience of auxiliary bearings improvements. These systems provide good damping for a "smooth landing" of the rotor in case of ailure in the control system, even in case of a flexible rotor rotating over its first free-free bending mode.

The figure shows a centrifugal compressor with flexible rotor (300 kg, 13,500 rpm) on which evidence of good landing behaviour has been made.

The shaft can cross its first free-free critical speed on the auxiliary bearings.

Fig. 8. Example of landing capability on auxiliary bearings: centrifugal compressor.

INDUSTRIAL APPLICATIONS

Many industrial applications in various and severe environmental conditions have been set into operation. They have been in operation for years and have shown their reliability.

Vacuum:	**Ultra-high vaccum pump**
	Neutron chopper
Extreme temperatures:	**Cryogenic turbo expander**
	Liquid helium cryogenic compressor
	Hot helium blower
Machining:	**Very high speed milling**
	Creep feed grinding
Turbomachines:	**Steam turbine**
	Centrifugal compressor
	Synopsis of turbomachines.

Ultra-high vacuum pump:

This design makes use of an internal stator. The external rotor (7 kg) rotates at 30,000 rpm. The turbomolecular pump enables to achieve a very high level of vacuum (10^{-11} Torr). It is well suited for vibration-sensitive applications such as electron microscopy and mass spectrometry. It is also absolutely oil-free.

Fig. 9. Ultra-high vacuum pump

Neutron chopper:

The principle of this laboratory equipment which works in vacuum is to chop a neutron flux by a rotating shutter.

The synchronisation of the shaft speed with the neutron source is essential. Therefore, a constant speed of 30,000 rpm is required ($\Delta\omega/\omega$ = 10^{-6}).

Thanks to the very low and constant electromagnetic friction torque of the AMB the speed control can be achieved.

Fig. 10. Neutron chopper

Cryogenic turbo expander:

This application has been shown before as an example of high load factor. We focus now on its industrial operation. As said before, it is used in air liquefaction plants. The temperature difference on this short shaft (900 mm) is very high: -190°C on the expander side, +110°C on the compressor side. The complicated lubrication system has been eliminated. The machine is now fully oil-free and can be started at any time without heating up. Therefore, it can follow very easily the production demands in liquid gases.

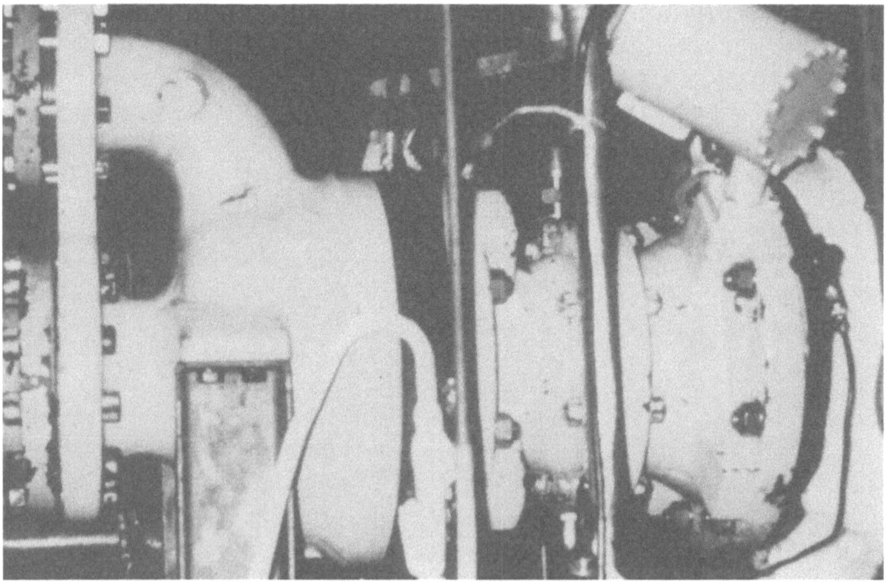

Fig. 11. Cryogenic turbo expander

Liquid helium cryogenic compressor:

Liquid helium (supercritical phase) at a temperature of 4 K requires a special compressor design. This compressor has a vertical shaft (6 kg) rotating at 36,000 rpm, on AMB. The heat transfer from the motor and AMB to the liquid helium is less than 1 W. This is of major importance. The efficiency of the whole refrigerating loop is much improved (1 W heat input means 1 kW energy consumption to cool it down).

Fig. 12. Liquid helium cryogenic compressor

Hot helium blower:

The hot helium blower is developed within the scope of the very high temperature nuclear reactor program. In these reactors the core is cooled by helium which goes to heat exchanger and brought back to the reactor by blowers.

The AMB of the blower presented here have been insulated with a special ceramic coating. They withstand a temperature of 450°C. Other AMB advantages for this application are the non-pollution of the helium by lubricant and the high reliability.

Fig. 13. Hot helium blower

Very high speed milling:

The very high speed milling is especially interesting in the aircraft industry where finished parts (such as frame, wings, integral structures,...) come out from solid parts with more than 90% chip removal.

The S2M electrospindles on AMB provide both very high speed and high power (from 30,000 rpm/ 35 kW up to 60,000 rpm/22 kW).

The milling machine shown here is equipped with a 45,000 rpm/ 30 kW electrospindle.

Fig. 14. Very high speed milling

Creep feed grinding:

The grinding machine presented here has been designed with a S2M electrospindle B15/1000 (60,000 rpm/22 kW).

Thanks to Cubical Bore Nitrid (CBN) tools, this grinding process may be now performed at very high speed (up to 300 m/s). A high productivity is reached together with high precision.

The machining is done in one pass and a width accuracy of 1 micron can be reached by the electronic displacement of the electrospindle rotor in the air gap.

Fig. 15. Creep feed grinding machine

Steam turbine:

The first steam turbine on AMB has been started by S2M and JMB in Japan (Mitsubishi Heavy Industries).

It is a new machine of 3 MW at 15,000 rpm, which has been designed in order to integrate all the new technological developments available today, and, among others, AMB. Thanks to them, the mechanical design has been simplified by the elimination of the lubrication system. Other major advantages are the fact that no vibrations are transmitted to the stator, and the fact that the user has at his disposal a continuous monitoring of the rotor dynamic without additional equipment.

Fig. 16. Steam turbine

Centrifugal compressor:

The picture presented here shows the same compressor before and after its retrofit with AMB and dry seals. The difference is obvious. All the accessories have disappeared.

It is a compressor of pipeline rotating at 11,500 rpm (4,100 kW).

The AMB brings the following advantages to the user of a centrifugal compressor:

- reduced losses,
- availibility (immediate starting of the machine, without lubrication system heating time),
- elimination of mechanical maintenance,
- elimination of vibrations,
- no pollution of the process gas by the lubricant...

Fig. 17. Centrifugal compressor

Synopsis of turbomachines:

More than 30 turbomachines are now running on AMB.

TURBOMACHINERY EQUIPPED WITH ACTIDYNE^{r} ACTIVE MAGNETIC BEARINGS

MACHINE TYPE	QUANTITY	START UP DATES	POWER RANGE kW	NOMINAL SPEED RANGE	APPLICATION
TEST COMPRESSORS	5	1980 - 1987	20 kW to 4,000 kW	10,000 rpm/ 16,000 rpm	- TEST STANDS
INDUSTRIAL COMPRESSORS	15	1985 - 1988	3,100 kW to 26,100 kW	5,000 rpm/ 14,500 rpm	- PIPELINES - REFINERIES
HERMETIC COMPRESSOR	1	1981	300 kW	18,000 rpm	- TEST STAND
MOTORIZED BLOWERS	4	1980 - 1987	5 to 200 kW	3,600 rpm/ 12,000 rpm	- TEST STANDS - HELIUM LOOP
GAS TURBINE	1	1988	12,000 kW	5,250 rpm	- COMPRESSOR DRIVE
STEAM TURBINE	1	1987	3,000 kW	15,000 rpm	- COMPRESSOR DRIVE
TURBO-GENERATOR	1	1986	5,000 kW	3,000 rpm	- ENERGY RECOVERY (BLAST FURNACE GAS)
TURBO EXPANDERS	3	1985 - 1988	300 kW to 900 kW	30,000 rpm/ 38,000 rpm	- AIR LIQUEFACTION

Fig. 18. Synopsis of turbomachines

Conclusion:

Thanks to its 20 years of experience and development, S2M has transformed the concept of AMB into an industrial reality, as easily demonstrated by its achievements to date.

PRESENT STATUS OF ACTIDYNE® ACTIVE MAGNETIC BEARINGS	
ROTATION SPEED	0 to 800,000 rpm
ROTOR DIAMETER	14 mm to 1.25 m (0.6 in to 50 in)
LOAD PER BEARING	up to 100,000 N* (22,000 lbs)
AMBIENT TEMPERATURE	-253°C to +450°C (-420°F to +840°F)
ENVIRONMENT	air, vacuum, helium, nitrogen, hydrocarbons
QUANTITY OF SUSPENSIONS MANUFACTURED	> 400 related to more than 30 different applications

*The present technology allows 300,000 N (66,000 lbs) per bearing.

The companies of the S2M group, thanks to their wealth of huma resources and their constant effort in research, development, and industrialization have provided the industry with highly performing economical and efficient AMB.

As our clients conclude:

"We choose Active Magnetic Bearings
For their high performance
And S2M for its experience."

Magnetic Bearing Turbomachinery Operating Experience

H. J. Moses, F. D. Pinckney and D. A. Weise

MAGNETIC BEARINGS, INC.
Radford, Virginia 24141 USA

Summary

The theory of active magnetic bearing technology utilizing electromagnetics and closed loop feedback control is presented. A system engineering process has been developed to apply active magnetic bearings to turbocompressors of various physical sizes, horsepower ratings and performance characteristics. This process is explained and a specific example reviewed. Characteristics of a number of North American machines designed using the system engineering methodology are tabulated. Future technology applications and enhancements are outlined.

Introduction

Volumes of research, development and analytical work can be identified relating to the history of magnetic bearings and suspension devices. The concept of magnetically suspending a rotor began to receive serious attention in the mid-nineteenth century. U.S. patents dating back to the early 1800's incorporated the use of "magnetic bearings" to support rotating shafts, reduce mechanical friction, and improve the efficiency of electrical meters. In 1842 Ernshaw analytically demonstrated the instability inherent in totally passive (permanent) magnetic bearing systems and presented the requirement of at least one active degree of freedom to achieve rotor stability.

In the years since then passive, active, and hybrid systems have been experimented with, each with its own advantages and disadvantages. While the basic theories involved have not significantly changed over the years, advancements in magnetic materials, servo circuitry, and power electronic components have provided many enhancements to the magnetic bearing technology. Today private corporations and institutes around the world are applying magnetic bearings in devices as diversified as centrifugal compressors, helium circulators, x-ray tubes, energy storage flywheels, high vacuum pumps, machine tool spindles, cryogenic compressors and space guidance. While the applications have been extremely varied, the technologies involved can be classified as passive, active, or hybrid systems. To date, the technology most successfully applied to large turbomachinery has been the fully active magnetic bearing system as described below.

Active Magnetic Bearing Operation

In theory, the principle is quite basic. An electromagnet will attract any piece of ferrous material. By using a stationary electromagnet (stator) and a rotating ferrous material (rotor) a shaft can be suspended in a magnetic field while maintaining accurate position under varying loads. This can be accomplished given a small space (air gap) between the stator and rotor and proper electronic control of the electromagnet. In the case of the active magnetic bearing, this concept is utilized for both radial and axial configurations. The bearing system described always operates in an attraction mode and never repulsion.

The radial and axial bearing rotors make use of a ferrous laminated sleeve and solid disc respectively. Applying ferrous rotor elements to the shaft allows the shaft material to be constructed from a nonmagnetic metal or composite material. While the radial bearing requires laminations due to the number of flux reversals during rotation, the axial rotor disc can be solid since the magnetic flux level is changing but the polarity is not.

As with any type of electromagnet, a wound field stator is required to produce a force output. Both the radial and axial bearing stators incorporate laminations to minimize stray losses and improve the bearing response time. The radial bearing stator is wound to provide four independently controllable quadrants for maximum rotor stability. The axial bearing, attracting the rotor in only one plane, requires the use of two stators, one on either side of the rotor disc, to provide double acting control.

Inductive position sensors are used to detect the exact radial and axial location of the shaft. Similar to the bearings, these sensors utilize a ferrous rotor and a wound field stator. As the air gap at the sensors changes with shaft disturbances, the inductance of the sensor also changes. It is this change in inductance with air gap variation that provides the position feedback signal required for closed loop servo control.

Figure 1, Radial and Axial Bearing Configuration, shows an isometric view of both a radial and double acting axial bearing with their associated position sensors.

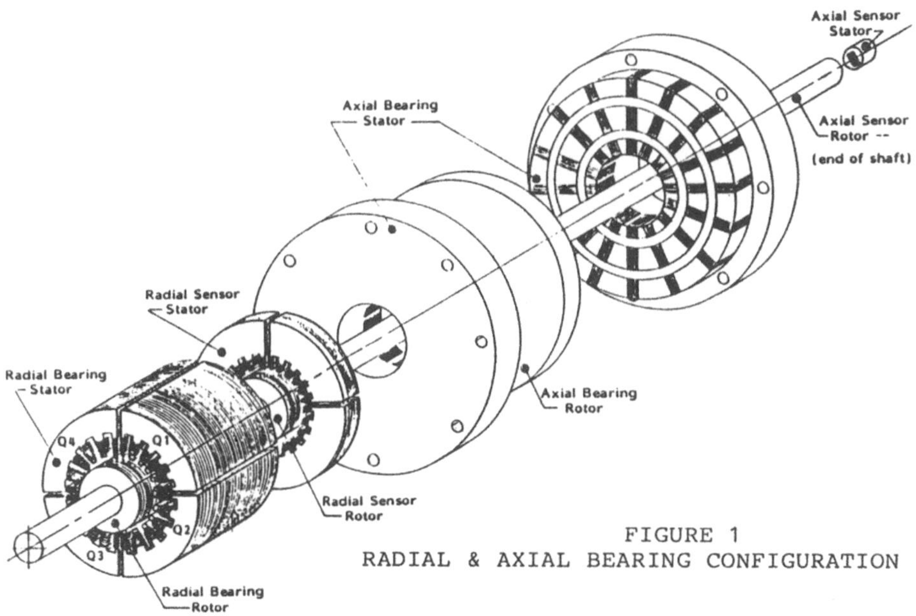

FIGURE 1
RADIAL & AXIAL BEARING CONFIGURATION

Control electronics are required to process the position signal and power the appropriate bearing coils. The exact shaft location is detected by the position sensors, and a DC voltage is generated which is related to rotor displacement. This DC voltage (where the shaft is) is compared to the position reference signal (where the shaft should be). Any difference between these two signals generates an error signal which is used to maintain control of the rotor. This signal is then amplified, filtered, and conditioned prior to commanding the specified power amplifier(s). Current is increased or decreased in the appropriate bearing coil(s) to maintain the rotor at equilibrium. Figure 2, Basic Control Loop Diagram, shows a block diagram of the closed loop servo control.

BASIC CONTROL LOOP DIAGRAM

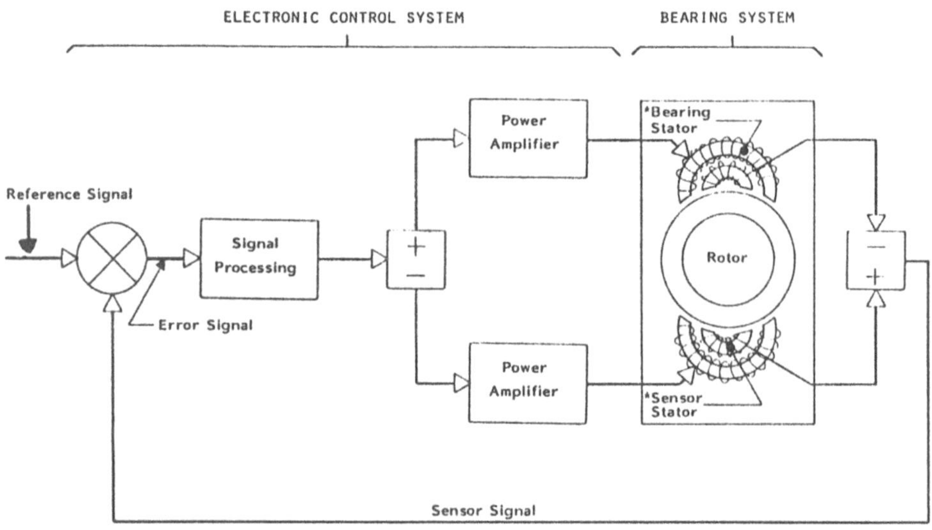

*Stators shown at 0° and 180° for Simplicity

FIGURE 2

System Engineering

A system engineering methodology has been developed to apply active magnetic bearings as described above to a specific turbomachine with a high degree of design success. Figure 3,

System Engineering Methodology, flow charts the process beginning with a machine performance specification.

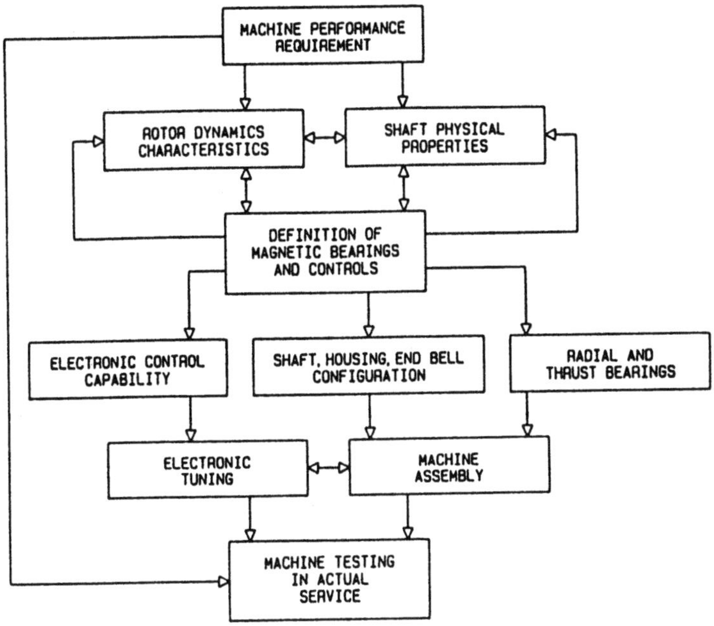

FIGURE 3
SYSTEM ENGINEERING METHODOLOGY

As shown by the steps in Figure 3, application of magnetic bearings, from conceptual layout through equipment start-up, requires a "system" rather than a "component" engineering approach. Physical configuration and operating environment, static and dynamic load requirements, and rotor dynamics must all be addressed to achieve successful implementation for a machine such as a centrifugal compressor.

Optimizing the physical configuration is an iterative process involving the machine (compressor) designer and the magnetic bearing designer. New options exist for the machine (compressor) designer such as bearings immersed in the process gas, larger shaft diameters at the bearing journals and more flexibility in locating bearings within the machine (compressor) case (i.e. midspan bearings, inboard or outboard thrust bearings). The bearing operating environment is usually established early in the design process; of particular concern is the nature of the fluid in contact with the bearing components and the operating temperatures. This is of paramount importance in design of the bearing hardware, especially winding and encapsulant selection.

Detailed consideration must be given to the static and dynamic loads the magnetic bearing will control. Magnetic bearings are not forgiving of gross overloads and it is often difficult to establish the load requirements, especially on a new machine. The magnetic bearing designer can manipulate several parameters (bus voltage, airgap, number of turns, control class) to optimize dynamic load capability but ultimately bearing active area and the magnetic material saturation flux density will limit the bearing load capability. For this reason it is often desirable to select a bearing with substantial reserve capacity. One approach is to design so that static levitation is achieved at about 1.0 Tesla flux density in the bearing airgap. This gives a margin of about two to one on static load capacity when using materials (silicon iron) with a 1.5 Tesla saturation flux density. This margin results from the magnetic force being proportional to the square of the magnetic flux density:

$$\left(\frac{1.5}{1.0}\right)^2 = 2.25 \qquad (1)$$

Rotor dynamics must be considered in parallel, and often dictate, the machine (compressor) physical configuration. The magnetic bearing controller governs the bearing dynamic characteristics of stiffness and damping which determines the rotor system dynamics. The machine (compressor) designer seeks to achieve an acceptable design based on critical speed margins, amplification factors and rotor stability while the magnetic bearing designer must meet minimum gain and phase margins over a broad bandwidth for control loop (and therefore rotor) stability. The "system approach" found most useful is to first generate an undamped critical speed map and mode shapes for the rotor based on a desired physical configuration. This requires detailed model data from the machine designer. The mode shapes may indicate problems with nodes near the bearing or between the bearing and sensor. Machine modifications or bearing design changes such as sensor location or even multiple sensors may be necessary. If mode shapes are acceptable, reasonable bearing stiffness characteristics may be estimated from the critical speed map. The controller gain and phase characteristics may then be tailored (within limits) to the application. Controller characteristics are often based on previous working designs or modifications of previous designs. The magnetic bearing designer must insure the proposed bearing characteristics are reasonable and achievable in practice from both control loop stability and bearing load capacity perspectives. A design for example that has a rotor running speed stiffness (gain) so high that the full dynamic load capacity of the bearing is commanded (by the controller) in a few microinches of rotor motion is probably unreasonable for this class of machines. A listing of the proposed bearing characteristics is then supplied to the machine designer to verify acceptable rotor response using magnetic bearings.

The final step in the "System Engineering" process is rotor levitation and control loop compensation with the actual machine (compressor). The magnetic bearing designer must then confirm that the proposed bearing characteristics have been achieved while also satisfying more fundamental control loop stability requirements.

A Specific Application

A specific application will now be reviewed to highlight some details of the system engineering involved in an actual machine. The application was a magnetic bearing retrofit of a single stage natural gas pipeline compressor. The machine had a 700 lb. rotor operating at a maximum speed of 11,700 rpm. The radial bearings had a maximum load capacity of 890 lb. per bearing quadrant. Static levitation was achieved at 0.8 Tesla flux density in a 0.020 inch airgap. The compressor had a conventional physical configuration with all bearings outside of the process gas and the thrust bearing inboard of the radial bearings. Operating speed was under the first shaft bending mode (3rd critical) as shown in Figure 4, Critical Speed Map. First and Third free-free (low bearing stiffness) mode shapes are shown in Figure 5 and Figure 6 respectively. The final controller gain and phase and resulting bearing stiffness and damping characteristics are shown in Figure 7. The bearing stiffness (k) and dynamic stiffness (K) are also plotted on the critical speed map (figure 4). The dynamic stiffness K is the vector sum of the (real) stiffness and the (imaginary) damping stiffness and is noted as "KMAG" in figure 7. Although a higher gain control loop (i.e. stiffer bearings) were initially proposed, it was found during site tuning to cause the shaft second mode to encroach on the operating speed range. The data shown reflects the actual bearing characteristics as measured during start-up at the site. A second machine, identical to the first unit, has since been commissioned. Controller and bearing characteristics of the two units are identical. The two units now have a total of more than 10,000 hours of operation.

Benefits of Active Magnetic Bearings

Application of active magnetic bearings depends not only on a successful implementation as described above but also on recognizing and quantifying the expected economic benefits.

Reasons for utilizing magnetic bearings in rotating machinery vary with each particular application, although many common threads are evident. Heavy equipment users, typically employing oil lubricated tilting pad bearings, see many advantages including efficiency and safety in eliminating the oil lubrication system. Such a system which utilizes external lube oil pumps, piping, reservoirs, and filters also contains costly elements to install and maintain. In many cases, more heavy equipment down time is attributable to failures in machinery subsystems than actual machinery failure itself.

Other users of magnetic bearings cite higher speeds, harsh environment operation and optimized rotor dynamic characteristics as reasons for using magnetic bearings. A typical weighing of factors for a pipeline compressor is shown in Figure 8, Magnetic Bearing Advantages.

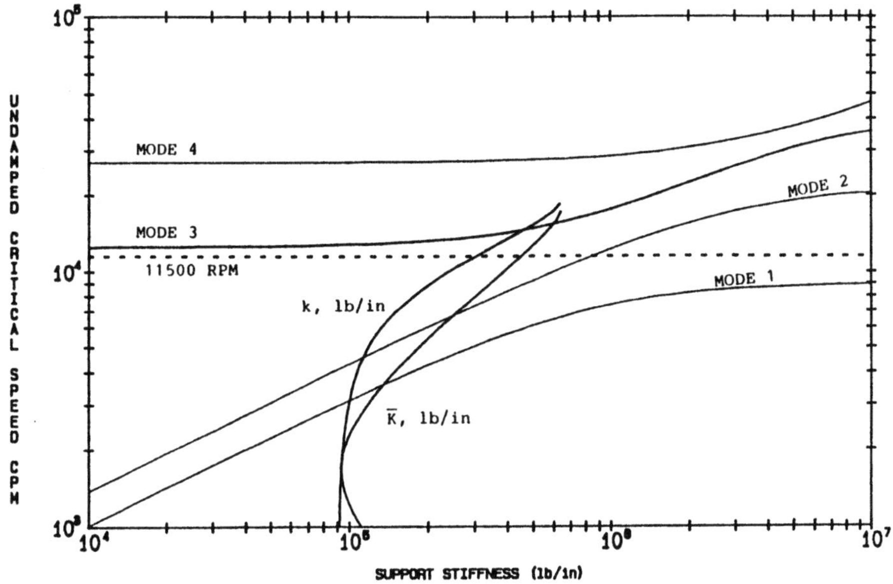

FIGURE 4
CRITICAL SPEED MAP

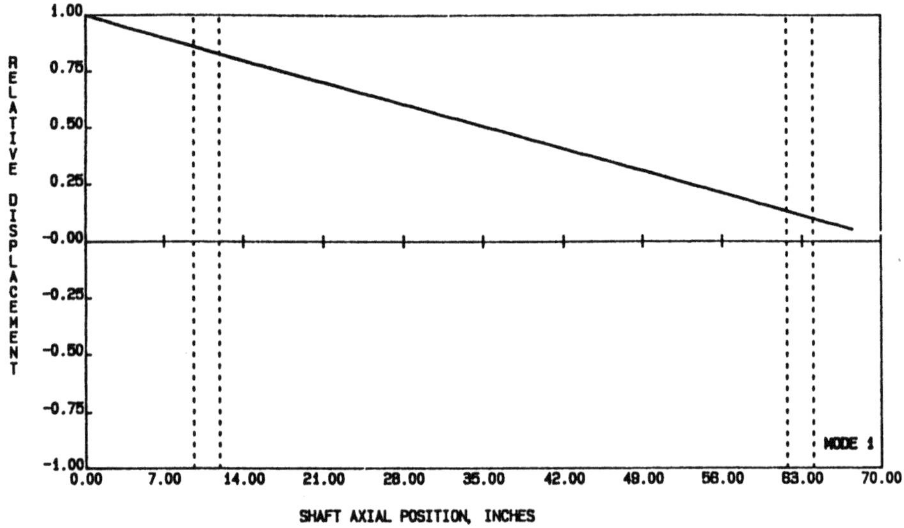

J59007
FDP 10-1-88

UNDAMPED NATURAL FREQUENCY (CPM): 1008
BEARING STIFFNESS (LB/IN): 1.0E4

FIGURE 5
MODE SHAPE

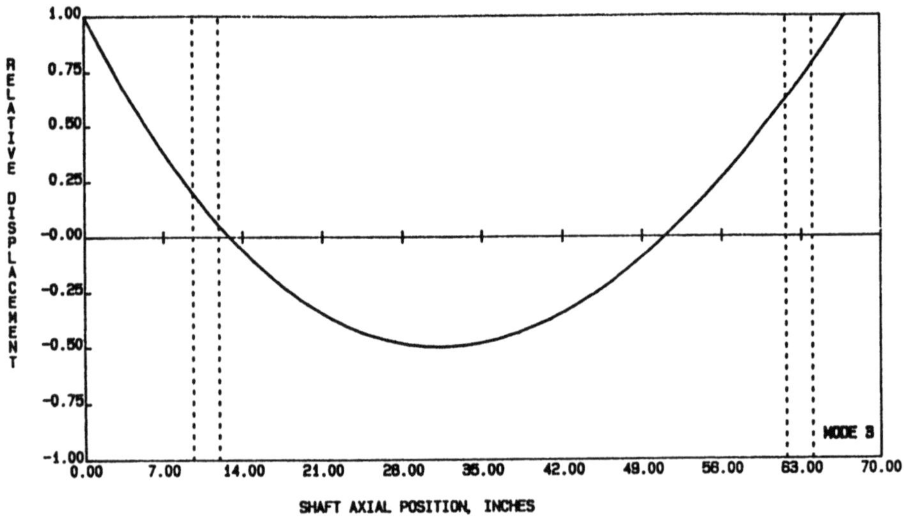

J59007
FDP 10-1-88

UNDAMPED NATURAL FREQUENCY (CPM): 12475
BEARING STIFFNESS (LB/IN): 1.0E4

FIGURE 6
MODE SHAPE

****
MAGNETIC BEARINGS INC
JOB #59007
BEARING CHARACTERISTICS
TUNED 11-86
SF=318265 K=345000LB/IN @ 150HZ
****

FREQ	GAIN	PHASE	RPM	K,LB/IN	C,LB*SEC/IN	KMAG
32.750	0.298	7.85	1965.00	0.9408E+05	0.6300E+02	0.9497E+05
35.110	0.304	12.33	2106.60	0.9461E+05	0.9375E+02	0.9685E+05
37.650	0.312	16.67	2259.00	0.9525E+05	0.1206E+03	0.9943E+05
40.370	0.323	20.83	2422.20	0.9599E+05	0.1440E+03	0.1027E+06
43.290	0.335	24.79	2597.40	0.9685E+05	0.1645E+03	0.1067E+06
46.420	0.350	28.53	2785.20	0.9787E+05	0.1824E+03	0.1114E+06
49.770	0.367	32.02	2986.20	0.9903E+05	0.1980E+03	0.1168E+06
53.370	0.387	35.27	3202.20	0.1005E+06	0.2119E+03	0.1230E+06
57.220	0.409	38.28	3433.20	0.1021E+06	0.2240E+03	0.1300E+06
61.360	0.433	41.02	3681.60	0.1040E+06	0.2347E+03	0.1378E+06
65.790	0.461	43.52	3947.40	0.1063E+06	0.2441E+03	0.1466E+06
70.550	0.491	45.77	4233.00	0.1090E+06	0.2526E+03	0.1562E+06
75.650	0.524	47.76	4539.00	0.1122E+06	0.2600E+03	0.1660E+06
81.110	0.561	49.51	4866.60	0.1160E+06	0.2666E+03	0.1786E+06
86.970	0.602	51.01	5218.20	0.1205E+06	0.2725E+03	0.1916E+06
93.260	0.647	52.24	5595.60	0.1260E+06	0.2777E+03	0.2058E+06
100.000	0.696	53.23	6000.00	0.1325E+06	0.2822E+03	0.2214E+06
107.200	0.749	53.95	6432.00	0.1403E+06	0.2862E+03	0.2384E+06
115.000	0.808	54.40	6900.00	0.1497E+06	0.2893E+03	0.2571E+06
123.300	0.872	54.56	7398.00	0.1609E+06	0.2919E+03	0.2775E+06
132.200	0.942	54.42	7932.00	0.1744E+06	0.2935E+03	0.2998E+06
141.700	1.018	53.98	8502.00	0.1905E+06	0.2943E+03	0.3240E+06
152.000	1.101	53.20	9120.00	0.2099E+06	0.2938E+03	0.3504E+06
163.000	1.190	52.05	9780.00	0.2329E+06	0.2916E+03	0.3787E+06
174.800	1.286	50.51	10488.00	0.2603E+06	0.2876E+03	0.4093E+06
187.400	1.388	48.53	11244.00	0.2925E+06	0.2811E+03	0.4418E+06
200.900	1.497	46.06	12054.00	0.3306E+06	0.2718E+03	0.4764E+06
215.400	1.610	43.01	12924.00	0.3747E+06	0.2583E+03	0.5124E+06
231.000	1.726	39.27	13860.00	0.4253E+06	0.2396E+03	0.5493E+06
247.700	1.840	34.66	14862.00	0.4817E+06	0.2140E+03	0.5856E+06
265.600	1.944	28.91	15936.00	0.5416E+06	0.1792E+03	0.6187E+06

FIGURE 7
TUNED BEARING CHARACTERISTICS

ADVANTAGES	WEIGHTING
-NO LUB SYSTEM REQUIRED	✓ ✓ ✓
-VERY LONG LIFE	✓ ✓ ✓
-MINIMAL MACHINERY VIBRATION	✓ ✓
-ACCURATE SHAFT POSITION	✓ ✓
-HIGH SPEED OPERATION	✓
-HARSH ENVIRONMENT OPERATION	✓ ✓ ✓
-NO PROCESS CONTAMINATION	✓ ✓ ✓
-REDUCED MAINTENANCE	✓ ✓
-HIGH RELIABILITY	✓ ✓
-ELECTRONIC SYSTEM INTERGRATION	✓ ✓

FIGURE 8: MAGNETIC BEARING ADVANTAGES

Current Experience

Having discussed the design process and reasons for incorporating active magnetic bearings, Figure 9, Turbomachinery Applications, shows the characteristics of a dozen magnetic bearing equipped machines which are either in service or planned for 1988 commissioning. Total operating time for this group is presently in excess of 30,000 hours. Based upon utilization estimates, it is anticipated that these machines will accumulate about 50,000 hours annually as the full set of units being constructed is placed in service. These turbomachines are owned by either Nova, an Alberta Corporation, Shell Canada Products Limited, or Trans Canada Pipelines and are located in Canada. The equipment was originally supplied by North American manufacturers either with magnetic bearings or retrofit modifications were performed to incorporate magnetic bearings into conventional units. Together, this grouping represents the largest concentration of turbomachinery operating on magnetic bearings anywhere in the world.

Experience to date continues to verify the system engineering process discussed above. Machine operation, once turned over to the operating organizations, has been trouble free. Reliable oil free operation with better efficiency, lower operating costs, and decreased vibration levels have all been achieved. Figure 9, Turbomachinery Applications, indicates a wide range of shaft rotor dynamics, machine sizes, loadings, and configurations. The design data base includes beam type and cantilever rotors, high thrust loadings, bearings operating directly in high temperature or corrosive process mediums, and repeat units of the same type. The latter are especially noteworthy in that they confirm the potential for lower cost standardized applications with predictable results.

MACHINE	ROTOR TYPE	SERVICE	DUTY	COMM	ROTOR WEIGHT LB	THRUST LOAD LB	SPEED RPM	JOURNAL DIAMETER	RATING HP	OPERATING HOURS (*)
CDP-230 COMP.	2 STAGE BEAM	PIPELINE METHANE	SEASONAL	1985	3200	12000	5250	10.6"	14650	9696
CDP-416 COMP.	4 STAGE BEAM	PIPELINE METHANE	SEASONAL	1986	280	3370	14500	3.7"	4150	9000
1B26 COMP.	1 STAGE BEAM	PIPELINE METHANE	CONTINUOUS	1986	780	4050	11000	6.5"	5540	8130
1B26 COMP.	1 STAGE BEAM	PIPELINE METHANE	CONTINUOUS	1987	780	4050	11000	6.5"	5540	2956
CBF-842 COMP.	8 STAGE BEAM	REFINERY WET GAS	CONTINUOUS	1987	1420	4590	10250	6.0"	4500	3900
3B37 COMP.	3 STAGE BEAM	REFINERY HYDROGEN	CONTINUOUS	1988	1050	4100	10230	6.0"	4500	---
RFB30 COMP.	1 STAGE OVERHUNG	PIPELINE METHANE	SEASONAL	1988	1730	15750	5200	10.0"	14250	---
5P2 COMP.	2 STAGE BEAM	PIPELINE METHANE	SEASONAL	1988	1500	10000	6800	6.0"	16600	---
5P2 COMP.	2 STAGE BEAM	PIPELINE METHANE	SEASONAL	1988	1500	10000	6800	6.0"	16600	---
5P2 COMP.	2 STATE BEAM	PIPELINE METHANE	SEASONAL	1988	1500	10000	6800	6.0"	16600	---
RF2BB30 COMP.	2 STATE BEAM	PIPELINE METHANE	SEASONAL	1988	2000	18000	5000	10.0"	29600	---
GT-51 TURBINE	1 STAGE OVERHUNG	PIPELINE	SEASONAL	1988	3530	24500	5250	8.3" 12.5"	14650	---

*AMB CABINET OPERATING HOURS AS OF APRIL 12, 1988

FIGURE 9
TURBOMACHINERY APPLICATIONS

Advanced Designs

Machines as detailed in Figure 9 have superior performance capabilities, but they do not fully exploit the possible improvements magnetic bearings can provide for turbomachinery. New machine designs are required to utilize the bearing and control capabilities in innovative ways. Turbomachinery performance depends heavily upon the dynamics of shaft and rotor design. Shaft suspension techniques have not changed markedly in decades, but magnetic bearings enable no contact operation at sustained high speeds with variable stiffness and damping coefficients and intelligent electronic control.

These characteristics can be applied to a number of turbomachinery types. Smaller compressors running at higher speeds with no oil system can save space on off shore platforms or in station modernization/upgrade situations. Higher efficiency steam turbines with no lubrication requirements are feasible. Gearboxes may be eliminated in some processes by directly matching driver and compressor speeds. Aircraft gas turbine thrust to weight improvements are possible through higher speeds and lighter weight structures. Canned pump life will be greatly enhanced by the ability of the magnetic bearing to perform in the process fluid. Fans and blowers will have bearings better able to survive harsh environments. Cascaded machinery strings for electrical generators or cogeneration can make use of the bearing capability as a transducer to damp out vibration.

Electronic control capabilities provide the features necessary to enable an intelligent bearing system. This intelligence can be applied in several ways. Machine performance can be altered adaptively depending on speeds or loads. Changes in operating conditions such as surge, unbalance, or seal wear can be recognized. Machine status, health maintenance trends, and internal bearing system self diagnosis data are all available within the electronic control system for local presentation or integration into control room functions.

<u>Technology Trends</u>

Technology advancements are proceeding in many areas applicable to further improving the performance and decreasing the cost of future magnetic bearing systems. Digital electronics utilizing custom chips and power semiconductors in hybrid packages will enable modular control systems to economically fulfill a wide variety of turbomachinery applications. Advances in magnetic materials and structures ultimately involving superconductivity will result in higher force density/smaller size bearings. Active magnetic bearing product enhancements will continue at a rapid pace because the basic technologies - computer simulation, control analysis, power electronics, electronic circuit packaging, and electro-magnetics are all strong areas of technology research.

Identification of Bearing and Rotor Parameters

Dynamic Response of Magnetically Supported Rotor

David W. Lewis, Professor
Eric Maslen, Graduate Research Assistant
Robert R. Humphris, Research Professor

Mechanical and Aerospace Engineering Department
University of Virginia
Charlottesville, Virginia 22901, USA

Summary

A theoretical analysis of the dynamic response of a rotor supported in magnetic bearings is presented. This predicted response is compared with the experimental response of a small rotor subjected to an impulse loading produced by a spring loaded solenoid. Theoretical and experimental results are compared for various stiffness and damping settings for magnetic bearings.

Nomenclature

C	damping matrix	K_d	rate (derivative) gain
f	exogeneous force	K_x	position stiffness
G	control transfer function	M	mass matrix
K	stiffness matrix	X	Laplace of position
K_p	proportional gain	τ	time constant
K_i	current stiffness	α,β	transfer function coefficient

Introduction

The stability of rotors may be affected significantly by use of bearings having stiffnesses that are unequal in mutually perpendicular directions, the theory of which was presented so eloquently by Smith [1]. Credance to the work by Smith came from experiments reported in 1924 by Newkirk [2] who suggested that the unsymmetrical flexibility of the bearings was the essential feature inducing stability even in the absence of damping in the bearings. At the time of this work, no convenient methods were available to implement controllable variations in the stiffness of a bearing in two orthogonal directions although bearing supports could be designed to produce unequal, but not adjustable bearing stiffnesses.

* This work was supported in part by the U.S. Army Research Office, NASA Lewis Research Center, and Sundstrand Corporation.

The use of magnetic bearings has included applications to very large machinery including high-speed multistage compressors [3]. Considerable details of the controls for magnetic bearings are now available for design and for proper applications [4]. Innumerable patents have been filed and countless papers reported in the field of magnetic bearings, some 200 of which were summarized by Humphris [5] in 1985. The reported works have probably doubled since that time as the potential for magnetic bearings has become better understood and accepted.

The option now exists through the implementation of digital controls [6] to design "smart" magnetic bearings to incorporate the findings reported decades ago. By changing the stiffnesses of the bearings in two mutually perpendicular directions as the speeds are traversed, rotors can be made more stable [7]. Simultaneously, varying the stiffnesses of magnetic bearings as a rotor speed changes offers the means to avoid running through any critical speeds as the "critical" speeds of the system will not be fixed values, but variable quantities dependent upon the controlled bearing stiffness and damping values.

As magnetic bearings become more commonplace, questions regarding some behavior details about them seem germane. This paper addresses one such detail, namely, "Can the dynamic response of a rotor in magnetic bearings be predicted theoretically when the rotor is subjected to an impulsive loading?" This is a first step in predicting the behavior of a rotor that suffers a sudden upset such as the loss of a blade in a compressor.

<u>Experimental Set-Up</u>

The arrangement used for the experimental impact testing is shown in Figure 1. The 12.7 mm (0.5 inch) diameter shaft is supported at each end by a magnetic bearing consisting of four electromagnets uniformly distributed radially about the shaft. Details of the controls and physical properties of this system have been previously reported [8]. The bearing span is 508 mm (20 in.) and a 0.8 Kg (1.8 lb) disk is located at the shaft center. A rubber coupling joins the rotor to a variable speed electric motor. Coupling properties are estimated at 17500 N/m stiffness and 8.80 N-sec/m damping. Eddy current position sensors are located vertically and horizontally at each end, displaced 30 mm from the magnetic bearing disks. Signals from these sensors are used, with the control electronics, to maintain the shaft in support and are also used as the transducers for observing the effects or results for this impact study.

A simple spring loaded solenoid, taken from an ordinary door-bell chime, is used as the impact mechanism. An electrolytic capacitor is initially charged to a given voltage.

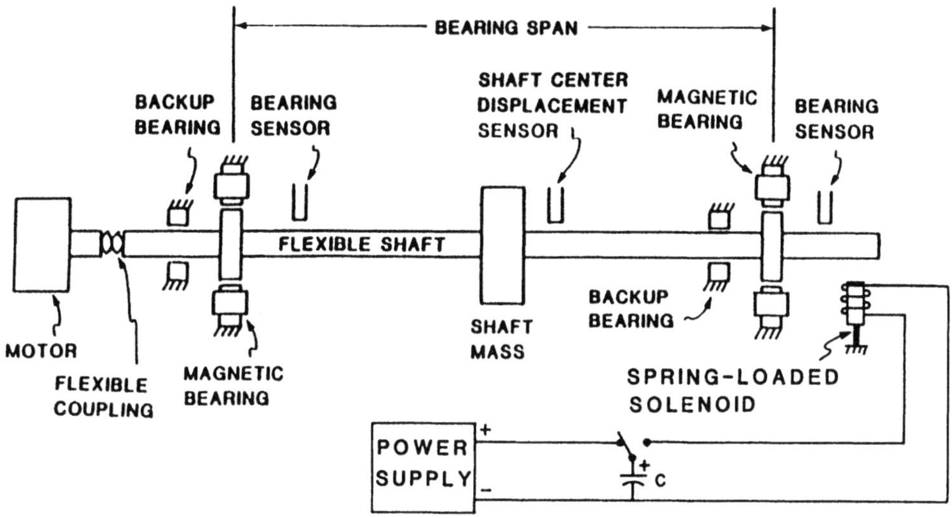

Figure 1. Experimental Schematic Diagram with Spring-Loaded Solenoid

When the push button switch is actuated, the stored charge is dumped through the spring loaded solenoid coil, causing the plunger to impact on the desired point of the shaft. The force of the impact is accurately controlled by varying the power supply voltage applied to the capacitor. Very repeatable impacts are obtained, as observed by a digital signal analyzer used to capture the transient behavior.

State Space Rotor/Bearing Model Development

To obtain a description of a rotor supported in magnetic bearings adequate for stability analysis and transient response simulation, a detailed model is required. Previously reported works have either assumed a simple second order plant model [7] or have examined higher order plant models using state space techniques, but neglected the high frequency limitations (poles) of the controller [9]. In this development, a general methodology is outlined for constructing a model which includes not only higher order plant dynamics and controller bandwidth limitations, but also permits accurate modeling of sensor/actuator collocation error. This last effect has been discussed fairly widely in the literature since 1977 [9,10].

Assuming that an undamped structure model for the rotor has been obtained of the form

$$M\ddot{\underline{x}} + K\underline{x} = \underline{f} \tag{1}$$

and that any mechanical bearings, seals, and gyroscopic effects can be modeled by linear stiffness and damping matrices K_b and C_b, respectively, the matrix differential equation can be expanded to the state space form

$$\frac{d}{dt}\begin{bmatrix} \underline{x} \\ \dot{\underline{x}} \end{bmatrix} = \left[\begin{array}{c|c} 0 & I \\ \hline -M^{-1}(K + K_b) & -M^{-1}C_b \end{array}\right]\begin{bmatrix} \underline{x} \\ \dot{\underline{x}} \end{bmatrix} + \begin{bmatrix} 0 \\ M^{-1} \end{bmatrix}\underline{f} \tag{2}$$

Magnetic bearings may be described by a linearized force law relating the bearing force to displacement of the journal and symmetric perturbation of the bearing coil currents [8]:

$$f_a = -K_x x_a - K_i i \tag{3}$$

The perturbation current, i, is related to the motion of node_s of the rotor by the linear control law

$$I = G(s)X_s \tag{4}$$

The bearing force is applied to node_a while the sensor providing the feedback control senses the position of node_s. If node_a and node_s are different (as they generally are), then collocation error is introduced.

To incorporate the bearing description of equation (3) into the state space description, the frequency domain equation governing the perturbation current must be transformed back to the time domain. Thus,

$$i^{(n)} + \beta_{n-1} i^{(n-1)} + \ldots + \beta_1 i^{(1)} + \beta_0 i = \alpha_m x_s^{(m)} + \ldots + \alpha_1 x_s^{(1)} + \alpha_0 x_s \tag{5}$$

in which the superscript (i) denotes the ith time derivative and the coefficients α_i and β_i are derived from the transfer function G(s):

$$G(s) = \frac{\alpha_m s^m + \ldots + \alpha_1 s + \alpha_0}{s^n + \beta_{n-1} s^{n-1} + \ldots + \beta_1 s + \beta_0} \tag{6}$$

In general, the numerator of G(s) will be of order higher than one, implying that equation (5) will involve higher derivatives of x_s than are available in the state vector. This problem is circumvented as G(s) can always be expanded, via partial fraction expansion, into a sum of transfer functions all of which are zero or first order in the numerator. In this manner, a bearing having a second or higher order transfer function numerator can be split into several separate bearings acting at the same point, each having first or zero order transfer function numerators.

So equation (5) can, without loss of generality, be written as:

$$i^{(n)} = \alpha_0 x_s + \alpha_1 x_s^{(1)} - \beta_0 i - \beta_1 i^{(1)} - \ldots - \beta_{n-1} i^{(n-1)} \tag{7}$$

Transforming to state space form,

$$\frac{d}{dt}\begin{bmatrix} x_s \\ x_s^{(1)} \\ i \\ i^{(1)} \\ \vdots \\ i^{(n-1)} \end{bmatrix} = \left[\begin{array}{cc|ccccc} 0 & 1 & & & & & \\ * & * & & & & & \\ \hline & & 0 & 1 & & & \\ & & 0 & 0 & 1 & & \\ & & & & & \ddots & \\ & & & & & & 1 \\ \alpha_0 & \alpha_1 & -\beta_0 & -\beta_1 & \ldots & & -\beta_{n-1} \end{array}\right] \begin{bmatrix} x_s \\ x_s^{(1)} \\ i \\ i^{(1)} \\ \vdots \\ i^{(n-1)} \end{bmatrix} \tag{8}$$

Equation (8) is incorporated into equation (1) by augmenting the state vector with the required current terms:

$$\frac{d}{dt}\left[\begin{array}{c} \underline{x} \\ \hline \dot{\underline{x}} \\ \hline i \\ i^{(1)} \\ \vdots \\ i^{(n-1)} \end{array}\right] = \left[\begin{array}{c|c|cccc} 0 & I & & & & \\ \hline -M^{-1}(K + K_b) & -M^{-1}C_b & & & & \\ \hline & & 0 & 1 & & \\ & & & & 1 & \\ & & & & & \ddots \\ & & & & & 1 \\ \alpha_0 & \alpha_1 & -\beta_0 & -\beta_1 & \ldots & -\beta_{n-1} \end{array}\right] \left[\begin{array}{c} \underline{x} \\ \hline \dot{\underline{x}} \\ \hline i \\ i^{(1)} \\ \vdots \\ i^{(n-1)} \end{array}\right] + \left[\begin{array}{c} 0 \\ \hline M^{-1} \\ \hline 0 \end{array}\right] f \tag{9}$$

Finally, equation (3) forms a feedback control law for the force term in equation (9):

$$f = \left[\begin{array}{ccccccc} 0\cdots0 & -K_x & 0\cdots0 & \vdots & \vdots & -K_i & 0 \cdots 0 \end{array}\right] \left[\begin{array}{c} \underline{x} \\ \hline \dot{\underline{x}} \\ \hline i \\ i^{(1)} \\ \vdots \\ i^{(n-1)} \end{array}\right] \tag{10}$$

An example will illustrate the assembly process. The dynamics of an axisymmetric nongyroscopic three mass rotor model is described by:

$$\begin{bmatrix} 2 & 0 & 0 \\ 0 & 4 & 0 \\ 0 & 0 & 1 \end{bmatrix} \ddot{\underline{x}} + \begin{bmatrix} 2 & -3 & 1 \\ -3 & 5 & -2 \\ 1 & -2 & 1 \end{bmatrix} \underline{x} = \underline{f} \tag{11}$$

There are two magnetic bearings, acting on stations 1 and 3. The bearing acting on station 1 has no collocation error, while that acting on station 3 senses station 2. They are described by:

$$f_1 = 5x_1 - 2i_1, \quad f_3 = 3x_3 - 3i_2 \tag{12}$$

The transfer functions relating the perturbation currents to the rotor motion are

$$G_1(s) = \frac{0.1s + 2}{s^3 + 4s^2 + 5s + 3}, \quad G_2(s) = \frac{5}{s^2 + 6s + 2} \tag{13}$$

After expanding to state space form, incorporating the bearing transfer functions, and implementing the feedback of equation (10), the system differential equation becomes

$$\frac{d}{dt}\begin{bmatrix} x_1 \\ x_2 \\ x_3 \\ \dot{x}_1 \\ \dot{x}_2 \\ \dot{x}_3 \\ i_1 \\ \dot{i}_1 \\ \ddot{i}_1 \\ i_2 \\ \dot{i}_2 \end{bmatrix} = \left[\begin{array}{ccc|ccc|ccc|cc} & & & 1 & & & & & & & \\ & & & & 1 & & & & & & \\ & & & & & 1 & & & & & \\ \hline 1.5 & 1.5 & -.5 & & & & -1 & & & & \\ .75 & -1.25 & .5 & & & & & & & & \\ -1 & 2 & 2 & & & & & & & -3 & \\ \hline & & & & & & & 1 & & & \\ & & & & & & & & 1 & & \\ 2 & & & .1 & & & -3 & -5 & -4 & & \\ \hline & & & & & & & & & & 1 \\ & 5 & & & & & & & & -2 & -6 \end{array}\right] \begin{bmatrix} x_1 \\ x_2 \\ x_3 \\ \dot{x}_1 \\ \dot{x}_2 \\ \dot{x}_3 \\ i_1 \\ \dot{i}_1 \\ \ddot{i}_1 \\ i_2 \\ \dot{i}_2 \end{bmatrix} \tag{14}$$

<u>Model of Laboratory Apparatus</u>

For the purposes of this investigation, a ten mass single plane model was used. The coupling is modeled as a simple bearing. All rotational degrees of freedom were removed from the model using static condensation.

Each magnetic bearing has eight poles. The stators are constructed of "soft" silicon magnet iron and are unlaminated. Hysteresis losses appear, on the basis of static measurements, to be substantial and are modeled as providing an added lag time constant of 1.0 msec in the control loop. The 1.0 msec value was determined by comparing the experimentally determined stability threshold with that predicted by the mathematical model corresponding to $K_p = 2.5$ and $K_d = 2.85$ with the motor coupling disconnected. This linear model of a highly nonlinear effect appears to be adequate for small motions of the shaft, but may introduce substantial error for large motions. The bearing coefficients are estimated (including losses due to leakage and fringing) as:

$$K_x = -36000 \text{ N/m} \quad , \quad K_i = 206 \text{ N/A} \tag{15}$$

The control algorithm employed is simple proportional-derivative (PD) feedback. The actual controllers used are adjustable, having a transfer function defined by

$$G_c(s) = 0.05 \frac{\alpha_1 K_p s^2 + (\alpha_2 K_p + \alpha_3 K_d) s + K_p}{(\tau_1 s+1)(\tau_2 s+1)(\tau_3 s+1)} \tag{16}$$

The various coefficients are reported in Table 1. The transfer functions of the controllers have been carefully documented in earlier works on the basis of experimental measurement [7,11]. Position sensing is accomplished with eddy current probes which have a sensing gain of 7900 V/m and a time constant of 56 μsec. The output of the controller drives a transconductance amplifier which converts the control voltage to the required current to drive the bearing coils. The amplifier has a transconductance of 0.5 A/V and a pair of poles with time constant of 220 μsec due to a well damped tank circuit in the output section required for amplifier stability.

Combining the transfer functions of the probe, controller, and amplifier, and adding the pole which models the actuator hysteresis, the overall control transfer function is found to be

$$G(s) = 197.5 \frac{\alpha_1 K_p s^2 + (\alpha_2 K_p + \alpha_3 K_d) s + K_p}{(\tau_1 s+1)(\tau_2 s+1)(\tau_3 s+1)(\tau_4 s+1)(\tau_5 s+1)(\tau_6 s+1)(\tau_7 s+1)} \tag{17}$$

Table 1. Controller Coefficients

Symbol	Value	Symbol	Value
α_1	2.0×10^{-8} sec^2	τ_3	47 μsec
α_2	0.320 msec	τ_4	56 μsec
α_3	2.20 msec	τ_5	220 μsec
τ_1	220 μsec	τ_6	220 μsec
τ_2	100 μsec	τ_7	1.0 msec

Impulse Response Simulation

There are several methods for simulating the impulse response of a system from a state space model. Among these are numerical techniques such as Runge Kutta or Euler time integration [12], direct analytic solutions such as Lukes' ABC algorithm [13], and frequency domain techniques based on the Laplace transform [14]. In this investigation, the direct analytic approach was adopted both for its speed in obtaining simulations and for the ease with which an impulse of finite duration can be represented. The actual time history of the impulse applied by the experimental apparatus was unavailable, but a 1.0 msec impulse was modeled.

Conclusions

Figure 2 shows the measured experimental and the predicted theoretical response of the rotor at the outboard sensor subjected to an impulse applied at the outboard end (end opposite motor). These data are for controller settings, $K_p = 2.5$ and $K_d = 4.5$ corresponding to proportional and damping gains of 117 N/mm and 0.79 N-s/mm, respectively. The general forms of the two curves agree as well as the fundamental frequencies (first three criticals are 2000, 3800, and 5500, experimentally measured and 2190, 4350, and 5768 rpm, theoretically predicted).

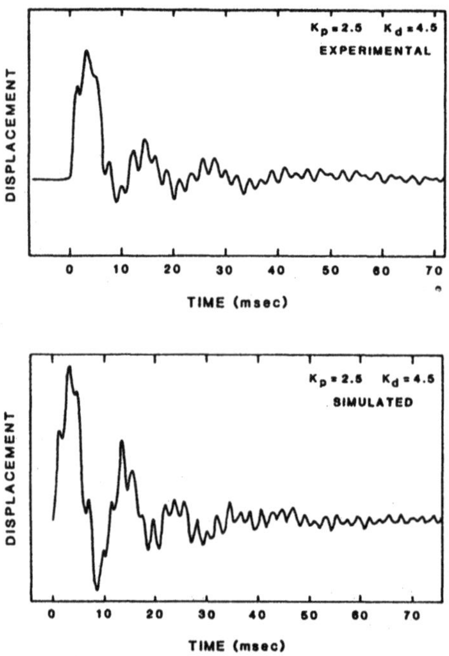

Figure 2. Experimental and Calculated Response with $K_d = 4.5$

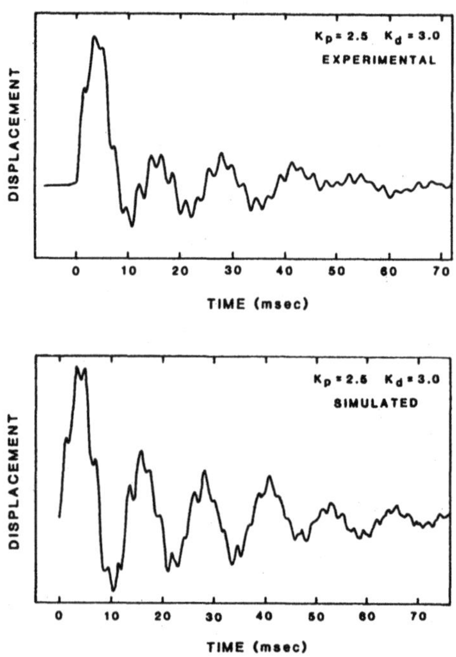

Figure 3. Experimental and Calculated Response with $K_d = 3.0$

Figure 3 shows results similar to those of Fig. 2 except with a reduced derivative gain of $K_d = 3.0$. The stability threshold for $K_p = 2.5$ is $K_d \simeq 2.4$, so the reduction from 4.5 to 3.0 would be expected to produce substantially reduced damping. This is supported by both the experimental and simulated results of figure 3 where, again, concurrance of theory and experiment is quite good.

This work illustrates that by having a good description of the control circuitry and by making a detailed model of the system, even relatively complex transient behavior can be predicted accurately.

References

1. Smith, D. M., "The Motion of a Rotor Carried by a Flexible Shaft in Flexible Bearings," Proc. of the Royal Society, Vol. 142, 1933, pp. 92-118.

2. Newkirk, B. L., "Shaft Whipping," Gen. Electric Review, Vol. 27, No. 3, pp. 169-178, March 1924.

3. Kirk, R. G., Hustak, J. F., and Schoeneck, K. A., "Active Magnetic Bearings Applied to Industrial Compressors," Magnetic Suspension Technology Workshop, NASA, Hampton, VA, 2-4 February 1988.

4. Schweitzer, G. and Lange, R., "Vibrations in Rotating Machinery," I. Mech. E. Conference Publication 1976-9, pp. 301-306.

5. Humphris, R. R., "A Bibliography of Magnetic Bearings," University of Virginia Report No. 643092/MAE85/336 (June 1985).

6. Keith, F. J., Williams, R. D., Allaire, P. E., and Schafer, R. M., "Digital Control of Magnetic Bearings Supporting A Multimass Flexible Rotor," Magnetic Technology Workshop, NASA, Hampton, VA, 2-4 Feb. 1988.

7. Humphris, R. R., Kelm, R. D., Lewis, D. W., and Allaire, P. E., "Effect of Control Algorithms on Magnetic Journal Bearing Properties," Oct. 1986, Vol. 108, J. of Engr. for Gas Turbines and Power, Trans. ASME, pp. 624-632.

8. Humphris, R. R., Allaire, P. E., and Lewis, D. W., "Design and Testing of Magnetic Bearings for Vibration Reduction," Proceedings of the 41st Meeting of the Mechanical Failures Group, Naval Air Test Center, Patuxent River, Maryland, October 28-30, 1986.

9. Schweitzer, G. and J. Salm, "Modelling and Control of a Flexible Rotor with Magnetic Bearings," Proceedings of the Conference on Vibrations in Rotating Machinery, IMECHE, York University, England, September 1984.

10. Balas, M. J., "Active Control of Flexible Systems," Dynamics and Control of Large Flexible Spacecraft, Blacksburg, AIAA Symposium, 1977.

11. Keith, F. J., "Digital Control System Design in Active Magnetic Bearings," Master's Thesis, University of Virginia, Charlottesville, May 1988.

12. Lapidus, L. and J. Seinfeld. Numerical Solution of Ordinary Differential Equations. Academic Press, New York, 1971.

13. Lukes, D. L. Differential Equations: Classical to Controlled. Academic Press, New York, 1982, pp. 47-63.

14. D'Azzo, J. J. and C. H. Houpis. Linear Control System Analysis and Design. McGraw-Hill Book Company, New York, 1981.

Identification and Optimization of Active Magnetic Bearing Systems Using Measured Nyquist Diagrams

Masujiro Hisatani

Machinery System Research Section, Tamano Laboratory
Mitsui Engineering and Shipbuilding
3-16-1, Tamahara, Tamano City, 706 JAPAN

Summary

An effective measurement and identification technique of the stability of active magnetic bearing systems is introduced. Frequency response is reduced into an open-loop transfer function and displayed as the Nyquist diagram. This shows the stability of the system both qualitatively and quantitatively. This technique is applied both for the axial direction, and for the radial direction with a strong gyroscopic effect. Measured Nyquist diagrams along the radial direction clearly show the divergence of eigenvalues caused by gyroscopic effect. A guide to optimize the system is indicated for the safe high speed rotation of the machine.

1. INTRODUCTION

The basic design of the control circuit of active magnetic bearing systems is simple: the main purpose is to overcome the negative stiffness of the rotor and afford an appropriate damping. In the case of a single degree of freedom (DOF) system, it is easily realized by PID control or phase-lead compensation. When a multi-DOF system has to be treated, it is described by state variable equations and techniques of either pole assignment or optimal regulator can be used.

However, actual tests often reveal the difference of the system characteristics with those of the design condition. It is usually worse than the designed one and sometimes the system becomes unstable. The causes of the system deterioration can be considered as frequency characteristics of the sensors, power amplifiers and control coils, incompletion of the compensation circuit, non-linearity of the magnetic attractive force characteristic, mechanical flexibility of the rotor, etc.

The most important "performance" of a control system is the stability. It is expressed by complex eigenvalues. A more elaborate model of the mechanical and electrical system than that of the design can be constructed to get complex eigenvalues closer to real values. However, it is difficult to model the frequency characteristic of every component accurately. Complex eigenvalues cannot be measured directly, but the performance can be measuered in the form of Nyquist or Bode diagrams. In order to investigate the actual system performance, impulse, step, or frequency response tests are often performed (1) (2) (3) and data are obtained in time or frequency domain. The obtained data are that of closed-loop and usually it is not easy to grasp the total system and get a good prospect to improve it adjusting parameters of the compensation circuit.

In this paper, an effective technique is introduced which measures and identifies the actual control system and directly shows the stability of the system as the Nyquist diagram.

2. EXPERIMENTAL APPARATUS

Fig.1 shows the vertical experimental apparatus of five-axis active control. A large hollow cylinder is connected to the top of a shaft and this causes a great gyroscopic effect. The rotor mass is about 3.4 Kg and the design rotational speed is 30,000 rpm. This is an elastic rotor with the first bending mode of 345 Hz. Radial displacement sensors of eddy current type are installed adjacent to radial coils. Sensor signals are input to analogue compensators and the outputs are sent to analogue power amplifiers of current feed-back type which energize control coils with currents proportional to the input signal up to a certain frequency range. A coordinate system with the origin at the center of the rotor mass is shown in Fig.2.

3. CONTROL SYSTEM OF AXIAL DIRECTION

Fig.3 is a block diagram of the control system and the measuring system of the axial direction, which is a single DOF system independent of multi-DOFs of radial directions. G_R and Gc are the transfer functions of the rotor and the compensation circuit respectively, k_s the sensitivity of the displacement sensor, R_{PA} the equivalent resistance of the power amplifier, and k_i the current stiffness of the control coils. A test signal e_t is added to the displacement signal e_z and these signals are input to the FFT analyzer to obtain a closed-loop transfer function which is expressed as :

$$G_{CL} \equiv e_z/e_t = -k_s G_R k_i/R_{PA} \cdot G_c(e_z + e_t)/e_t$$

where e_z and e_t are assumed to be Laplace transformed.

The open-loop transfer function of the system is

$$G_{OL} \equiv k_s G_R k_i/R_{PA} \cdot G_c$$

Therefore G_{CL} and G_{OL} have the following relation.

$$G_{OL} = -G_{CL}/(1+G_{CL}) \tag{1}$$

The analyzed result of G_{CL} is transferred from FFT to the personal computer and the open-loop transfer function G_{OL} is calculated using equation (1). This is displayed in a complex plane as a Nyquist diagram with frequency as a parameter. Stability is inspected by the turning mode of the curve around the point (-1,0).

Now G_{OL} is proportional to Gc, the compensator. Therefore, when the stability of the system is not sufficient and the gain and the phase of Gc has to be modified, it can be done more easily using the Nyquist diagram of G_{OL} rather than using the closed-loop G_{CL} which is not directly proportional to Gc.

Some examples of measured Nyquist diagrams are shown in Fig.4. Frequencies in Hz are indicated on the curves. Curves No.1 and 2 represent designed PD (proportional and differential) gains but with and without I(integral) action respectively. Both curves turn around the point (-1,0) looking to the left, so the system is stable with phase margins about 18 to 20 degrees. The initial part (near DC) of curve No.1 extends greatly upward and indicates enlarged static stiffness due to integral action.

However, the phase margin is about two degrees less than the non-integral curve of No.2. The basic characteristics of the curve has been explained in [4].

In order to express the system stability directly, the frequency and damping ratio of the eigenvalues is identified as follows. The components of the system are approximated simply as:

$$G_R(s)=1/(m s^2-k_d)$$
$$G_c(s)=k_P+k_D s+k_I/s$$

where s represents Laplace operator, m the rotor mass, and k_d the displacement stiffness. A rigid rotor and an ideal PID control with gains k_P, k_I and k_D respectively are assumed. The open-loop transfer function is then expressed as:

$$G_{OL}(s)=k_s k_i/R_{PA}\cdot(k_P+k_D s+k_I/s)/(m s^2-k_d)$$

Let $s=j\omega$, then $G_{OL}(j\omega)$ approximates the measured diagrams in Fig.4. A curve-fitting least square method was used to identify the eigen-frequency and the damping ratio which is lead from the characteristic equation $1+G_{OL}=0$. TABLE 1 summarizes measured and identified results. The damping ratio of the original setting (No.1) is less than expected. Therefore the differential gain was almost doubled (No.3) and a satisfactory value for the damping ratio was obtained. A larger value of the differential gain brings noise and high frequency oscillation. The optimization of the system can be achieved through a trade-off of these effects using Nyquist diagrams.

4.CONTROL SYSTEM OF RADIAL DIRECTION

The equation of motion for a rigid rotor can be stated by four radial DOFs, x, y, θ_x, and θ_y as follows:

$$\left.\begin{aligned} m\ddot{x} &= k_d x + k_i i_x \\ m\ddot{y} &= k_d y + k_i i_y \\ J_r\ddot{\theta}_x + J_a\omega_R\dot{\theta}_y &= k_{dt}\theta_x + k_{it} i_{tx} \\ J_r\ddot{\theta}_y - J_a\omega_R\dot{\theta}_x &= k_{dt}\theta_y + k_{it} i_{ty} \end{aligned}\right\} \qquad (2)$$

where Jr and Ja are the moment of inertia about the radial and axial directions respectively, ω_R the angular rotational speed, and i the control current.

The equations above indicates that the translations x and y are essentially independent and can be treated as single DOFs as well as z. On the other hand, rotational motion θ_x and θ_y are coupled with each other by the gyroscopic terms when the rotor is rotating. Thus the rotational motion is treated as one of two DOFs as shown by a block diagram of Fig.5. Here two kinds of compensation circuits are used : G_{MC}, main compensation and G_{XC}, cross compensation. $J_a \omega_R s$ represents gyroscopic terms.

Although the control system now seems much more complicated, the same simple measuring technique can be applied with a test signal e_t input to θ_y system, and the following transfer functions concerning θ_y are obtained:

$$G_{CL} \equiv \frac{e_{\theta y}}{e_t} = -\frac{G_{OL\theta}}{1+G_{OL\theta}} \Big/ \left[1+\left\{\frac{G_R(G_{XC}'+J_a\omega_R s)}{1+G_{OL\theta}}\right\}^2\right] \quad (3)$$

$$G_{OL} \equiv -G_{CL}/(1+G_{CL}) = G_{OL\theta} \Big/ \left[1+\frac{G_R{}^2(G_{XC}'+J_a\omega_R s)^2}{1+G_{OL\theta}}\right] \quad (4)$$

where

$$G_{OL\theta} = k_s k_i / R_{PA} \cdot G_R G_{MC}$$
$$G_{XC}' = k_s k_i / R_{PA} \cdot G_{XC}$$

The second term of the denominator represents the gyroscopic and cross compensation effect. Eq.(3) and (4) shows that, on the contrary to the simple case of axial direction, the control characteristics of the radial rotational motion changes greatly as the rotational speed of the rotor increases. This is clearly indicated by the measured data of Fig.6. In this case, PI control was used in the cross compensation G_{XC} . The lower frequency range of the diagrams indicates the effect of P control of G_{XC} which stabilizes precession, the lower eigenvalue diverged by the gyroscopic effect. At higher frequency range over 100 Hz the curve approaches (-1,0) as the rotational speed increases, indicating the decrease of stability of nutation, the higher branch of the eigenvalue.

This inclination of instability is caused by the phase lag of the main compensation circuit due to various kinds of filters used. Fig.7 shows simulated curves corresponding to Fig.6, with low pass filters included in G_{MC} in eq.(4). The stabilization of precession and the destabilization of nutation are expressed. D control of the cross compensation must be useful for stabilizing nutation. But, in this case, it causes the onset of the oscillation of higher bending mode. Therefore a notch filter in the main compensation was used with a successful result shown in Fig.8, where f_N is the center frequency of the notch filter and k_N is the proportion of full filtering.

The eigenvalues of the bending mode are also diverged by the gyroscopic effect. Fig.6 shows the divergence of the first bending mode, which is originally located at 345 Hz, and diverges into 320 Hz and 410 Hz at 10,000 rpm. Care should be taken if these branched eigenvalues approach instability.

The measuring technique stated above guarantees a reliable observation of the system stability as the rotational speed increases, and ensures a safe high-speed rotation of the machine. Realistically suitable variable-gain control method will be realized using the result of optimization obtained. For elastic rotors this measuring technique could be expanded into a more precise form considering the variation of the transfer function with space and utilizing more information from additional displacement sensors.

5. CONCLUSION

(1) A simple and effective measuring technique was introduced which identifies the actual performance of the control system of active magnetic bearing systems and enables its optimization. The result is displayed as a Nyquist diagram.

(2) The method is applicable not only for a simple case of single DOF but also for multi DOFs having coupled terms such as gyroscopic terms. The measured Nyquist diagram clearly indicates the gyroscopic effect which diverges system eigenvalues of both the rigid mode and of the bending mode.

Acknowledgement

The author wishes to express his appreciation to Mr. S. Nakata and Mr. K. Nakamura for their assistance, and is grateful to Mitsui Engineering and Shipbuilding for the permission to present this paper.

References

1. Humphris,R.R.,Allaire,P.E.,et al:Effect of Control Algorithms on Magnetic Journal Bearing Properties. J. of Eng. for Gas Turbines and Power Vol.108,No.4 October 1986

2. Bradfield,C.D.,Roberts,J.B.,Karunendiran,R.:Performance of an electromagnetic bearing for the vibration control of a supercritical shaft. PIME Vol.201,No.C3,1987

3. Hisatani,M., Inoue,Y., Mitsui,J.: Development of digitally controlled magnetic bearing. Bulletin of JSME Vol.29 No.247 January 1986

4. Kamerbeek,E.M.H.: Magnetic bearings. Philips Tech.Rev.41, 1983/84,No.11/12

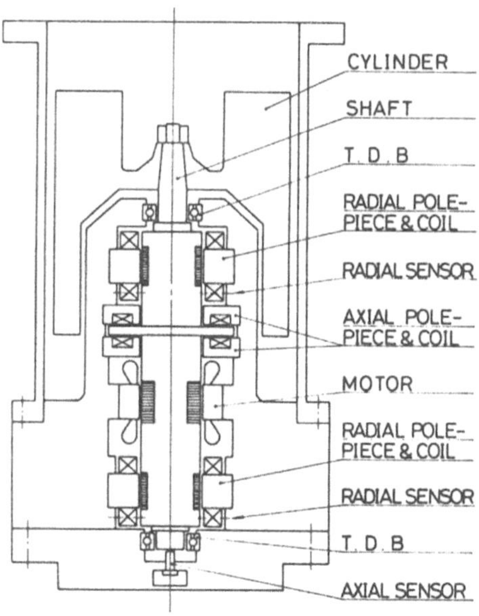

Fig.1 Experimental apparatus

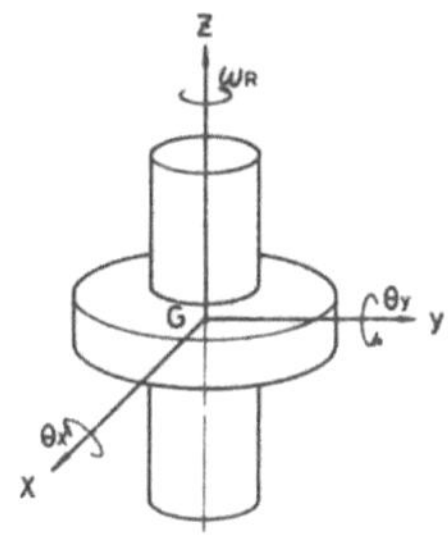

Fig.2 Co-ordinate system

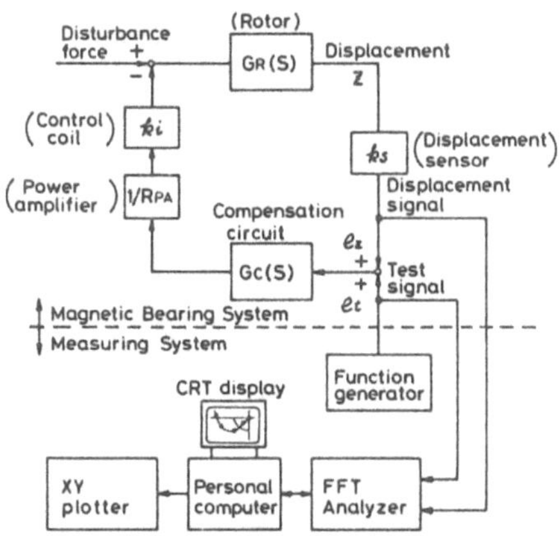

Fig.3 Control system & Measuring System (axial direction)

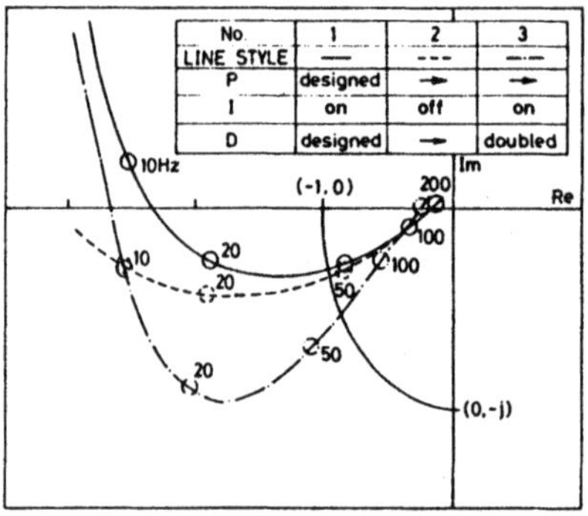

Fig.4 Measured Nyquist diagram (axial direction)

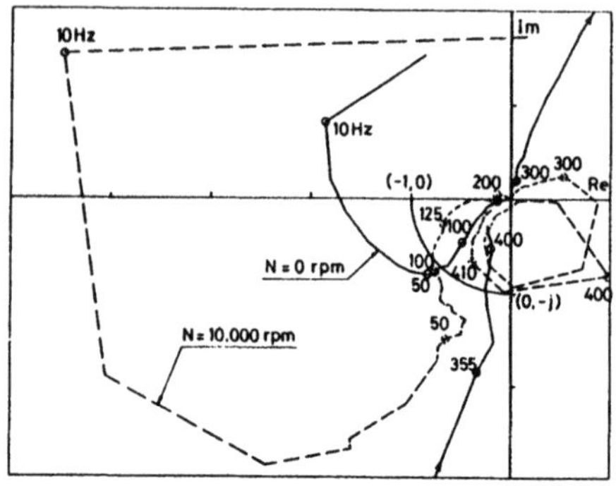

Fig.6 Measured Nyquist diagram of θ_y direction

No.	1	2	3
Measured Phase margin	18°	20	30
Identified eigenfrequency damping ratio	42Hz 0.24	42 0.26	37 0.58

TABLE 1 Identification of eigenvalues

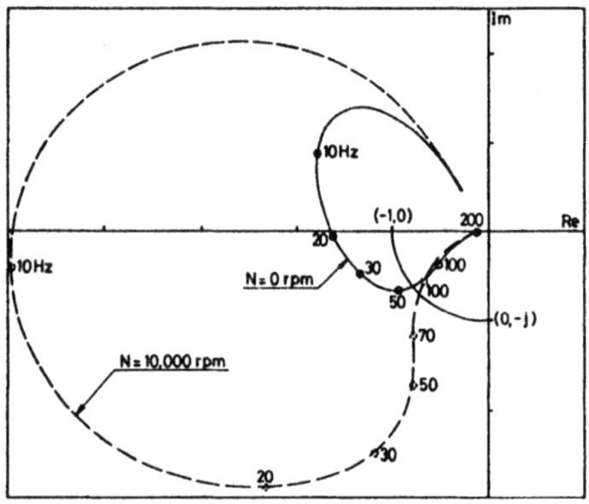

Fig.7 Simulated Nyquist diagram of θ_y direction

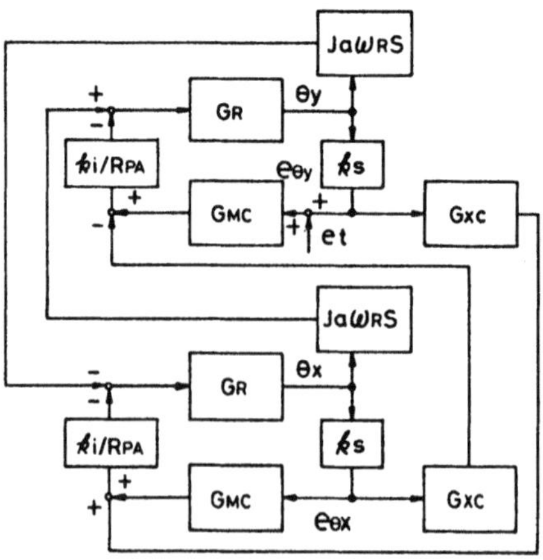

Fig.5 Control system of radial rotational motion

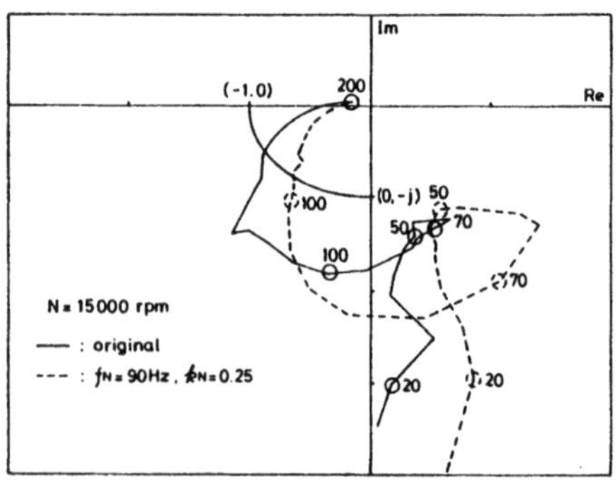

Fig.8 Effect of notch filter on the stability of nutation

New Test Techniques Using Magnetic Bearings

Heinz ULBRICH

Institute B of Mechanics
Dept. of Mechanical Engineering
Technical University, Munich

Summary
The appearance of a multitude of excitation mechanisms in rotor systems under working conditions is a well known fact. A magnetic bearing is used to experimentally generate these effects. It not only allows a perfect simulation of the real excitation forces, but in addition enables the generation of arbitrary test forces in order to evaluate the effectiveness of active controlled supports of rotors. The magnetic bearing along with its electronic components will be presented as a force generator. By the use of a computer, a fully automated test run with arbitrary excitation mechanisms can be realized. Subsequently, possible applications of magnetic bearings for research purposes will be demonstrated experimentally. Two simple examples will be outlined more in detail.

Introduction

The advantages of magnetic bearings in comparison to conventional bearings are becoming increasingly interesting for industrial applications in such fields as space or vacuum technology, and as bearings in tooling machines, in turbomachinery, in centrifuges, etc., i.e. /1, 2, 3, 4/.
The use of magnetic bearings as "active" machine elements is not only of interest as an optimal bearing for rotating machinery. They are also well suited for producing realistic simulations of specific excitation forces. Aside from thus enabling significant reductions in the amount of experimental efforts for fundamental studies, a separation of the various effects is possible. On the actual machinery, in contrast, the sum of all occuring effects is present at all times. In connection with controlled rotor systems, a practical means of testing control concepts is given. These points are at the center of the following discourse. First, the magnetic bearing along with its electronic components will be presented as a force generator.

Later, the "active" machine element is integrated into an existing test facility at a location chosen with regard to control conditions. Theoretical correlations will be briefly outlined. Subsequently, possible applications of magnetic bearings for research purposes will be discussed in conjunction with experimental results attained on the test facility.

Magnetic bearing as force generator

The complete configuration of the active element including its necessary electronic components is schematically shown in Fig. 1. It consists mainly of the magnetic bearing with four elec-

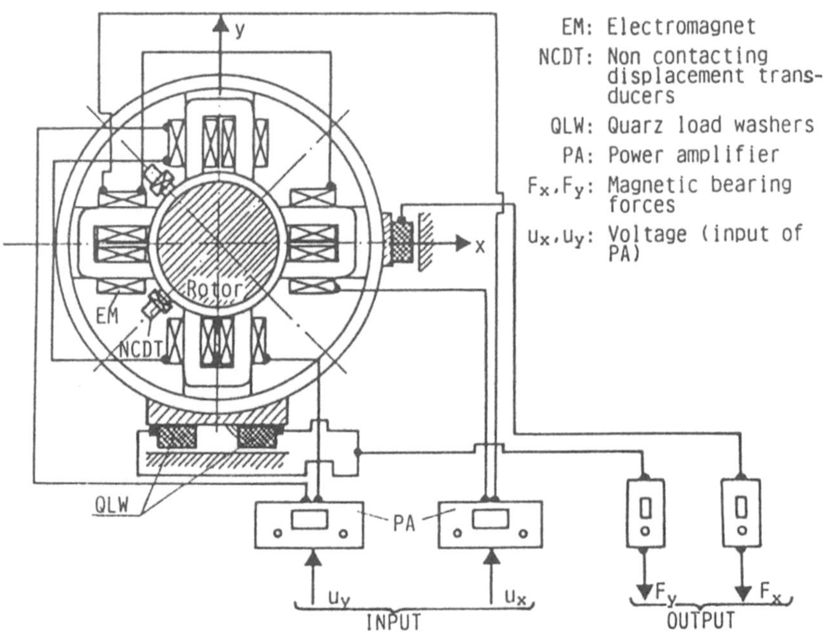

Fig. 1: Schema of the magnetic bearing including the sensor and electronic device

tromagnets (EM) switched in a differential arrangement, which produces a linear I/O-characteristic. The operation of the power amplifiers (PA) as current source yields the following equation between the input voltages u_x, u_y and the resulting forces F_x, F_y:

$$\begin{bmatrix} F_x \\ F_y \end{bmatrix} = \begin{bmatrix} k_u & 0 \\ 0 & k_u \end{bmatrix} \begin{bmatrix} u_x \\ u_y \end{bmatrix} + \begin{bmatrix} k & 0 \\ 0 & k \end{bmatrix} \begin{bmatrix} x \\ y \end{bmatrix} , \tag{1}$$

where k_u represents the force-voltage-factor, k the force-displacement-factor, and x, y the radial displacements of the rotor in the respective directions.

The rotor motion can be measured by built-in non contacting displacement transducers. Forces on the rotor are simultaneously determined in x-and y-directions by quarz load washers. In order to decouple the forces into the two measurement directions, special axial ball guides are used. This statically and kinematically defined assembly permits the application of defined forces into the rotor at the location of the magnetic bearing. Naturally, all conceiveable force functions, including nonlinear types, are possible.

Experience has shown that a targeted experimental investigation can only be conducted with a reasonable amount of effort, if a preceding theoretical study of the system has been completed. The modelling of a rotor system as a hybrid multibody system (HMBS) has proven to be useful, especially in respect to including active components. The HMBS-modell contains rigid bodies(e.g. bearing units) and elastic subsystems (e.g. rotors, blades), as determined by specific needs /5/.

System formulation

The setting-up of the equations of motion for such HMBS is best carried out by a computer. A very effective method is the direct evaluation of the principle of d'Alembert (/6/). In order to separate the distributed coordinates, one makes use of the product method by Ritz:

$$v(z,t) = \mathbf{v}(z)^T \mathbf{c}_v(t), \tag{2}$$

where $\mathbf{v}(z)$ is the vector of permitted shape functions (natural modes of vibration of the nonrotating rotor) and $\mathbf{c}_v(t)$ is the vector of the matching time functions. This leads to the equations of motion in their usual form:

$$\mathbf{M}\ddot{\mathbf{q}}(t) + \mathbf{P}\dot{\mathbf{q}}(t) + \mathbf{Q}\mathbf{q}(t) = \sum_{i=1}^{k} \mathbf{h}_i(t,\mathbf{q},\dot{\mathbf{q}}) , \tag{3}$$

where $\mathbf{q} \in \mathbb{R}^f$ represents the displacement vector, with $(\dot{})$ the time derivative, f the number of degrees of freedom which is the sum of f_r and f_e, f_r are the degrees of freedom of rigid bodies, and f_e are the degrees of freedom of elastic bodies, $\mathbf{M}$ is the fxf mass matrix, $\mathbf{P}$ is the fxf matrix of forces proportional to velocities, $\mathbf{Q}$ is the fxf matrix of forces proportional to displacements, $\mathbf{h}_i$ is the vector of i-th external force which is dependent on the acting excitation forces (i.e.: unbalance, internal dissipative effects, fluid forces etc.).

In addition to excitation forces that are due to the nature of the system, almost arbitrary further excitations can be achieved with an appropriate location of the magnetic bearing. This shall be demonstrated by two examples.

First example: non-conservative forces

In the presence of non-conservative forces, such as originate from unsymmetrical flow in turbomachinery, unstable vibrations can be induced (/7/). This typ of force appears in (3) on the right hand side. It can be written in the form

$$\mathbf{h}_i = \int_0^L \mathbf{J}^T \mathbf{f}_i \, \delta(z-z_i)\,dz, \quad 0 \leqslant z_i \leqslant L \,, \tag{4}$$

where $\mathbf{J}$ is the Jacobian matrix of translation (see ref. /5/), L is the length of the elastic structure $\delta(z-z_i)$ is the Dirac function, with which the location where the force acts is taken into account. The non-conservative force $\mathbf{f}_i$ is given by

$$\mathbf{f}_i = \begin{bmatrix} 0 & -q \\ q & 0 \end{bmatrix} \begin{bmatrix} u(z_i,t) \\ v(z_i,t) \end{bmatrix} \,, \tag{5}$$

where $u(z_i,t)$ and $v(z_i,t)$ represent the radial displacements of the rotor at the point z_i in the respective directions. The degree of influence this force has on the natural vibrations can be judged by the elements of the Jacobian matrix $\mathbf{J}$ (controllability). These elements assume the value of the amplitudes of the natural modes of vibration at the point z_i (shape function, eq. (2)). Those natural modes of vibration possessing a mode at this point are not influenced by the force.

A simulation of the effects of non-conservative forces by the magnetic bearing can be formulated in the following manner:

$$\mathbf{f}_i = \begin{bmatrix} F_x \\ F_y \end{bmatrix} = \begin{bmatrix} k_u & 0 \\ 0 & k_u \end{bmatrix}\begin{bmatrix} u_x \\ u_y \end{bmatrix} + \begin{bmatrix} k & 0 \\ 0 & k \end{bmatrix}\begin{bmatrix} x \\ y \end{bmatrix} = \begin{bmatrix} 0 & -q \\ q & 0 \end{bmatrix}\begin{bmatrix} x \\ y \end{bmatrix}, \tag{6}$$

where $x = u(z_i,t)$ and $y = v(z_i,t)$. To satisfy equation (6) the control vector is determined as

$$\begin{bmatrix} u_x \\ u_y \end{bmatrix} = \begin{bmatrix} k_{11} & k_{12} \\ k_{21} & k_{22} \end{bmatrix}\begin{bmatrix} x \\ y \end{bmatrix}, \tag{7}$$

and the amplification factors assume following values:

$$k_{11} = k_{22} = -k/k_u \; ; \quad k_{21} = -k_{12} = -q/k_u \; . \tag{8}$$

Second Example: test forces (here as step-function)

The step-function excitation is often used to examine the dynamic behavior of systems. It also can be easily generated and accurately reproduced by the magnetic bearing. The equation for a step-function force in x-direction is given as follows:

$$\mathbf{f}_j = \begin{bmatrix} F_x \\ F_y \end{bmatrix} = \begin{bmatrix} k_u & 0 \\ 0 & k_u \end{bmatrix}\begin{bmatrix} k_{11} & k_{12} \\ k_{21} & k_{22} \end{bmatrix}\begin{bmatrix} x-x_o \\ y \end{bmatrix} + \begin{bmatrix} k & 0 \\ 0 & k \end{bmatrix}\begin{bmatrix} x \\ y \end{bmatrix} = \begin{bmatrix} Q_x \\ 0 \end{bmatrix}, \tag{9}$$

where the amplification factors assume the values

$$k_{12} = k_{21} = 0, \quad k_{11} = k_{22} = -k/k_u \wedge x_o = Q_x/k \; . \tag{10}$$

These two examples indicate the numerous of applications that a magnetic bearing permits in simulating excitation forces. Furthermore, the excitation forces can be generated, individually or as the sum of various effects.

Test facility

Fig. 2 shows the facility used in examining such possible applications.

Fig. 2: Test rig (without electronics and oil supply system)

The facility mainly consists of the, elastic rotor structure ① , the bearing units (ball- or journal bearings) ② , the sensors ③, the actuators ④, the oil supply system (not shown), and the electronics (not shown).

Two types of actuators are used. Forces can be applied directly onto the rotor through the magnetic bearing or indirectly via the bearing housings through electromagnetic actuators. They can act as control forces or as any conceiveable kind of excitation force (e.g. an impulse, periodic force, circulatory force, fluid induced force, force due to bearing damage, foundation vibration (earthquake simulation), etc.). Furthermore, special electronics were developed for the adaptation of analog control voltages with the aid of a computer. All measurement and computer systems and components are interconnected by corresponding data buses (IEC). A fully automated test operation is thereby possible.

Test results

The results shown in Figs. 3 and 4 were gained on the facility. Fig. 3a depicts the locus diagram (radial displacement) of the rotor and in Fig. 3b the supplementary locus diagram of the effective non-conservative (circulatory) force produced by the magnetic bearing. One can recognize the equal sense of direction for both paths of locus diagrams resulting in an increasing amplitude and transverse force and thus evidencing the correlation given by eq. (6). In Fig. 4 the magnetic bearing served as test force generator. The test force was used for controller testing purposes in the regarded form of an impulse force. The control forces act upon the rotor and are generated by electromagnetic actuators via the journal bearing housing. A destinct improvement of system damping by the use of the controllers can be ascertained. In comparing Figs. 4a and 4b one

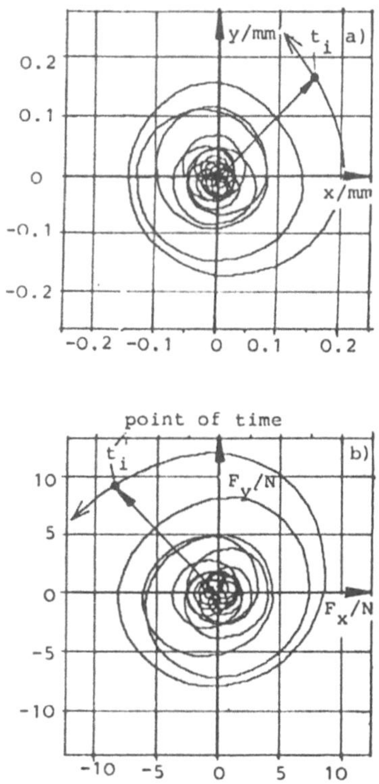

Fig.3: Locus diagrams with regard to the magnetic bearing location
a) displacement of the rotor
b) acting magnetic bearing force

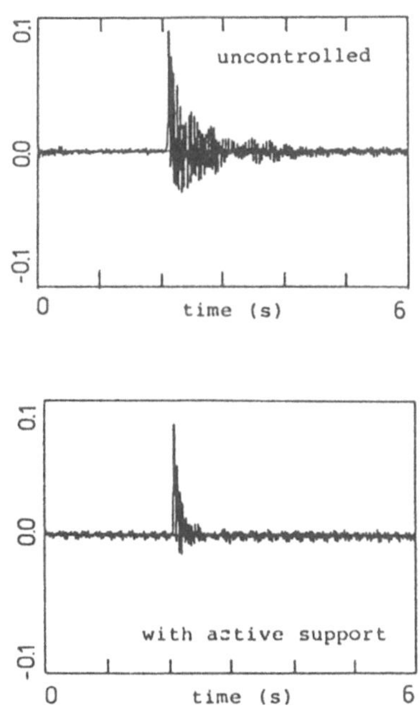

Fig.4: Transient behavior of system (impulse force produced by the magnetic bearing)

also recognizes the reproduceable amplitude of the transverse impulse generated by the magnetic bearing.

Conclusions

The use of magnetic bearings in rotor systems is not only advantageous as a substitute for conventional bearings and/or a vibration controller. They also open new possibilities in the field of test techniques. Realistic excitation processes can be simulated. This often means a considerable reduction in the scope of necessary experimental efforts. In addition, magnetic bearings are very well suited for testing the dynamic behavior of rotor systems by creating reproduceable disturbance forces. In connection with controlled rotor systems a means of testing various controllers is given. And through the use of a computer, a fully automatic operation of test cycles with various excitations is possible.

References

1. Innerhofer, G.; Hammer, J.: Low-Cost Magnetic Bearing Reaction Wheel. Proc. of the IFAC Symp., Oxford, England, 1980.

2. Ulbrich, H.; Schweitzer, G.; Bauser, E.: A Rotor Supported without Contact - Theory and Application. 5th World Congress on Theory of Machines and Mechanisms, Proc., Montreal, 1979.

3. Liard, G.: Active Magnetlager - ein Schlüssel zu höheren Spindeldrehzahlen. Kugellager-Z, 56, 213, 1982, pp. 8-13.

4. Habermann, H.; Liard, G.: An Active Bearing System. Tribology Intern., 1980, pp. 85-89.

5. Ulbrich, H.: Dynamik und Regelung von Rotorsystemen. Fortschr.- Ber. VDI-Z, Reihe 11, Nr. 86, 1986.

6. Bremer, H.: Kinetik starr-elastischer Mehrkörpersysteme. Fortschr.- Ber. VDI-Z, Reihe 11, Nr. 53, 1983.

7. Ebner, F.L.: Stability of Clearance - Excited Turborotors with External Anisotropy. ASME-Paper 85-DET, 1985.

Identification of Rotordynamic Parameters on a Test Stand with Active Magnetic Bearings

N.G. WAGNER; W.D. PIETRUSZKA

Mannesmann Demag, and University-GH-Duisburg
Duisburg, Germany

Summary

A rotor test stand using active magnetic bearings and a digital control unit to identify system parameters is presented. On the basis of a discrete-time system description, reference is made to a basic control circuit and a feasible identification method for on-line processing. First experimental results in terms of basic control circuit design and identification are discussed.

1. Introduction

Because of their physical properties and design, active magnetic bearings can perform further functions beyond that of supporting the shaft. Techniques of active vibration control [6, 7] and identifying parameters can be used particularly in conjunction with a digital control system. The identification of rotor-dynamic parameters in a test stand utilising active magnetic bearings allows to improve the reliability of rotor-dynamic calculations.

This is of major importance for turbo-compressor rotors as the trend towards ever increasing power density leads to the design of very slender rotors which are run far above their first natural bending frequency. Only with sufficiently precise knowledge of the bearing characteristics and the magnitude of the various excitation sources can stable operation be ensured through an adequate rotor-dynamic design [4, 5]. This applies to machines both with conventional bearings and with magnetic bearings.

The prediction of excitation by fluid forces originating from the inter-stage and shaft seals require results from experiments [3]. Realistic experiments can be conducted on a test stand with active magnetic bearings serving as measuring unit. The objective of this paper is to present a rotor test stand for that purpose and a possible method of parameter identification.

The approach to be shown can be generalised and transferred to other rotor systems. On the one hand, it enables the controller parameters to be adapted with varying operating conditions such that an optimum rotor behaviour can always be achieved, and on the other it allows to match magnetic bearing units to unknown rotor systems at the start-up phase.

2. Test Stand

The use of active magnetic bearings allows flow tests under realistic conditions which prevail in this stand in contrast with conventional test stands, in particular:

a) Magnetic bearings may be installed in the pressure chamber. This allows tests at a *high pressure level* without shaft seals which would affect the dynamics of the entire system. Apart from the known magnetic forces and fluid forces, there are no further disturbing forces acting on the rotor if the shaft is well balanced.

b) Owing to the possibility of presetting test functions, magnetic bearings allow the simulation of a great variety of *real operating conditions*.

The test set-up employed is shown schematically in figure 1. In view of the algorithms to be processed a digital signal processing unit was developed because of the greater flexibility it offers. The core is a 16 bit signal processor with an on-board multiplier [7]. The assembler programs are set up in a host computer. The evaluation module (EVM) provides the connection between the host computer and the signal processor

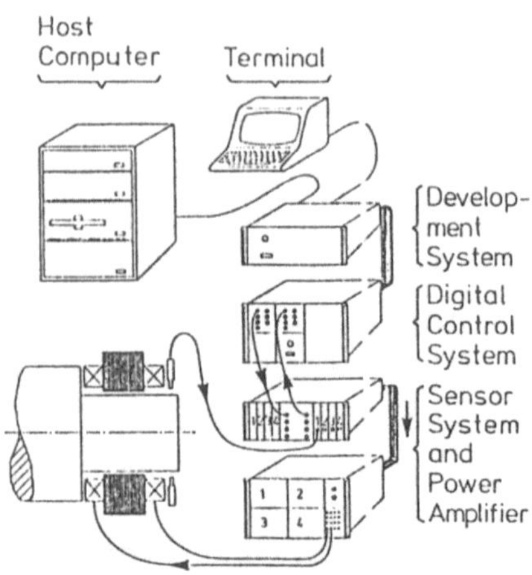

Fig. 1: Digital Signal Processing

and includes, inter alia, the assembler and debug options. The EVM processor and the signal processor work in a master/slave configuration. An analog interface with 7 input/output channels was developed at the Duisburg University as input/output of the signal processor [7]. Inductive sensors are used to measure distances and the power amplifiers are pulsed DC types.

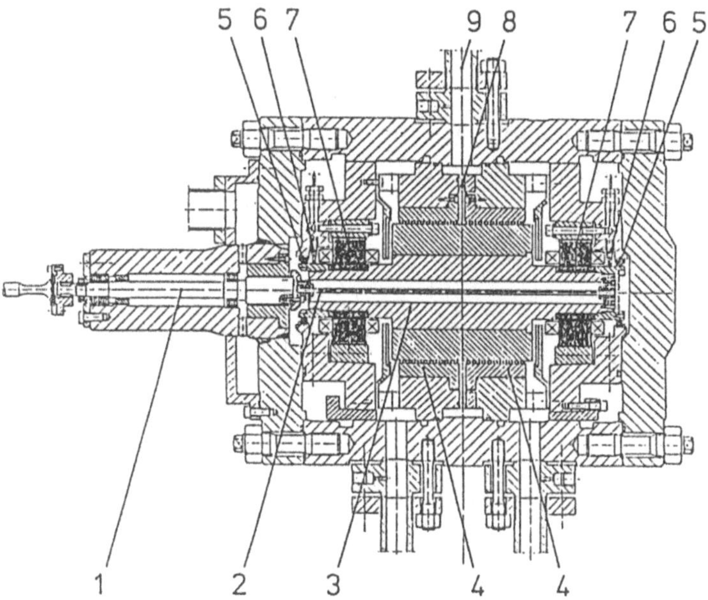

Fig. 2: Test stand with active magnetic bearings

Figure 2 is a sectional view of the test stand. The drive is conventional by a variable-speed motor via a gear unit. Since the drive is not integral with the stand, disturbing moments originating from the drive are avoided and the test rotor is simple and symmetric. The dynamic separation of the test rotor (3) from the drive (1) is through a most bendable and torsionally elastic bar (2) introduced through the hollow rotor. The figure shows a labyrinth seal being tested. The symmetric arrangement of two identical labyrinths (4) provides balanced thrust. The test gas is admitted in the casing centre (9) and accelerated tangentially by swirl rings (8). The magnetic bearings (7) are jointly installed with the auxiliary bearings (5) and sensors (6) in a housing and are radially adjustable. The electrical lines to the sensors and bearings leave the housing through pre-stressed glass penetrations.

3. Equations of Motion

The test shaft shown in figure 2 is projected to the model of the rigid rotor according to figure 3. The axial motion need not be considered because of being restrained by the torsion bar. The radial rotor motions are described by the coordinates of centre of gravity x, y, and cardan angles α, β. Magnetic forces F_r and F_l act in the right and in the left bearing plane. The flexural stiffness of the torsion bar is simulated by a substitute spring with stiffness factor k_a at the right shaft end. Furthermore a static unbalance $F = m\ e\ \Omega^2\ \exp(j\Omega t)$ and a dynamic unbalance

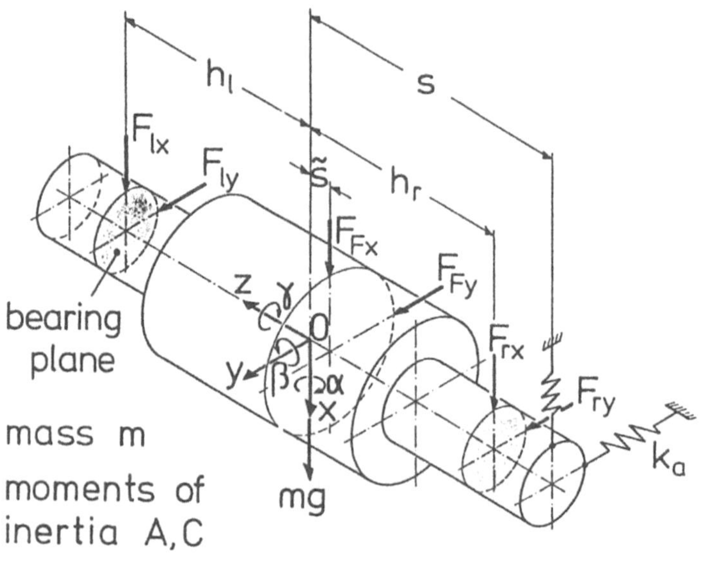

Fig.3: Model of rigid rotor, coordinate system

$M = E\,\Omega^2 \exp(j\Omega t)$ has to be considered ($j = \sqrt{-1}$). The resulting fluid forces ideally act at the rotor centre where because of inevitable unsymmetry, axial shift $\tilde{s}$ of that force has to be considered. In the event of little vibration x, y of the rotor about the centre these fluid forces can be taken as proportional to deflection and velocity [3], with coefficients d_R, d_Q, k_R, k_Q which are unknown at this stage:

$$\begin{bmatrix} F_{Fx} \\ F_{Fy} \end{bmatrix} = -\begin{bmatrix} d_R & d_Q \\ -d_Q & d_R \end{bmatrix} \begin{bmatrix} \dot{x} \\ \dot{y} \end{bmatrix} - \begin{bmatrix} k_R & k_Q \\ -k_Q & k_R \end{bmatrix} \begin{bmatrix} x \\ y \end{bmatrix} \qquad (1)$$

The angles α, β of rotor incline are not considered in equation (1) since disturbing moments which may occur are compensated, which results in the rotor performing pure motions of translation.

In the case of minor motions about the centre the magnetic forces for the left and the right bearing (index l, r) can be linearised [2, 8, 9]. Therefore the bearing force can be expressed as

$$\begin{bmatrix} F_x \\ F_y \end{bmatrix}_{l,r} = k_s \begin{bmatrix} x_L \\ y_L \end{bmatrix}_{l,r} + k_q \begin{bmatrix} i_x \\ i_y \end{bmatrix}_{l,r} \qquad (2)$$

where x_L, y_L bearing coordinates; i_x, i_y bearing currents, and $k_s = \mu_o\, N^2\, A_L\, I_o^2/s_o^3$, $k_q = 0{,}5\, \mu_o\, N^2\, A_L\, I_o/s_o^2$ with $\mu_o = 4\pi\, 10^{-7}$ Vs/Am, N number of windings; A_L pole surface; I_o magnetic biasing current; s_o nominal air gap.

Employing the complex variables $z = x + j\,y$ and $\varphi = \beta - j\,\alpha$ and the complex bearing currents in the left and the right bearing $u_l = i_{lx} + j\, i_{ly}$, $u_r = i_{rx} + j\, i_{ry}$, the linear motion equations, taking account of the gyroscopic influence, can be expressed as:

$$\begin{bmatrix} m & 0 \\ 0 & A \end{bmatrix} \begin{bmatrix} \ddot{z} \\ \ddot{\varphi} \end{bmatrix} + \begin{bmatrix} D_{11} & 0 \\ \tilde{s}\, D_{11} & D_{22} \end{bmatrix} \begin{bmatrix} \dot{z} \\ \dot{\varphi} \end{bmatrix} + \begin{bmatrix} C_{11} & C_{12} \\ C_{21} & C_{22} \end{bmatrix} \begin{bmatrix} z \\ \varphi \end{bmatrix} =$$

$$= \begin{bmatrix} me \\ E \end{bmatrix} \Omega^2 \exp(j\Omega t) + k_q \begin{bmatrix} 1 & 1 \\ h_l & -h_r \end{bmatrix} \begin{bmatrix} u_l \\ u_r \end{bmatrix} + \begin{bmatrix} mg \\ o \end{bmatrix} \qquad (3)$$

with the following expressions (cf. Fig. 3):

$$D_{11} = 2(d_R - jd_Q) \qquad C_{12} = -k_a s - k_s(h_l - h_r)$$

$$D_{22} = -j\Omega C \qquad C_{21} = -k_a s - k_s(h_l - h_r) + 2k_R\tilde{s} - j2k_Q\tilde{s}$$

$$C_{11} = k_a + 2(k_R - k_s) - j2k_Q \qquad C_{22} = k_a s^2 - k_s(h_r^2 - h_l^2)$$

Generally, motions of the centre of gravity and incline z, φ occur as combined motions. In the present case (cf. Fig. 2) the two bearings are at the same distance from the centre of gravity, i.e. $h_l = h_r$. The equations can be completely separated if the fluid forces act exactly at the rotor centre ($\tilde{s} = 0$) and the flexural stiffness of the torsion bar is negligible ($k_a = 0$). We shall use these simplifications in our further approach.

Owing to the digital signal processing method we use, the system works on the sampling principle with the result that the analysis is continued in a discrete-time manner. The conversion of the continuous-time representation (3) into the discrete-time representation is accomplished using the transition matrix [1]. For the complex motion of centre of gravity we employ a derived difference equation which only contains the deflections involved:

$$z(k) + a_1 z(k-1) + a_2 z(k-2) = b_1 u(k-1) + b_2 u(k-2) + w_G + w_o + w_s s(k-1) \quad (4)$$

with the complex system parameters a_1, a_2, b_1, b_2, into which enter, inter alia, the fluid forces, with the control variable $u(k) = u_l(k) + u_r(k)$, variable of unbalance w_s (time function: $s(k-1) = \exp(j\Omega(k-1)T)$) as well as weight w_G and an additional constant disturbance w_o. If we now introduce backward shift operator q^{-1}, equation (4) can be transformed into

$$\left[1 + A(q^{-1})\right] z(k) = B(q^{-1}) u(k) + w_G + w_o + w_s s(k-1) \quad (5)$$

using the polynomials

$$A(q^{-1}) = a_1 q^{-1} + a_2 q^{-2}; \quad B(q^{-1}) = b_1 q^{-1} + b_2 q^{-2}$$

4. Basic Control Circuit

If the fluid forces are not considered now, the known parameters in $A_o(q^{-1})$, $B_o(q^{-1})$ can be used, with $\Omega = \Omega_A$ as given design angular velocity, to present the control law

$$u_o(k) = -\frac{1}{P(q^{-1})}\left[Q(q^{-1})z(k)+u_G\right] \tag{6}$$

for the system free of unbalance

$$\left[1+A_o(q^{-1})\right] z(k) = B_o(q^{-1})u_o(k)+w_G \tag{7}$$

In equation (6) P and Q stand for polynomials

$$P(q^{-1}) = 1+p_1\, q^{-1}; \qquad Q(q^{-1}) = q_o+q_1\, q^{-1} \tag{8}$$

with initially unknown coefficients for matching to the rotor dynamics desired. The compensation of weight is by means of u_G.

In the simplest case the controller coefficients in (6) are determined by placing poles z_i (i=1,2,3) for the control circuit taken from (6) and (7). This leads to coefficient comparison [1] of the form

$$P(z^{-1})\left[1+A_o(z^{-1})\right] + B_o(z^{-1})\, Q(z^{-1}) = \prod_{i=1}^{3}(z-z_i) \tag{9}$$

with single-valued coefficients for the basic control circuit.

5. Identification of Parameters

Figure 4 is a typical representation of the parameter identification procedure.

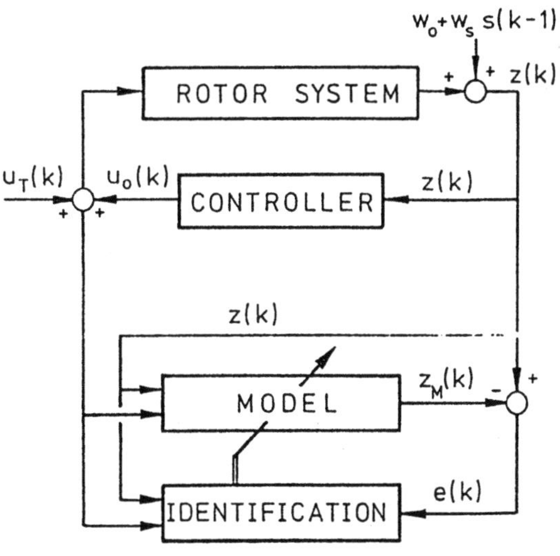

Fig. 4: Typical procedure of parameter identification

This diagram shows the rotor system with parameter vector

$$\underline{p}^T = \left[a_1, a_2, b_1, b_2, w_o, w_s\right] \tag{10}$$

connected in parallel with an equally structured deterministic model including external disturbance with adjustable parameters

$$\underline{p}_M^T(k) = \left[a_{M1}(k), a_{M2}(k), b_{M1}(k), b_{M2}(k), w_{Mo}(k), w_{Ms}(k) \right] \qquad (11)$$

Error

$$e(k) = z(k) - z_M(k) = \left[\underline{p} - \underline{p}_M(k)\right]^T \underline{\Phi}(k-1) = \underline{\Delta p}^T(k)\, \underline{\Phi}(k-1) \qquad (12)$$

with data vector

$$\underline{\Phi}^T(k-1) = \left[-z(k-1), -z(k-2);\ u(k-1),\ u(k-2);\ 1;\ s(k-1) \right] \qquad (13)$$

tends towards zero if

$$\underline{p}_M(k) \longrightarrow \underline{p} \qquad \text{for } k \longrightarrow \infty$$

In contrast with (6), the variable in (13) is to be extended by the test signal $u_T(k)$:

$$u(k) = u_o(k) + u_T(k).$$

On the basis of error (12) there are numerous methods which can be employed to draft a suitable on-line identification algorithm. Since the hardware and numeric expenditure must be within reasonable limits, currently two approaches can be highlighted: one is the well-proved least squares method which is most advantageous in that it provides rapid convergence with well selected initial values. However, its disadvantage is the still high numeric expenditure for quick on-line identification. On the other hand, the minimisation of e(k) using the gradient method has proved to be successful. This approach ist characterised by a low numeric expenditure. Its disadvantage is the slow convergence and the related greater susceptibility to disturbance. Despite these disadvantages we are using the gradient method for error minimisation.

Taking the real error function

$$J(\underline{p}_M(k)) = \frac{1}{2}\, e(k)\, \overline{e}(k) \qquad (14)$$

$$= \frac{1}{2}\, \underline{\Phi}^T(k-1)\, \underline{\Delta p}(k)\, \overline{\underline{\Delta p}}^T(k)\, \overline{\underline{\Phi}}(k-1),$$

in which the superlined values represent the conjugate complex numbers,and the related gradient

$$\text{grad}(J(\underline{p}_M(k))) = \overline{\underline{\Phi}}(k-1)\, e(k) \qquad (15)$$

as a basis, we obtain the identification algorithm

$$\underline{p}_M(k+1) = \underline{p}_M(k) + \underline{G}(k)\ \underline{\bar{\Phi}}(k-1)\ e(k). \tag{16}$$

For the hermitic weighting matrix $\underline{G}(k) = \underline{\bar{G}}^T(k)$ we write:

$$\underline{G}(k) = \frac{c_0}{\underline{\bar{\Phi}}^T(k-1)\ \underline{G}_0\ \underline{\Phi}(k-1)}\ \underline{G}_0 \qquad 0 < c_0 < 2 \tag{17}$$

with the constant diagonal matrix $\underline{G}_0$.

If the system-parameters are known, the disturbances can be identified separately.

Figure 5 gives an exemplary result of a numeric simulation of error $e(k)$ and identification variable $a_{M1}(k)$ varying as a function of time. We see the error initially decreasing relatively fast. As mentioned above, the residual adaptive variation of $a_{M1}(k)$ is slow because of the small residual error. This behaviour causes some difficulties during processing with the aid of a micro computer and requires the selection of suitable scales.

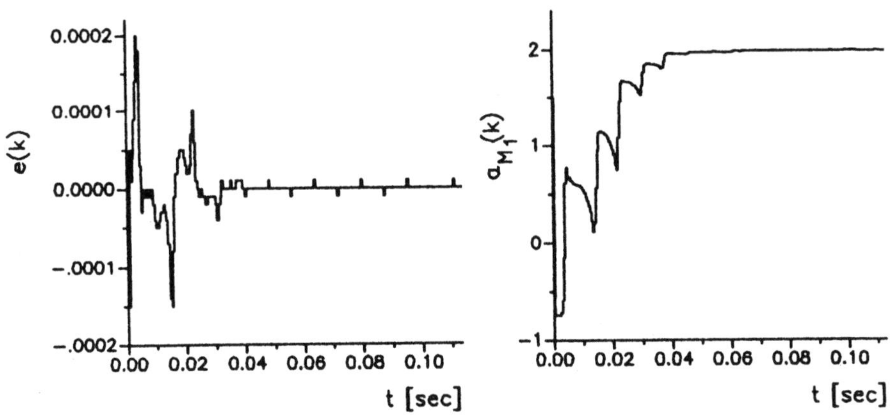

Fig. 5: Numeric simulation of the identification process

6. Experimental Results

At first we are discussing the behaviour of the basic control circuit. The problem faced when designing a digital control circuit is to correctly tune the controller coefficients and the sampling time with due consideration of the permanent disturbance of the measuring signal by noise. Most of this fine adjustment cannot be accomplished but by way of experiment. The test rotor showed a favourable running behaviour with 500 µs sampling time and poles placed between $z_i = 0.70 \ldots 0.85$ $(i=1,\ldots,6)$. However, here the static stiffness is low resulting in the weight having to be compensated by constant superimposition of u_G in (6) or the static stiffness being increased through an integrating portion.

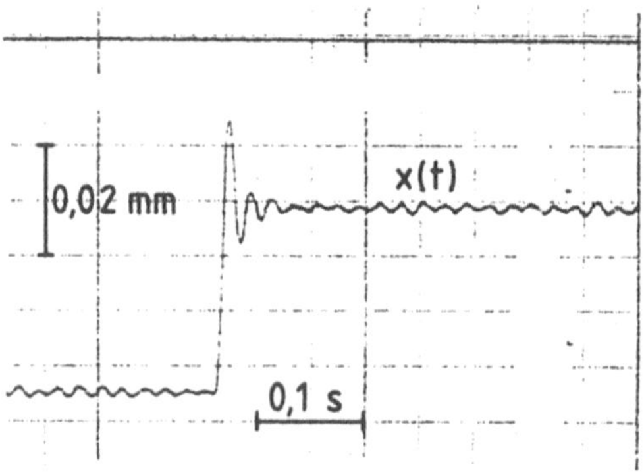

Fig. 6: Step response. Sampling time T=500 µs

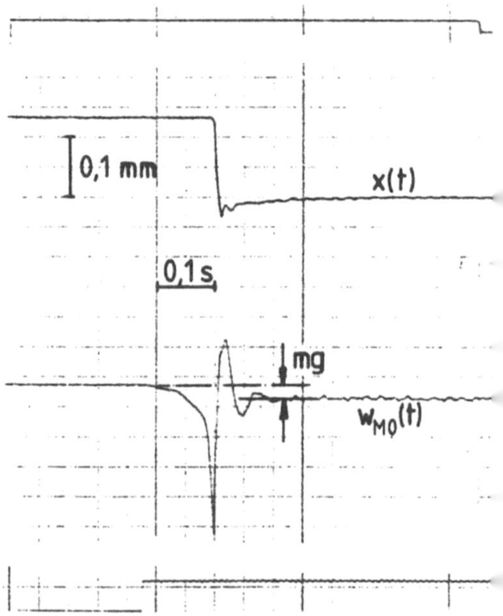

Fig. 7: Parameter identification

Figure 6 shows the deflection of centre of gravity x(t) when a reference signal is changed abruptly, the weight being compensated by u_G in (6) instead of an integrating portion. We readily see a good settling time to the new final condition. The identification algorithm is to be judged in the first analysis on the basis of the determination of the weight parameter w_0 in (4) and (10). Figure 7 is based on the system with the basic control circuit according to (6) $(u_G=0)$ and shows the identification process as a function of time giving deflection signal x(t) and parameter $w_{M0}(t)$. The identification process is completed after 0.2 seconds, and the identification time and the signal overshoot can be influenced by selecting a different weighting coefficient c_0 in (17).

References

/1/ Ackermann, J.: Abtastregelung, Springer, Berlin, 1983

/2/ Bleuler, H.: Decentralized Control for Magnetic Rotor Bearing Systems, Diss. ETH Zürich, 1984

/3/ Childs, D., Scharrer, J.: Experimental Rotor Dynamic Coefficient Results for Teeth-on-rotor and Teeth-on-stator Labyrinth Gas Seals, ASME-Paper 86-GT-12, Gas Turbine Conference and Exhibition, Düsseldorf, 1986

/4/ Fulton, J.W.: The Decision to Full Load Test a High Pressure Centrifugal Compressor in its Module Prior to Tow-out, Inst.Mech.Eng. Conf.Publ. 1984-2, The Hague, March 1984

/5/ Wang Xi-Xuan, Gu Jin-chu, Shen Qin-gen, Huan Yong Li, Zhu Lan-sheng: Some Field Experiences with Subsynchronous Vibration in Centrifugal Compressors, Proceedings of the Fifth Workshop on Rotordynamic Instability Problems in High-performance Turbomachinery, Texas A & M University, College Station, Texas, May 16-18, 1988

/6/ Pietruszka, W.D., Wagner, N.: Aktive Beeinflussung des Schwingungsverhaltens eines magnetisch gelagerten Rotors, VDI-Berichte Nr. 456, 1982

/7/ Pietruszka, W.D.: Zeitdiskrete Schwingungsbeeinflussung magnetisch gelagerter Rotoren mit Mikrorechnern, VDI-Berichte Nr. 603, 1986

/8/ Traxler, A.: Eigenschaften und Auslegung von berührungsfreien elektromagnetischen Lagern, Diss. ETH Zürich, 1985

/9/ Ulbrich, H.: Entwurf und Realisierung einer berührungsfreien Magnetlagerung für ein Rotorsystem, Diss. TU München, 1979

Unbalance and Disturbance Control

Seismic Test on Turbo-Molecular Pumps Levitated by Active Magnetic Bearing

Yukio MURAI[1], Katsuhide WATANABE[2], Youichi KANEMITSU[2]

1: EBARA Corporation, Fujisawa-shi, Japan
2: EBARA Research Co., Ltd., Fujisawa-shi, Japan

Summary
To evaluate the behavior of a rotating shaft undergoing seismic vibration, two types of turbo-molecular pumps were installed on a vibration table. One of the tested pumps has unique 4-axis control magnetic bearings that utilize both electromagnets and permanent magnets and features excellent zero power control, while the order has 5-axis control magnetic bearings that control all five degrees of freedom except for the rotating component around the shaft. Three types of seismic waves, El-centro, Taft, and Hachinohe and their time responses on a building floor were produced on the vibration table. The displacement of the shaft under vibrating conditions was less than 60 μmp-p in each test, thus it was confirmed that active magnetic bearings do actually possess sufficient bearing stiffness.

1. Introduction

A turbo-molecular pump (TMP) is an axial flow compressor which operates in high and ultra-high vacuum conditions. In recent years, the demand for oil contamination-free and vibration-free vacuum pumps has increased rapidly in the semiconductor industry in order to achieve high density integration of semiconductors. The TMPs levitated by active magnetic bearings (AMB) are attaching much attention as a new pump to meet these requirements. Because of their weakness in terms of bearing stiffness compared with ball bearings, it is very important that their stability be demonstrated with respect to external vibration forces such as seismic waves. To evaluate the behavior of the rotating shaft undergoing seismic vibration, two types of TMPs were installed and shaken on a vibration table. The followings are the results of the test and an introduction for the pump tested and their magnetic bearings.

2. Test pumps and their magnetic bearings

The following two pumps were tested;

Table 1

Pump model	KP18	KP13
Type of magnetic bearings	5-axis control	4-axis control
Operating Speed (mim^{-1})	45,000	32,000
Pumping Speed (ℓ/s)	310	620
Greatest Pressure (Torr)	10^{-10} order	10^{-10} order
Weight (kg)	10	27
Size (mm)	φ170 x ℓ272	φ250 x ℓ412

i) TMP with 5-axis control magnetic bearings

Fig. 1 shows the construction of a pump using 5-axis control magnetic bearings. The support system has the same construction as the conventional one with ball bearings used in the TMP. The two ball bearings which support the rotating shaft are replaced with two radial magnetic bearings and an axial magnetic bearing. By means of these magnetic bearings, all five degrees of freedom of shaft motion except for the rotating component around the shaft are actively controlled. The schematic construction of this bearing system is illustrated in Fig. 2.

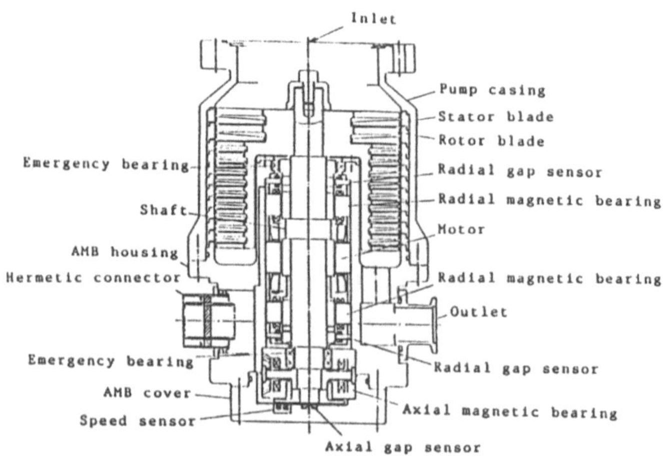

Fig. 1 Sectional drawing of TMP (KP18)

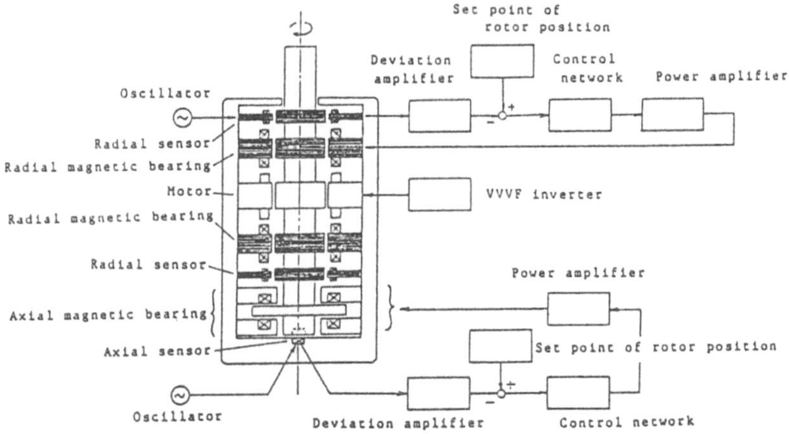

Fig. 2 Structure of 5-axis control AMB

As shown in Fig. 1 and 2, non-contacting inductive sensors that detect the rotor displacement are installed near each magnetic bearing. The rotor assembly is controlled to maintain its center position by regulating the attraction force of each magnetic bearing according to the rotor displacement signal from these sensors, and through the feed-back loop of the control circuit. The tested pump is designed so that the rated speed of the shaft is less than the bending mode frequency of the shaft.

Fig. 3 shows the result of a natural frequency analysis of the shaft. The bending mode frequency is denoted as "order 6".

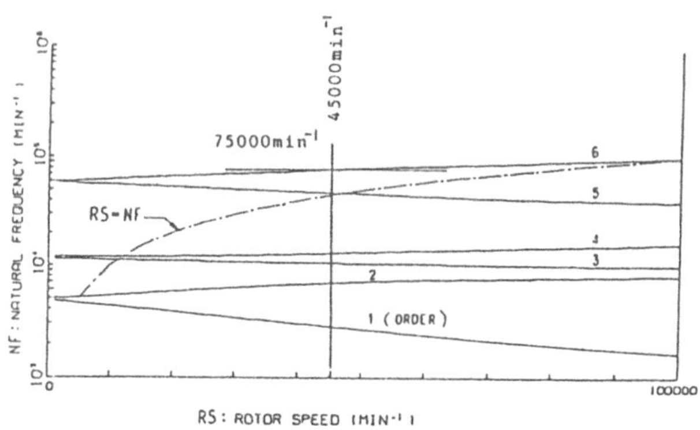

Fig. 3 Natural frequency analysis

ii) TMP with 4-axis control magnetic bearings

Fig. 4 shows the construction of a pump with 4-axis control magnetic bearings. The arrangement of the magnetic bearings is similar to the 5-axis type except that the axial magnetic bearing is not employed here. A schematic drawing of a bearing unit is shown in Fig. 5.

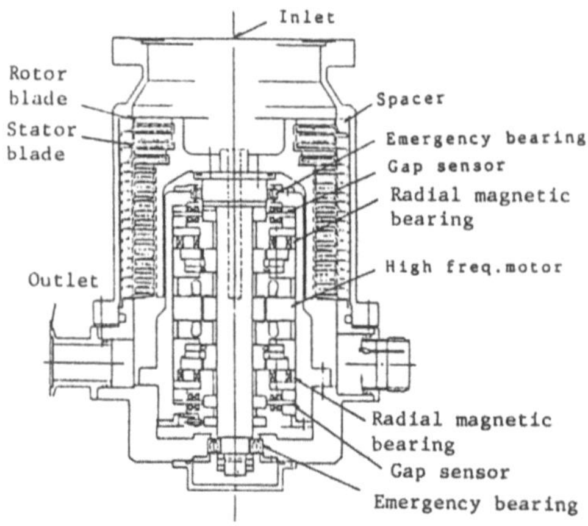

Fig. 4 Sectional drawing of TMP (KP13)

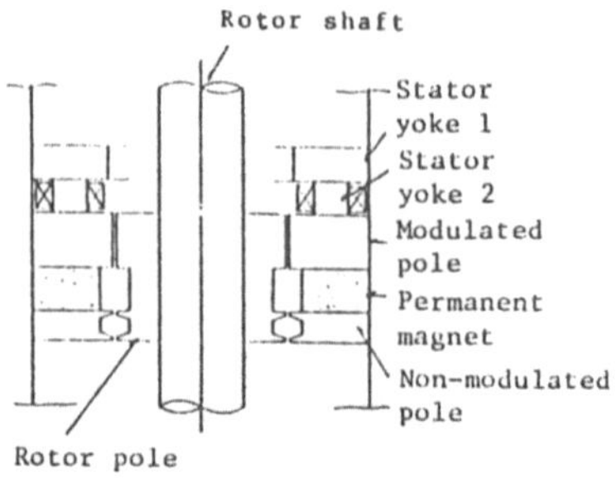

Fig. 5 Schematic of AMB Unit

This bearing unit is a radial control magnetic bearing in which passive stability is provided in the axial direction by the use of permanent magnets and electromagnets. No active control is needed in the axial direction, because axial stability is obtained by the use of bias flux through the tooth profile in the non-modulated magnetic pole. This type of bearing unit requires less electric power than the other bearing which use continuous current through a coil to establish bias flux, because bias flux is supplied from the permanent magnet.

Fig. 6 shows the construction of the entire bearing system.

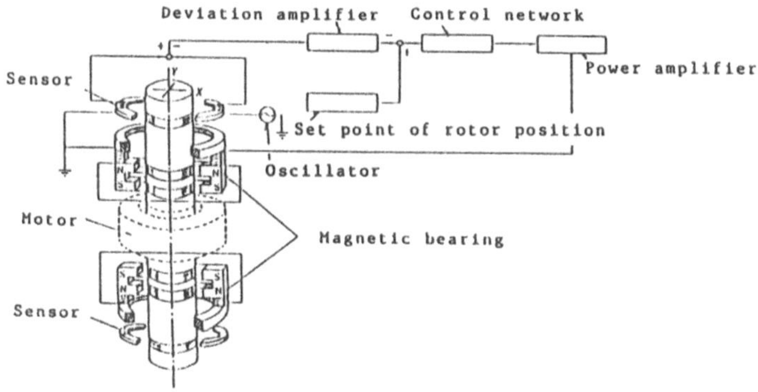

Fig. 6 Schematic of an active magnetic bearing

3. Seismic tests

1) Testing arrangement

Fig. 7 is a schematic diagram of the testing equipment. Two pumps were mounted together on a vibration table, and vibrated both horizontally and vertically in each test by changing their mounting positions. Accelerometers were set on the pump casing and on the vibration table as illustrated in the figure. The displacement of the rotating shaft was measured by the sensors of the magnetic bearings. All data recorded were converted, to digital form and analyzed by a computer. Table 2 describes the test items and the methods used.

Table 2

Test item	Testing method
1. Seismic response test	Pumps were vibrated directly with three different seismic waves (El-centro, Taft and Hachinohe), and the acceleration at the pump casing and the displacement of shaft were measured.
2. Test with seismic response of a model building	Pumps were vibrated with response waves of a model building under seismic vibration. El-centro, Taft and Hachinohe were used as seismic waves. The natural frequency of the model building was 2.5 Hz. The acceleration at the pump casing and the displacement of the shaft were measured.

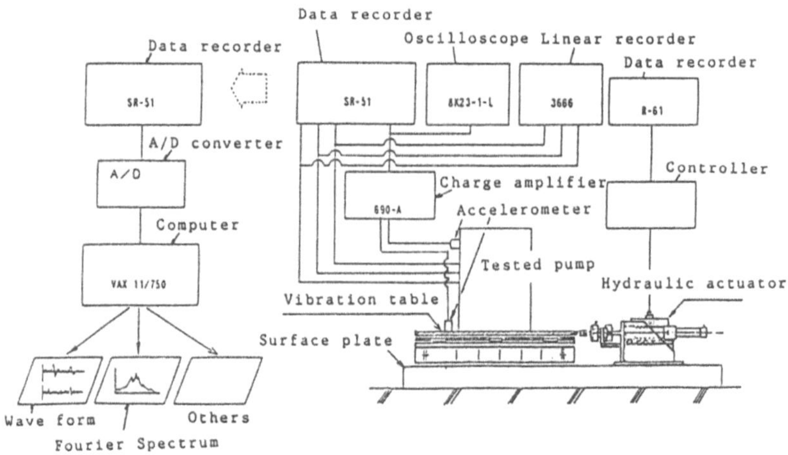

Fig. 7 Block diagram for measurement and analysis

2) Test results

i) Seismic response of the pumps

a. Radial direction

Pump model	Acceleration at pump casing	Displacement of shaft
KP18	187 - 385 gal	0.0084 - 0.0500 mm
KP13	200 - 457 gal	0.0036 - 0.0448 mm

b. Axial direction

Pump model	Acceleration at pump casing	Displacement of shaft
KP18	235 - 378 gal	0.0184 - 0.0260 mm
KP13	199 - 407 gal	No axial gap sensor

ii) Test with seismic response of a model building

a. Radial direction

Pump model	Acceleration at pump casing	Displacement of shaft
KP18	467 - 1052 gal	0.0200 - 0.0568 mm
KP13	516 - 1027 gal	0.0048 - 0.0236 mm

b. Axial direction

Pump model	Acceleration at pump casing	Displacement of shaft
KP18	235 - 378 gal	0.0196 - 0.0476 mm
KP13	199 - 407 gal	No axial gap sensor

Tested pumps operated in each test without touching the emergency bearings. Figs. 8 and 9 show examples of shaft displacement during this test.

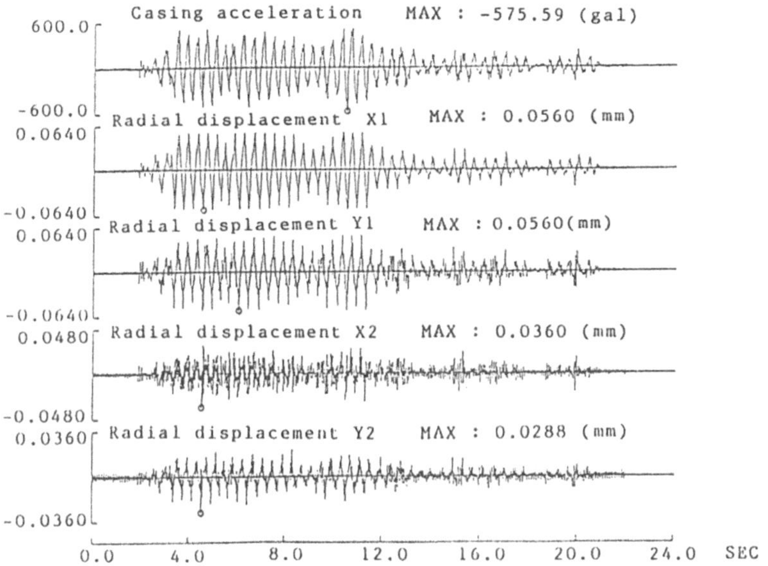

Seismic wave : El-centro NS, response of a model building
Pump : KP18

Fig. 8 Radial displacement of shaft under seismic vibration

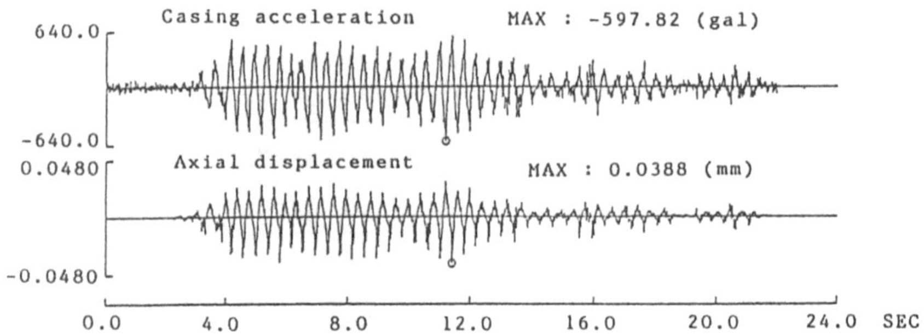

Seismic wave : El-centro NS, response of a model building
Pump : KP18

Fig. 9 Axial displacement of shaft under vibration

Acknowledgement

The authors gratefully ackowledge guidance on the development of magnetic bearings from Prof. C. Murakami of the Tokyo Metropolitan Institute of Technology and Mr. Nakajima of the National Aerospace Laboratory.

They also wish to express their thanks to Mr. Yano of the Kajima Institute of Construction Technology who collected most of the data and carried out the computer analysis of section 3.

A Suppression Method of Conical Motion of an Axi-Symmetrical Spinning Rigid Rotor Suspended by Magnetic Bearings

Chikara MURAKAMI* and Ichiju SATOH#

*Tokyo Metropolitan Institute of Technology, 6-6 Asahigaoka, Hino-shi, Tokyo 191, JAPAN. #Graduate Student of Tokyo Denki University.

Summary

New definitions of nutation and precession are proposed, which are usuful not only for physical understanding but also for intuitive interpretation of the optimal control of axi-symmetrical spinning rigid bodies. At first it is shown that the spin axis does coning motion about the instantaneous angular momentum vector with constant rotational speed (nutation angular velocity) regardless of its coning angle.

Next, steady state coning motions, which have been called as steady state precessional motion, are discussed based on angular momentum vector and it is shown that they have two modes; precessional mode and nutational one. Lastly, based on angular momentum vector, a method of control for conical motion of a spinning rigid rotor supported by a magnetic bearing system is derived and it has the same control structure as the optimal control which has been derived as a regulator problem by referenced paper.

Introduction : There were some confusion in the field of attitude dynamics and control of spacecraft at the beginning of the space age because of inadequate definitions of precession and nutation which are found in textbooks of classical dynamics. Moreover, there is no clear interpretation of dynamical behavior in the field of magnetic bearings. In this paper, new definitions of precession and nutation are proposed, which are useful not not only for physical understanding but also for intuitive interpretation of the optimal control of axi-symmetrical spinning rigid bodies. And then it is proved that the spin axis rotates about the instantaneous total angular momentum vector with constant nutational velocity regardless of its magnitude of coning motion using small cone angle approximation. Next, steady state coning motions, which have been called as steady state precessional motion, are discussed based on angular momentum vector and it is shown that they have two modes; precessional mode and nutational one. Lastly, based on angular momentum vector, a method of control for whirling motion of a spinning rigid rotor supported by a magnetic bearing system is derived and it has the same control structure as the optimal control which has been derived as a regulator problem by a referenced paper [2].

1. New Definition of Nutation and Precession, and Coning Motion

In textbooks of classical dynamics, precession is defined as horizontal motions of spinning axis, whereas nutation is defined as vertical ones. Here, the new definitions which are the same as the ones of attitude control field of spacecraft are as follows :

Precession: motion of the total angular momentum vector(H, invisible) of the rotor due to outer torques(T).

Nutation: coning motion of the spinning axis(S) about H.

We cannot see the total angular momentum vector(H), but the motion of the spinning axis(S) is nearly equal to the precessional motion when there is no nutation, i.e. $H \fallingdotseq S$. This will be shown in the next chapter.

Regardless of the both motions, a general and fundamental equation of motion is

$$dH/dt = T. \tag{1}$$

Therefore, H depends on outer torque(T) alone. On the other hand, S draws a cone about the instantaneous axis of H with constant rotational speed (nutational angular velicity)

$$h = \sigma\,\omega/\cos C = H/I_d \tag{2}$$

seen from inertial space, regardless of cone angle, where $\sigma = I_s/I_d$ (ratio of moment of inertia about S, I_s, to the one about the diametrical axis, I_d), ω is S-component of spinning velocity, and C is a half of cone angle. This coning motion of S in the case where T is restoring torque which is in proportion to the small tilting angle of S, θ, will be proved at the next chapter. It is well known that Eq.(2) is constant regardless of C even when C is not small, provided that H is unchanged or T = 0.

2. Equations of Motion

In the following, equations of motion of the rotor suspended by a magnetic bearing system will be derived, where $T = K\theta$, K is restoring spring constant which has the opposite sign to the spinning top on the ground.

As an inertia coordinate system, it is supporsed that mass center of the rotor is the origin, O, a vertical upside axis is Z, and two perpendicular axes are X and Y, respectively. As an moving coordinate system which expresses the attitude of the rotor, a non-spinning one is adopted, where the origin is mass center of the rotor, two axes, x and y, which are nearly horizontal and not fixed to the rotor, are near to X and Y axes, respectively, and z axis is the spinning principal axis, S.

Small tilting angle of S is θ, which is decomposed to x- and y-component as

θ_x and θ_y. Angular velocity vector of the non-spinning coordinate is expressed as $(\theta_x', \theta_y', 0)^T$, where $()^T$ notation is matrix transposition, and ['] means time derivative of a variable. Similarly, the angular momentum vector H is expressed as $(I_d\theta_x', I_d\theta_y', I_s\omega)^T$. If there is no external torque except for restoring one, then, T is expressed as follows:

$$T = -K\ (\theta_x,\ \theta_y,\ 0)^T. \tag{3}$$

Using these expressions, Eq(1) changes to the following component equations:

$$\theta_x'' + h\theta_y' + k\theta_x = 0 \tag{4}$$

$$\theta_y'' - h\theta_x' + k\theta_y = 0 \tag{5}$$

$$\omega = \text{const.} \tag{6}$$

where $k=K/I_d$. In the following, as only x- and y-component are important, we regard θ as a complex number, and express it using underlined notation:

$$\underline{\theta} = \theta_x + i\theta_y, \qquad (i=\sqrt{(-1)}) \tag{7}$$

Therefore, Eqs.(4) and (5) are combined and change into one equation:

$$\underline{\theta}'' - ih\underline{\theta}' + k\underline{\theta} = 0 \tag{8}$$

Equation (8) was derived as Euler's equation. To prove validity of Chapter 1's hypothesis, starting from this hypothesis, let us try to derive Eq.(8) geometrically.

As shown in Fig.1, using complex number plane, suppose $\underline{H}$ be positioned small angle $\underline{m}$ to the origin O, and $\underline{S}$ be positioned small angle $\underline{\theta}$ to O and small angle $\underline{c}$ to $\underline{H}$. Therefore,

$$\underline{\theta} = \underline{m} + \underline{c} \tag{9}$$

$$\underline{c} = \underline{c_0}e^{iht} \tag{10}$$

where $\underline{c_0}$ is an initial value of $\underline{c}$, meaning rotation of $\underline{c_0}$ about $\underline{H}$ with constant rotational speed of h. As the restoring torque is perpendicular to $\underline{\theta}$,

$$\underline{dm} = \underline{dH}/H = -iK\underline{\theta}dt/H \tag{11}$$

$$\begin{aligned} d\underline{c} &= (\underline{c_0} - \underline{dm})e^{ihdt} - \underline{c_0} \\ &\fallingdotseq (\underline{c_0} - \underline{dm})(1 + ihdt) - \underline{c_0} \\ &= i\underline{c_0}hdt - \underline{dm} \end{aligned} \tag{12}$$

$$d\underline{\theta} = \underline{dm} + d\underline{c} = i\underline{c_0}hdt. \tag{13}$$

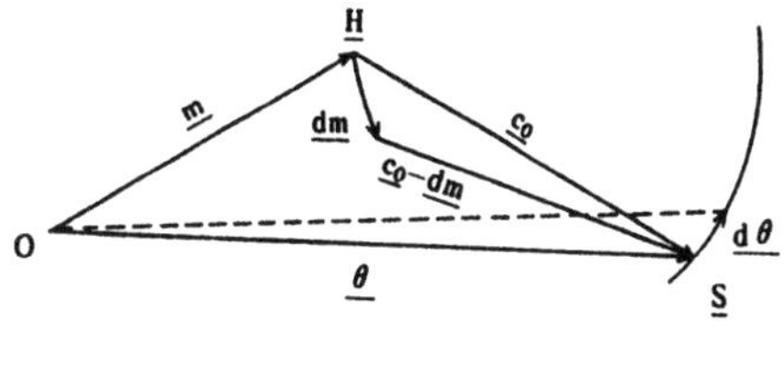

Fig.1. Delineation for formation of differential equation

Equation (13) is a first order one, and even if we change $\underline{c_0}$ to $\underline{c}$, generality is not lost. Substituting Eq.(9) into this one, and once more differentiation gives us:

$$\underline{\theta}'' - ih\underline{\theta}' + ih\underline{m}' = 0. \tag{14}$$

Using Eqs.(11) and (2) to this one, we can get Eq.(8). Thus, the validity of Chapter 1's hypothesis has been proved.

3. Approximate General Solution of Equation of Motion

Position of S, i.e. $\underline{S}$ or $\underline{\theta}$ is expressed as approximate general solution of Eq.(8). Using subscript $_0$ as initial values, and supposing $h^2 \gg k$, $\underline{\theta}$ becomes:

$$\underline{\theta} = \underline{S} \equiv \underline{\theta}p + \underline{\theta}n \tag{15}$$

$$\underline{\theta}p \fallingdotseq (\underline{\theta}_0 + i\underline{\theta}_0'/h)\, e^{-i(k/h)t} \tag{16}$$

$$\underline{\theta}n \fallingdotseq (-i\underline{\theta}_0'/h)\, e^{i(h+(k/h))t} \tag{17}$$

From this result, general solution of $\underline{S}$ consists of a fast coning motion (nutation mode) and a slow one (precession mode). In magnetic bearings, the both modes have opposite rotational direction, whereas in spinning tops on the ground, they have the same direction because of negative value of k.

Next, let us get an expression of H motion. Radial component of H is $I_d\theta'$ whose angle expression is $I_d\theta'/H$. Phase of θ' advances to θ by $\pi/2$. Therefore, position of H on the complex number plane is $i\underline{\theta}'/h$ (see Fig.2). Differentiating Eqs.(15)~(17), we get following approximate expressions:

$$i\underline{\theta}'/h = \underline{H} - \underline{S} \fallingdotseq (k/h^2)\underline{\theta}p - (1+(k/h^2))\underline{\theta}n$$

$$\fallingdotseq -\underline{\theta}n \tag{18}$$

Adding Eq.(15) to the above expression, $\underline{H}$ is expressed as

$$\underline{H} \fallingdotseq (1+(k/h^2))\underline{\theta}p - (k/h^2)\underline{\theta}n \tag{19}$$

$$\fallingdotseq \underline{\theta p}.$$

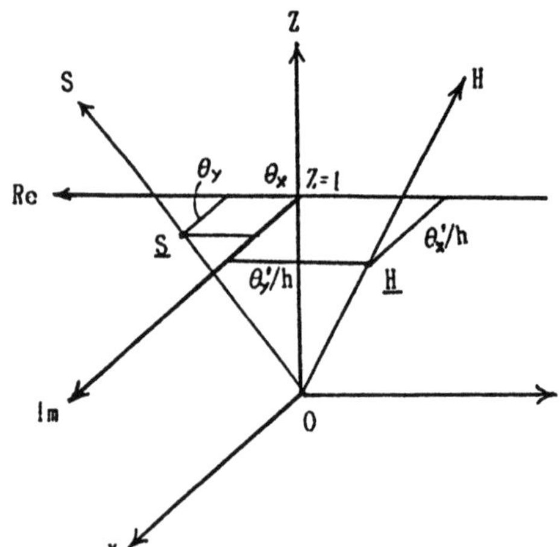

Fig.2. Relationship between complex number plane and inertia coordinate system

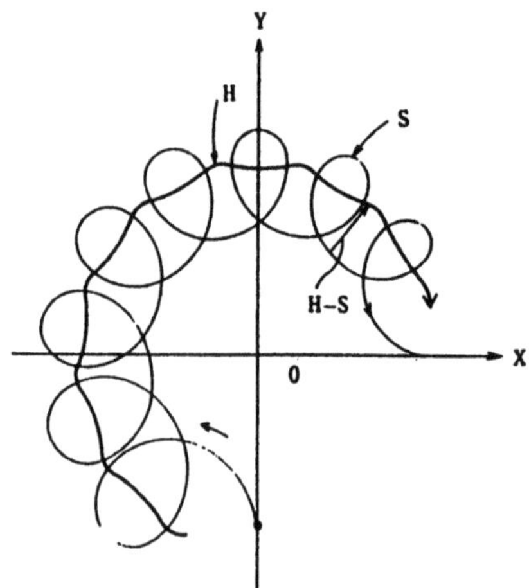

Fig.3. General coning motion of spin axis

From the results mentioned above, $\underline{\theta p}$ expresses approximate motions of H, whereas $\underline{\theta n}$ expresses approximate radius of circular motion whose center is H. Figure 3 is a simulation example of a general case where both modes are present and are rather decaying.

4. Steady State Coning Motion

In case of spinning spacecraft, k=0, and $\underline{\theta p}$ is fastend to a constant point, and pure nutations are observed. In case of spinning tops on the ground, k≠0, and both $\underline{\theta p}$ and $\underline{\theta n}$ are present. However, at high spinning speed, $\underline{\theta n}$ damps out quickly due to friction of ground, and pure but rather decaying $\underline{\theta p}$ is observed. If there is only one mode and no damping, the motion of S is a pure circular one. Nutation is fast anti-clockwise circular motions, whereas precession is slow clockwise (anti-clockwise in case of ground spinning tops) circular ones. In textbooks of classical dynamics, either motion is treatized as fast or slow steady precessional motion.

Motion of $\underline{H}$ in the steady circular motions is described here a little. In the steady precessional motion, the second term of Eq(19) is zero, and $\underline{S}=\underline{\theta p}$, therefore, $\underline{H}$ draws a circle a bit outside (inside, in case of spinning tops) of $\underline{S}$ by (k/h^2) times of $\underline{S}$. It would be said that both $\underline{S}$ and $\underline{H}$ move getting near and parallel with each other. Fig. 4 is an example of steady precessional motion. In steady nutational motion, $\underline{H}$ consists of only the second term of Eq. (19). Therefore, $\underline{H}$ positions opposite side of $\underline{S}$ to the origin, and draws a small circle. Fig. 5 is an example of steady nutational motion.

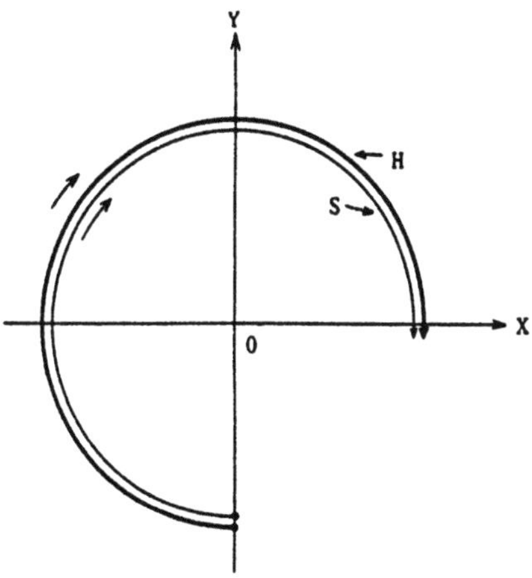

Fig. 4. Steady state precessional motion

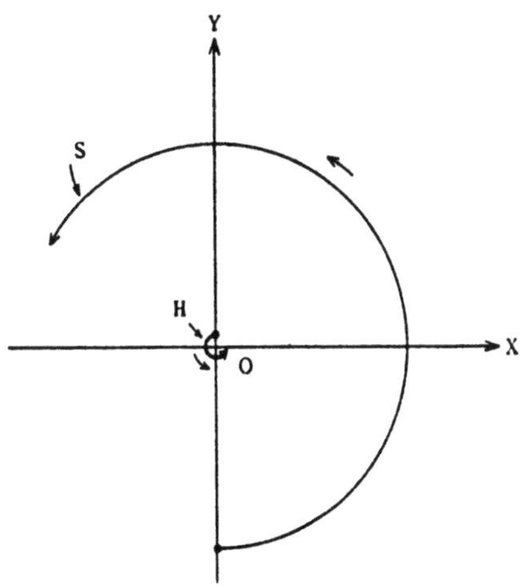

Fig. 5. Steady state nutational motion

5. Application to Suppression of Whirling Motion

Whirling motion of the rigid rotor suspended by magnetic bearings consists of two above-mentioned modes. To suppress the whirling motion, it is intuitively more easily understood to move $\underline{H}$ than to move $\underline{S}$. Because, $\underline{H}$ is governed by Eq. (1) or first order differential equation, and it has been known that $\underline{S}$ rotates about $\underline{H}$ at constant rotational speed. On the other hand, if we do not utilize directly Eq. (1), we must use the following Eq. (20) which is second order and consists of Eq. (8) and newly added control torqe T :

$$\underline{\theta}'' - i h \underline{\theta}' + k \underline{\theta} = \underline{T}/I_d \equiv \underline{u} \quad (20)$$

provided that $\underline{T}$ does not include the restoring torque. Thus, it is by far easier to move H than to try to directly move S.

For suppression of nutation, H should be moved near to S. This is a very common method in the field of attitude control of spacecraft. It is very easy to suppress precession; to bring $\underline{H}$ to the origin is sufficient. To summing up, both nutation and precession modes are suppressed at the same time, if $\underline{H}$ is brought to the origin while being moved near to $\underline{S}$. For this purpose, a concrete feedback method will be shown in the follow.

At first, to control $\underline{\theta_p}$, as is known from Eq. (16), $\underline{\theta}$ and $\underline{\theta}'$ should be fedback. On the other hand, from Eq. (18), to control nutation or $-\underline{\theta_n}$, only $\underline{\theta}'$ should be fedback. Fortunately, in both modes control, signs of θ' which should be fedback, are identical.

If we regard the magnetic bearing system as a two-gimbal servo system, each gimbal servo system has feedback loop of its own rotational position signal and its rate signal in general control. However, as $\underline{\theta}$ and $\underline{\theta}'$ are two-dimensional vectors, each gimbal servo system must include also the other one's sensor signals into its feedback loop. This is called inter-axis cross coupling feedback [1]. Returning to Eq. (16), as is already shown in Fig. 2, phase of $\underline{\theta}'$ advances to $\underline{\theta}$ by $\pi/2$, or in other words, phase of complex number plane lags by $\pi/2$ to the inertia frame, therefore, $\underline{\theta}'$ should be fedback directly to each servo. On the other hand, $\underline{\theta}$ should be delayed by $\pi/2$ and then be fedback, or specifically, θ_y and $-\theta_x$ should be fedback to x and y servos, respectively. After all, rate signals should be fedback independently, and position ones in mutually coupled mode. This feedback structure is identical to the one obtained as the optimal regulator problem of the magnetic bearing systems by referenced paper [2].

6. Simulation Examples of Control

Here, supposing that control torque $\underline{U}$ includes restoring torque also, $\underline{U}$ is generally expressed as the next equation:

$$\underline{U} = -K\underline{\theta} - (K_n + K_{p1})\,\underline{\theta'} + i\,K_{p2}\underline{\theta} \tag{21}$$

where K_n is nutation control gain, and K_{p1} and K_{p2} are precession control ones. In the following three examples, all initial conditions are the same as Fig. 3. Figure 6 is the case of approximate precession control where only the last term of Eq. (21) is used for H. Really, H approaches to the origin, but H and S diverge rapidly due to omission of K_{p1}, changing to a diverging nutation mode. Figure 7 is the case of nutation suppression where the third term of Eq. (21) is omitted (usual gimbal servo). Nutation is rapidly decayed and S and H are approaching each other, and a pure but rather decaying precession mode is observed. Figure 8 is the case where all terms of Eq. (21) are fed-back, and two modes decay rapidly at a time.

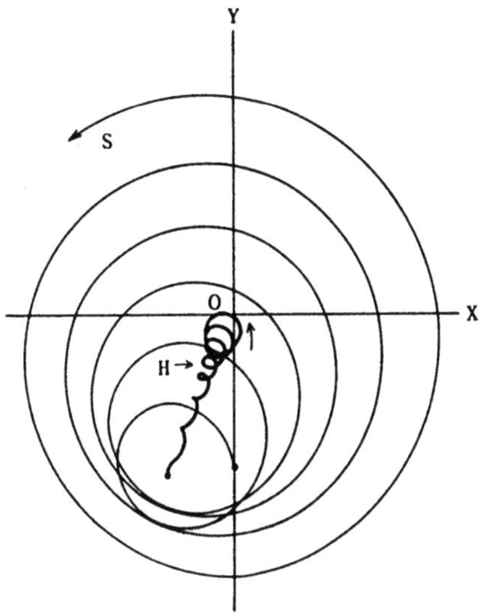

Fig. 6. Approximate precession control

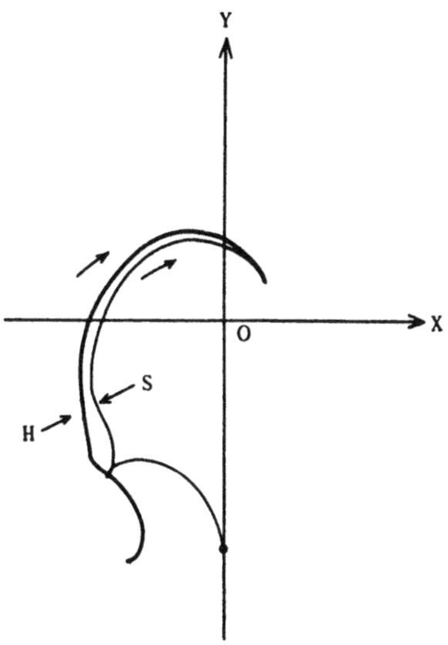

Fig. 7. Suppression of nutation alone (usual gimbal servo)

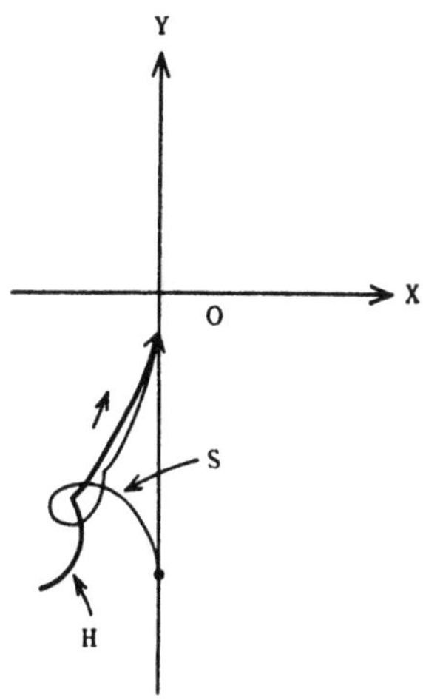

Fig. 8. Suppression of both precession and nutation

7. Conclusions.

(1) New definitions of precession and nutation are by far rational than conventional ones.

(2) Spin axis rotates about the instantaneous total angular momentum vector with a constant nutational velocity regardless of its radius of rotation.

(3) In suppression of whirling motion, it is more intuitively understandable to move the angular momentum vector than to move the spin axix directly.

(4) At high spinning velocity of the rotor, usual gimbal servo systems have little effect for suppression of precessional mode.

(5) Inter axis cross coupling feedback of angular position signals is very usufull for suppression of precessional mode because it brings the angular momentum vector to the origin.

REFERENCES

1. G. Heimbold "Impact of Magnetic Bearing Rotor Design on Satellite Nutational Stability", J. Guidance, vol. 7, no. 3, May/June 1984, pp. 279 285.

2. T. Mizuno and T. Higuchi "Design of the Control System of Totally Active Magnetic Bearings" International Symp. on Design and Systems, July 11 13, 1984, Tokyo. pp. 534 539.

A Design of Robust Servo Controllers for an Unbalance Vibration in Magnetic Bearing Systems

F. MATSUMURA, M. FUJITA, and C. OIDA

Department of Electrical and Computer Engineering,
Kanazawa University, 2-40-20 Kodatsuno, Kanazawa 920, Japan

Summary

This paper deals with the problem of an unbalance vibration in magnetic bearings. An efficient damping control scheme for the vibration is proposed. First, the state equation of a magnetic bearing system in which unbalance of a rotor is taken into consideration is shown. Next, the control system achieving the rotation around the geometrical axis is proposed. Here we formulate the control problem within a framework of the so-called Robust Servomechanism problem, and we design the control law using the Selective Turn-over method. Finally, the effectiveness of the control system is actually confirmed by several experimental results.

1. INTRODUCTION

In order that a rotor supported by magnetic bearings rotates around a certain axis smoothly, the rotor must be symmetric to the axis. But there exist residual unbalance on the rotor. By this unbalance, a shaft vibration is arisen when the rotor rotates at high speed. In this paper, based on modern control theory, we design a control system achieving a reduction of the unbalance vibration.

The following is discussed for a magnetic bearing combining radial control and thrust control [1], however, it can be applied to general magnetic bearing systems also.

2. PROBLEM STATEMENT

The state equation and the output equation of the magnetic bearing system in which an unbalance of a rotor is taken into consideration can be written as follows.

$$\dot{x} = Ax + Bu + D_1 w_1 + D_2 w_2 \tag{1}$$
$$y = Cx \tag{2}$$

where w_1 is a vector of sinusoidal disturbances caused by unbalance and w_2 is a vector of constant disturbance to this system. Each disturbance is not included in the equation given in [1]. These disturbances are arranged as follows.

$$\dot{w} = Fw \tag{3}$$

Above equations show that when a rotor rotates around the geometrical axis, sinusoidal disturbances arise a shaft vibration.

3. DESIGN OF ROBUST CONTROLLER

We introduce robust control [2] in this system, with which our problem can be solved. This system is augmented with a necessary number of servocompensator, which is described as the following equation,

$$\dot{x}_c = A_c x_c + B_c e \tag{4}$$

in cascade with each input line. The equation of the augmented system is as follows.

$$\begin{pmatrix} \dot{x} \\ \dot{x}_c \end{pmatrix} = \begin{pmatrix} A & 0 \\ B_c C & A_c \end{pmatrix} \begin{pmatrix} x \\ x_c \end{pmatrix} + \begin{pmatrix} B \\ 0 \end{pmatrix} u \tag{5}$$

where

$$A_c = \text{block diag}[F, F, \ldots, F]$$
$$B_c = \text{block diag}[K, K, \ldots, K]$$

$$F = \begin{pmatrix} 0 & p & 0 \\ -p & 0 & 0 \\ 0 & 0 & 0 \end{pmatrix} \qquad K = \begin{pmatrix} 1 \\ 0 \\ 1 \end{pmatrix} \qquad \text{p: angular frequency}$$

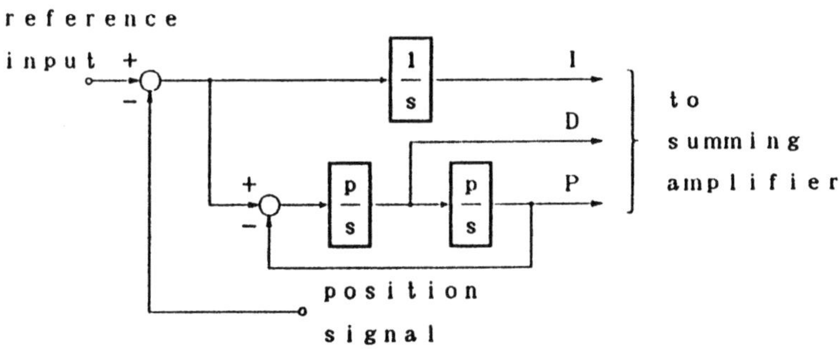

Fig.1. servocompensator

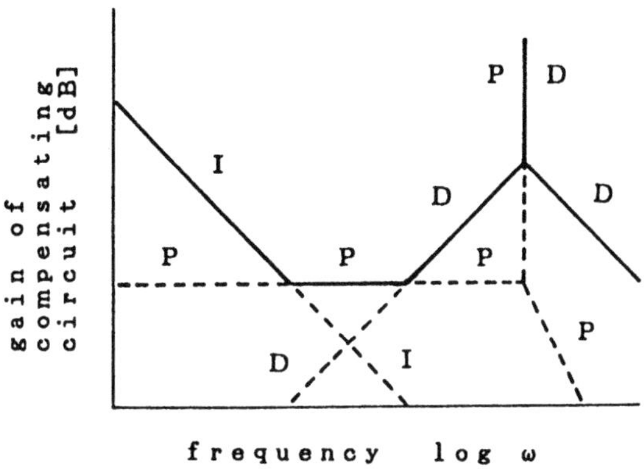

Fig.2. Frequency characteristic

The structure of servocompensator is shown in Fig. 1. This compensator consists of an integrator corresponding to internal model of step type disturbance and an oscillator corresponding to internal model of sinusoidal disturbances. Fig. 2 shows the frequency characteristic of this compensator. The line P shows the response of second order system with sharp resonance, the line D the differential of it and the line I the response of the integrator. The synthesized response is shown by solid line. This line and inertia stiffness make the frequency characteristic of stiffness. A sharp peak is realized at resonance angular frequency p.

Designing a compensator behaves according to variable rotational speed, sinusoidal disturbances can be reduced in a comprehensive range.

4. STABILIZATION

For the stabilizing of the augmented system (5), we use the Selective Turn-over method, with which we can allocate all the closed loop poles in the desirable region. As Fig. 3, any poles on the right space of the line can be folded back symmetrically with respect to an arbitrary straight line ($\mathrm{Re}\lambda = -k$) parallel to the imaginary axis. Selecting the line reasonably, imaginary part of eigenvalue of closed loop system can be specified arbitrarily. Allocated whole poles in stable space, it is proved that a derived control system is optimal regulator.

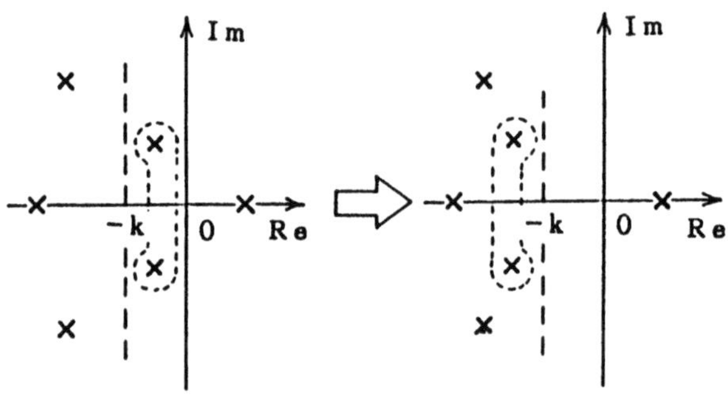

Fig.3. Selective Turn-over method

We stabilized this system according to the following procedure on the assumption that the rotational speed is 1900rpm.

(1) We select a line ($\mathrm{Re}\lambda = -10$) for the poles of an original system and an integrator. In earlier study [1], stability have been affirmed by this design.

(2) Since the poles of vibration is far from real axis, a little movement of poles makes feedback gain large. This leads a saturation of a control circuit. Considering this circumstance, we select a line ($\mathrm{Re}\lambda = -1$) for the poles of an oscillator.

The relationship of eigenvalues between an open loop system and its corresponding closed loop system is shown in Fig. 4. The pole ($\lambda = 0.0$) is a pole of an integrator, the poles ($\lambda = 0.0 \pm j199$) are those of an oscillator, the others are those of a magnetic bearing system.

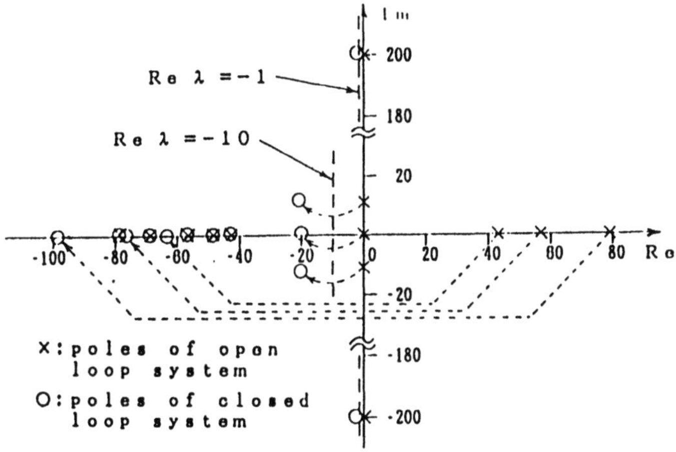

Fig.4. Poles of open and closed loop systems

5. EXPERIMENTAL MACHINE

The structure of the experimental machine is shown in Fig.5. There are a cage type induction motor at the center, electromagnets and radial position sensors on both sides, a dummy weight on the left side, and an axial position sensor on the right side.

Due to the staggered arrangement, radial and thrust direction control are achieved by only four pairs of magnet.

The rotor is 72mm in diameter at the electromagnets and the motor, 540mm in total length, 6.64kg in weight, and 4mm in gap length. The gap is wider than that of usual machines because only the rotor might be put in a vacuum with a glass tube. That is, this machine was designed for the purpose of an rotating anode X-ray tube.

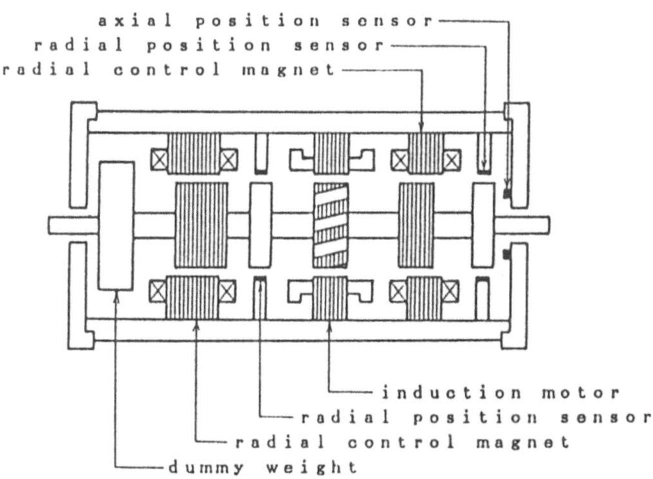

Fig.5. Structure of experimental machine

6. EXPERIMENTAL RESULTS

The following results are characteristics of the system designed for 1900rpm.

Fig.6 shows a displacement of the shaft (radial horizontal direction) rotating at 1900rpm. The case unbalance is not considered is also shown. As is evident from Fig.6, proposed control system reduces the unbalance vibration effectively.

Fig.7 is one of the characteristics changing a rotational speed from 1600 to 2000rpm. As an example, the case of 1900rpm is shown. It is similar to the preceding result and at the other rotational speed are so. This reveals that the unbalance vibration is reduced efficiently according to the changing rotational speed.

Fig.8 shows transient responses for the 3N step type disturbance force in a radial horizontal direction. The result shows that the control system achieves asymptotic disturbance rejection. This means that proposed control theorem enables us to achieve the reduction of unbalance vibration and the stability.

Fig.9 shows a frequency characteristics of stiffness for radial disturbance force. The stiffness at rotational frequency and nearby 0Hz is extremely high. This reveals that both step type and sinusoidal disturbances are almost eliminated.

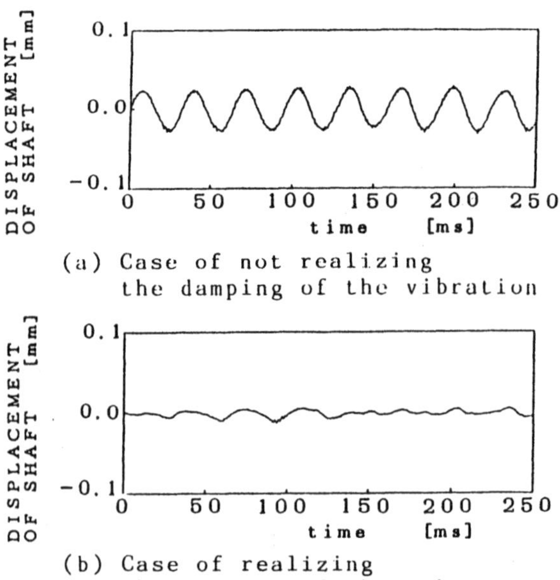

Fig.6. Displacement of the shaft

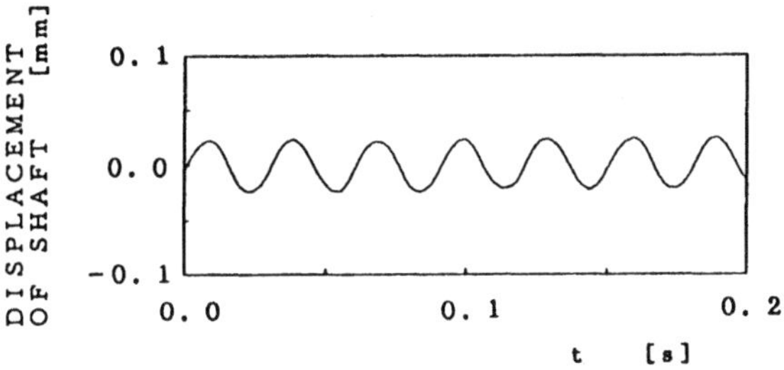

(a) Case of not realizing
the damping of the vibration

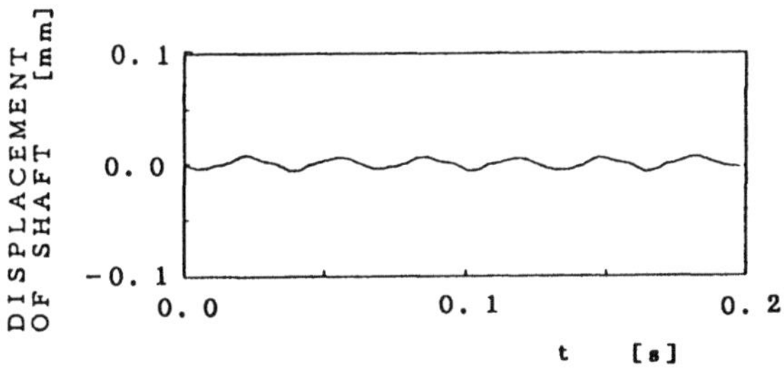

(b) case of realizing
the damping of the vibration

Fig.7. Displacement of the shaft

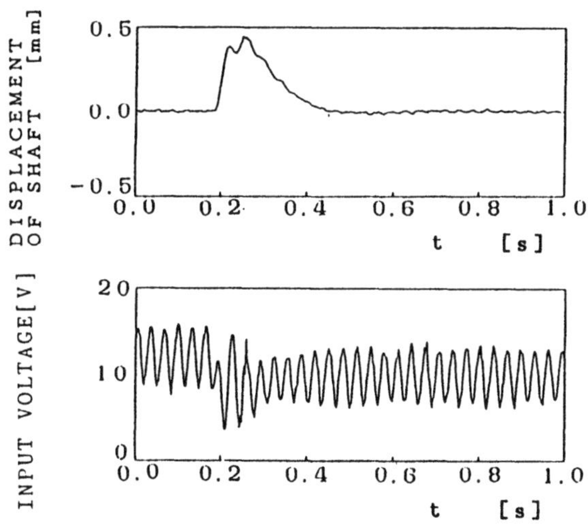

Fig.8. Response for step type disturbance

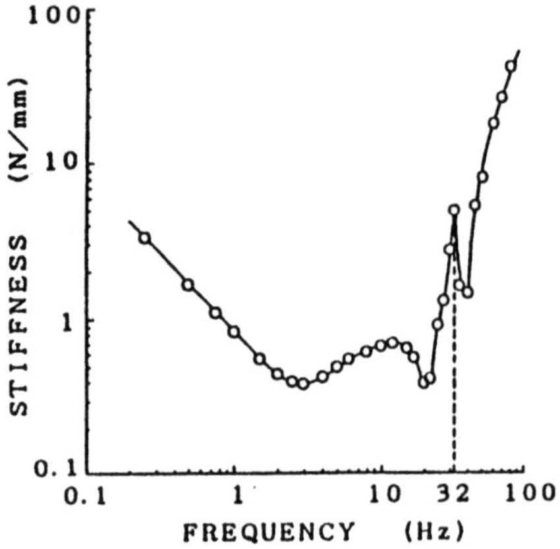

Fig.9. Frequency characteristic of stiffness

7. CONCLUSION

In order to reduce the unbalance vibration of a magnetic bearing systems, Robust Servo controller has been designed. The application of the Selective Turn-over method enables us to achieve the reduction of vibration and the stability. The effectiveness of the proposed controller has been demonstrated with several experimental results.

ACKNOWLEDGMENT

The authors are grateful to Mr.Ohkawa for his typing.

REFERENCES

1. F. Matsumura; M. Fujita; C. Oida: Theory and Experiment of Magnetic Bearing Combining Radial Control and Thrust Control. IEEE Trans. on Magnetics, Mag-23 (1987) 2581-2583.

2. E. J. Davison; A. Goldenberg: Robust Control of a Servomechanism Problem: The Servo Compensator. Automatica, 11 (1975) 461-471.

Balancing Measurement System Using Magnetic Bearings

Toshiro Higuchi

Institute of Industrial Science, University of Tokyo, Tokyo

Takeshi Mizuno

Faculty of Engineering, Saitama University, Urawa

summary

The principles and features of a measurement system with magnetic bearings for providing information for balancing a rotor are described. The proposed system uses magnetic bearings not only as a device supporting a rotor but also as a device measuring and compensating unbalance forces. In the system the total forces acting on a rotor can be balanced by controlling the bearing forces actively; measuring and locating unbalance on the rotor are performed in such states. Some experiments of measurement are carried out by using a magnetic bearing spindle as a single-plane balancing machine. The obtained results show the feasibility of a measurement system for balancing a rotor based upon the proposed principles of measuring.

1 Introduction

The magnetic bearing can suspend a rotor without any mechanical contact and lubrication. Since an active magnetic bearing can control force acting on the rotor dynamically, it can have additional functions which have not been achieved by other bearings. A control method, by which a magnetic bearing can suspend an unbalanced rotor without whirling, has been presented based upon the theory of output regulation with internal stability[1]. In the designed control system centrifugal forces due to unbalance are estimated by an observer and cancelled by electromagnetic force during rotation. When the position of a rotor is regulated to rotate without whirling, the total forces acting on the rotor are balanced; information which is necessary to balance the rotor can be obtained in a direct way from the controlled electromagnetic forces of the bearing. In this paper this regulation technique is applied to measuring and locating unbalance on rotor actually.

2 Description of the Test Instrument

2.1 Structure of the Instrument Figure 1 shows the section view of the test instrument used for investigations of measuring and compensating unbalance forces. It consists of an axial magnetic bearing, two radial magnetic bearings (ⓐ, ⓑ), a motor-stator and a rotor. The rotor, to which an object of balancing will be attached, is suspended by the magnetic bearings. A radial magnetic bearing has four electromagnets which are set in four equally spaced positions in the bearing plane and an inductive-type device for detecting radial movements of the rotor. The motor is three-phase induction-type and driven by an inverter which can vary the supply frequency.

2.2 Dynamic Modelling The attractive force of each magnet of the radial bearings , which is represented as F_n (n=1,...,8), can be approximately given by the following equation:

$$F_n=K_n(I_n/D_n)^2 \qquad n=1,\dots,8 \qquad (1)$$

where

K_n:coefficient of the magnet

I_n:excited current of the magnet ($=\bar{I}_n+i_n$)

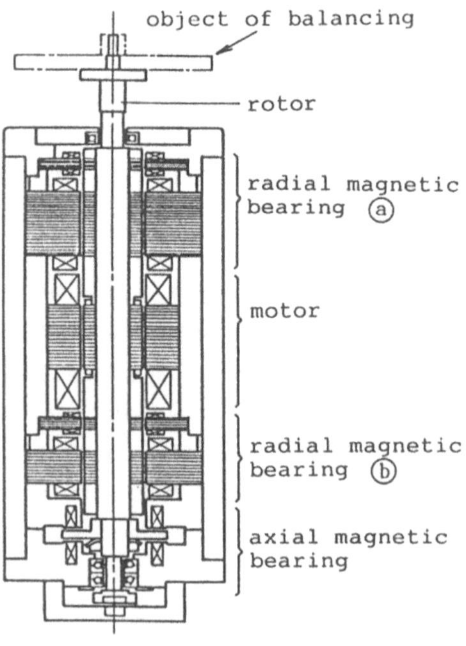

Fig.1 Structure of experimental setup

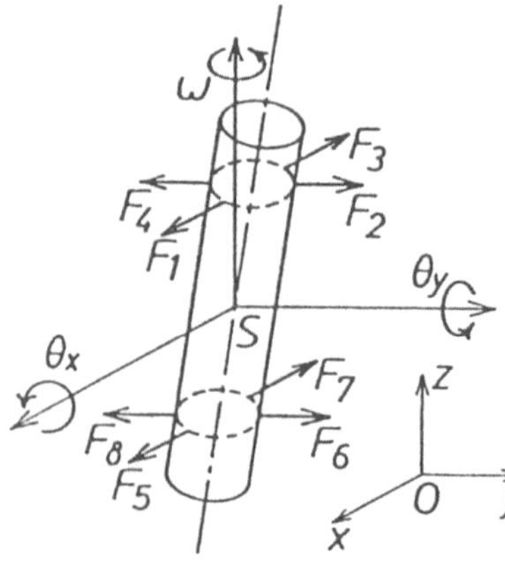

Fig.2 Coordinate axes and forces acting on the rotor

$\bar{I}_n, i_n$:stationary and incremental components of I_n

D_n:gap between rotor and pole of the magnet ($=\bar{D}_n+d_n$)

$\bar{D}_n, d_n$:stationary and incremental components of D_n

For small motions about the stationary, the magnetic force can be approximated by the linear relation:

$$F_n=\bar{F}_n-G_n d_n+H_n i_n \qquad n=1,\dots,8 \tag{2}$$

where

$$\bar{F}_n=K_n(\bar{I}_n/\bar{D}_n)^2, \quad Gn=2\bar{I}_n{}^2/\bar{D}_n{}^3, \quad H_n=2\bar{I}_n/\bar{D}_n{}^2$$

A coordinate frame **O**-xyz and the direction of F_n are shown in Fig.1. In nominal rotor position, the center of the rotor **S** is at the coordinate origin **O** and the center line agrees with z-axis. The stationary currents are set to realize the equilibrium conditions:

$$\bar{F}_1-\bar{F}_3+\bar{F}_5-\bar{F}_7=0 \tag{3}$$

$$\bar{F}_2-\bar{F}_4+\bar{F}_6-\bar{F}_8=0 \tag{4}$$

$$-(\bar{F}_2-\bar{F}_4)l_1+(\bar{F}_6-\bar{F}_8)l_2=0 \tag{5}$$

$$(\bar{F}_1-\bar{F}_3)l_1-(\bar{F}_5-\bar{F}_7)l_2=0 \tag{6}$$

and decoupling conditions:

$$(G_1+G_3)l_1-(G_5+G_7)l_2=0 \tag{7}$$

$$(G_2+G_4)l_1-(G_6+G_8)l_2=0 \tag{8}$$

where

l_1, l_2:distance between the center of the rotor **S** and the magnets of the radial bearing ⓐ and ⓑ

Then the equations of motion are given by

$$m\ddot{x}_s-(G_1+G_3+G_5+G_7)x_s=H_1 i_1-H_3 i_3+H_5 i_5-H_7 i_7+m\varepsilon\omega^2\cos(\omega t+\alpha) \tag{9}$$

$$m\ddot{y}_s-(G_2+G_4+G_6+G_8)y_s=H_2 i_2-H_4 i_4+H_6 i_6-H_8 i_8+m\varepsilon\omega^2\sin(\omega t+\alpha) \tag{10}$$

$$I_r\ddot{\theta}_x+I_a\omega\dot{\theta}_y-((G_2+G_4)l_1^2+(G_6+G_8)l_2^2)\theta_x$$
$$=-(H_2 i_2-H_4 i_4)l_1+(H_6 i_6-H_8 i_8)l_2+(I_r-I_a)\tau\omega^2\cos(\omega t+\beta) \tag{11}$$

$$I_r\ddot{\theta}_y-I_a\omega\dot{\theta}_x-((G_1+G_3)l_1^2+(G_5+G_7)l_2^2)\theta_y$$
$$=(H_1 i_1-H_3 i_3)l_1-(H_5 i_5-H_7 i_7)l_2+(I_r-I_a)\tau\omega^2\sin(\omega t+\beta) \tag{12}$$

where

x_s, y_s:displacements of the rotor in x and y directions

θ_x, θ_y:angular displacements of rotor axis about x and y axes

m:mass of the rotor

I_a, I_r:polar and transverse mass moments of inertia of the rotor

α, β:parameters on angular location of static and dynamic unbalance

ε, τ:amount of static and dynamic unbalance

ω:angular velocity of the rotor

From eqs.(9),...,(12) the whole system can be divided to a subsystem related to translation and a subsystem related to rotation. As a result the dynamics of the magnetic bearing system is expressed by a set of equations of the type:

$$\dot{\mathbf{x}}(t)=\mathbf{A}\mathbf{x}(t)+\mathbf{B}\mathbf{u}(t)+\mathbf{D}\mathbf{w}(t) \tag{13}$$

$$\dot{\mathbf{w}}(t)=\mathbf{E}\mathbf{w}(t) \tag{14}$$

where

$$\mathbf{x}=[x_1\ \dot{x}_1\ x_2\ \dot{x}_2]^t\ ,\quad \mathbf{u}=[u_1\ u_2]^t\ ,\quad \mathbf{w}=[w_1\ w_2]^t$$

$$\mathbf{A}=\begin{bmatrix}0 & 1 & 0 & 0\\ a & 0 & 0 & c\omega\\ 0 & 0 & 0 & 1\\ 0 & -c\omega & a & 0\end{bmatrix}\quad \mathbf{B}=\begin{bmatrix}0 & 0\\ b & 0\\ 0 & 0\\ 0 & b\end{bmatrix}\quad \mathbf{D}=\begin{bmatrix}0 & 0\\ 1 & 0\\ 0 & 0\\ 0 & 1\end{bmatrix}\quad \mathbf{E}=\begin{bmatrix}0 & -\omega\\ \omega & 0\end{bmatrix}$$

3.Compensation and Measurement of Rotor Unbalance

3.1 Compensation of Unbalance Forces The measuring method proposed in this paper are based upon a control system developed by the authors[1]. The ways of designing the control system are

(1) The system is stabilized by a state feedback since an attractive magnetic bearing system is inherently unstable without control.

(2) Since centrifugal forces due to unbalance cannot be detected in an usual magnetic bearing system, an observer is constructed in order to estimate unbalance forces.

(3) Using the output of the observer the effects of rotor unbalance are cancelled by controlling incremental currents of the electromagnets.

The resultant control input is

$$\mathbf{u}(t)=\mathbf{P}_1\mathbf{x}(t)-\hat{\mathbf{w}}(t)/b \tag{15}$$

where $\mathbf{P}_1$ is selected to stabilize a closed-loop system:

$$\dot{\mathbf{x}}(t)=(\mathbf{A}+\mathbf{B}\mathbf{P}_1)\mathbf{x}(t) \tag{16}$$

and $\hat{\mathbf{w}}(t)$ is the output of the observer which converges to $\mathbf{w}(t)$:

$$\lim_{t=\infty} \hat{\mathbf{w}}(t)=\mathbf{w}(t) \tag{17}$$

Then it is concluded that

$$\lim_{t=\infty} \mathbf{x}(t)=0 \tag{18}$$

It means that whirling motions due to unbalance disappear. The details are described in ref.1.

3.2 Measurement of Rotor Unbalance When the position of the rotor is regulated, the bearing forces are cancelling the unbalance forces. Necessary data for balancing the rotor, can be obtained in a direct way from the resultant bearing forces. This measuring method is to be said as a dynamic measurement with a closed-loop system, while a common method of measuring for balancing is a measurement with an open-loop system.

In an usual centrifugal balancing machine, measurements are performed at a speed far below or far above the resonant frequency of the rotor support system because it is difficult to measure accurately near the resonant frequency. In contrast with common balancing machines, test speed can be selected independently of the natural frequency of the machine in a proposed measurement system. Therefore it is possible to balance a rotor at its actual operation speed.

4.Experiments

4.1 Methods Figure 3 shows a photograph of the test instrument; its section view has been shown in Chapter 2. This instrument, which was designed as a spindle for grinding test, is used as a single-plane vertical-type balancing machine.

A test-object of balancing which is shown in Fig.4, is concentrically mounted to the head of the suspended rotor. A weight can be attached to the object at equally spaced positions in a face of the object for introduction of known unbalance. The angular position of unbalance is measured counter-clockwise from the reference position which is pointed by an arrow in Fig.4. A reflective-type photo-sensor detects a black mark which is pasted at the reference position and generates a reference signal.

As is mentioned above, when the position of the rotor is regulated, the bearing forces are identical with corrective forces to balance total forces acting on the rotor. For a direct measurement of bearing force, however, proper sensing devices, as piezoelectric or strain-gage tansducers, must be incoorporated in the instrument. On the other hand an incremental component of the current is proportional to the corrective force(see eq.(2)) and can be measured without special sensing devices. Thus current monitoring is used to determine the amplitude and location of unbalance in the experiments.

4.2 Results. The stationary motions of a rotating unbalanced rotor are shown in Fig.5; the angular velocity of the rotor is 314 rad/s. In the figures x_a and x_b refer to the outputs of the displacement sensor of the radial bearing ⓐ. When the magnetic bearing system is only stabilized, the rotor is whirling with amplitude of about 4.5 μm (see (a)). As contrasted with the case (a), the whirling motion becomes remarkably small when the compensation for unbalance is applied (see (b)). The results show that even an unbalanced rotor behaves like a balanced rotor when it is supported by magnetic bearings with the unbalance compensation controller. The wave-

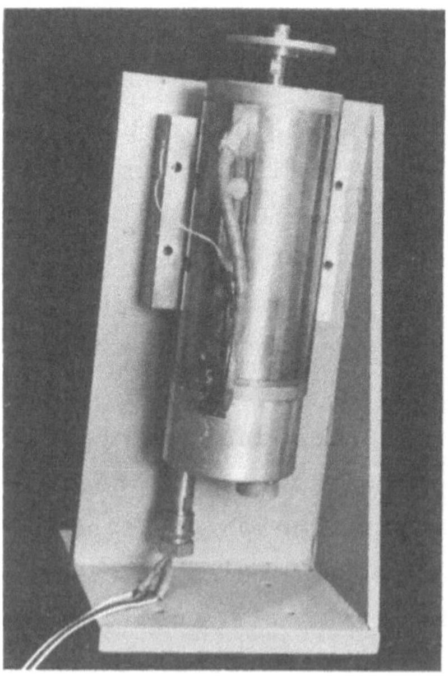

Fig.3 Outlook of experimental instrument

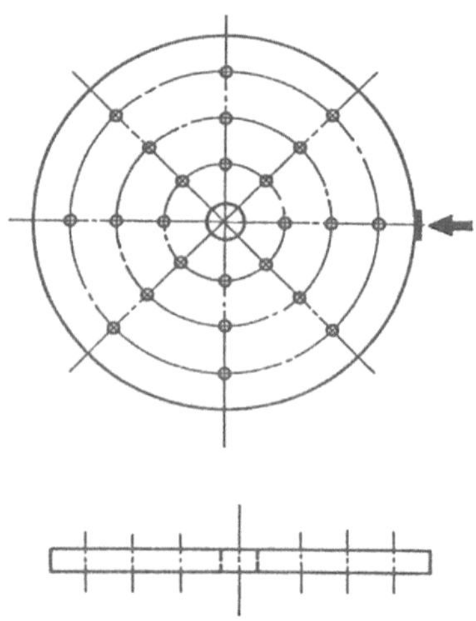

Fig.4 Example of test-object

forms of the incremental current i_3 with the compensation for unbalance applied are shown in Fig.6; a weight is attached to the test-object (see (b)) or not (see (a)). The amount and position of unbalance are determined from the amplitude and the phase relation to the reference signal of the current. The net unbalance introduced by a weight is obtained by comparing the measurement results in case(a) and (b). This procedure of measuring can reject the influences of initial unbalance of the suspended rotor and misalignment of the test-object.

The results of measurements are shown schematically in Fig.7; the position of each circle indicates the actual or estimated location of an attached weight and its area is proportional to the mass of the weight. The actual mass of an unbalance weight is 0.25g. The rotor is driven to rotate in (a)counter-clockwise

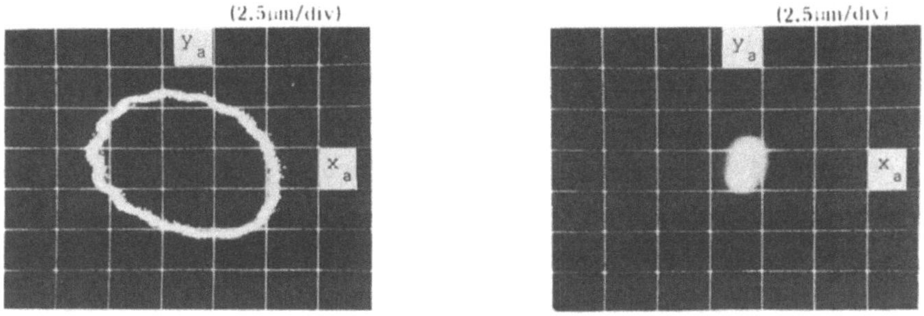

(a)without the compensation for unbalance

(b)with the compensation for unbalance

Fig.5 Stationary motions of an unbalanced rotor suspended by the magnetic berings (scale: 2.5 μm/div)

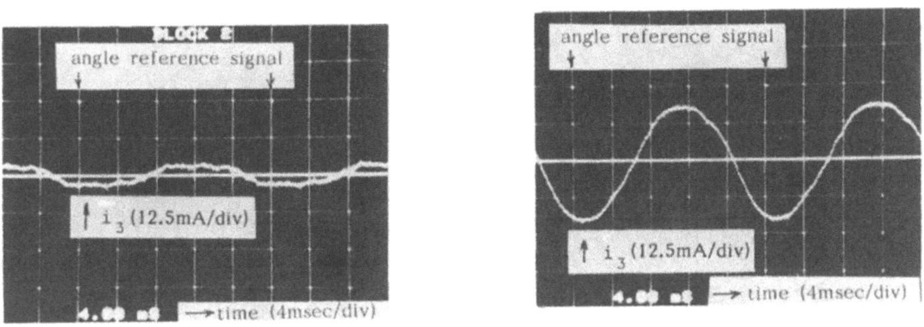

(a)without an unbalance weight

(b)with an unbalance weight attached to the test-rotor

Fig.6 Current and angle reference signal when the compensation for unabalnce works

(ω=314rad/s) and (b)clockwise (ω=-314rad/s) directions.

It is seen from Fig.7 that the estimated amount and location coincide with the actual ones to some degree. The measured amount of unbalance is, however, about 1.6 times as great as the actual one. The main source of error is probably the parameter estimation error of H_n and can be eliminated by a proper calibration. It is also seen from Fig.7 that each estimated location tends to be shifted to the direction of the rotor rotation from the correct one. This tendency is probably due to the delay-time of the photo-sensor generating a reference signal. Using a faster-response detector or averaging the located angles which are measured in clockwise and counter-clockwise rotating, more accurate location of unbalance will be obtained.

5 Conclusions

The principles of a measurement system with magnetic bearings for providing information for performing balancing operations have been presented. The experiments demonstrate the possibility of developing a new-type balancing machine with magnetic bearings.

References

1. Mizuno, T.; Higuchi, T.: Compensation for Unbalance in Magnetic Bearing System (in Japanese). Trans. Society of Instrument and Control Engineers, 20 (1984) 1095-1101.

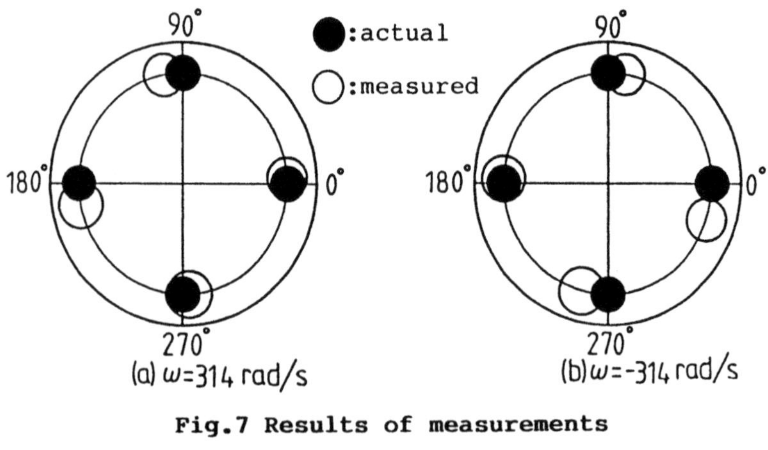

Fig.7 Results of measurements

Digital Control

Modeling for Flexible Mechanical Systems

O. Matsushita*, H Bleuler**, T. Sugaya*, R. Kaneko***

* Mechanical Engineering Research Laboratory, Hitachi Ltd., 502 Kandatsu-machi, Tsuchiura-shi, Ibaraki, Japan 300

** Institute of Mechanics, Swiss Federal Institute of Technology, CH 8092 Zürich, Switzerland

*** Hitachi Works, Hitachi Ltd., 1-1 Saiwaicho 3 chome, Hitachi-shi, Ibaraki, Japan 317

Outline

An alternative modelling technique is proposed in place of the well known modal reduction method for mechanical system modelling. The method, called **quasi-modal reduction**, belongs to the category of mode synthesis methods and represents a special way of substructuring.

Its main advantage is the **adaptability** of the reduced mechanical model to varying boundary conditions, as they are often encountered in control design for mechatronics systems. A mass-spring model of the reduced mechanical system can be derived for flexible mechanical systems.

As an application example, a possible control layout method, a Luenberger observer modified for mechanical vibrational systems, is indicated.

1. Introduction

The quasi-modal method will be presented with an example of a flexible rotor system supported by a contact-free active electromagnetic bearing. The method is suited for modelling and control layout of a quite general class of mechanical vibrational systems.

For this purpose, the FEM equations commonly used for vibration simulations must be simplified. The major drawback of the modal method is the inaccuracy of the results for varying boundary conditions, as the mode shapes depend themselves on a certain boundary condition. The controller in the process of being designed determines these boundary conditions and hence the mode shapes.

The method proposed here uses two types of modes independent of the actual boundary conditions: The modes for constrained (i.e. fixed) boundary conditions are combined with deflection modes generated by unit inputs at the boundary conditions.

2. The Quasi-Modal Method Demonstrated with a Simple Example

2.1 FEM-Model and Equations of Motion

A simple flexible rotor is supported by an ideal hinge and a magnetic bearing as shown in Fig. 1:

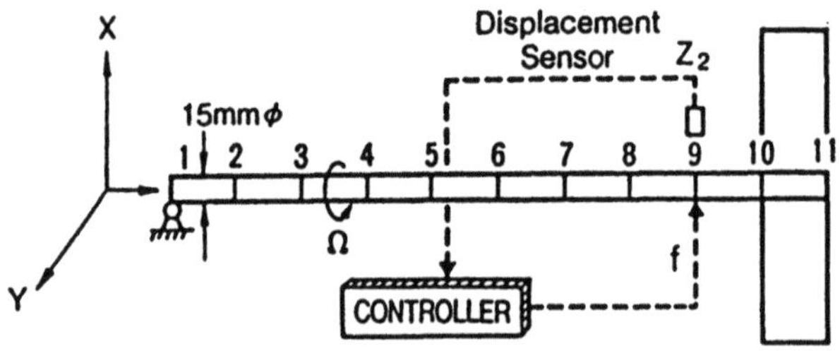

Fig. 1: Flexible rotor example. The shaft vibration is measured with a gap sensor placed at the magnetic bearing (collocation). The FEM-model uses the 10 shaft elements shown. This 40-degrees-of-freedom model will be used as the reference model.

The equations of motion are derived from the finite element model. The complex notation $z = x + iy$ (with the radial directions x and y) is applied, as usual in rotor dynamics.

The displacement vector z is divided into "external" displacements z_2 and "internal" displacements z_1. "External" degrees of freedom are at bearing and sensor locations, "internal" ones are not accessible. The equations of motion can then be represented as follows:

$$\begin{bmatrix} M_1 & 0 \\ 0 & M_2 \end{bmatrix} \begin{bmatrix} \ddot{z}_1 \\ \ddot{z}_2 \end{bmatrix} + \begin{bmatrix} K_{11} & K_{12} \\ K_{21} & K_{22} \end{bmatrix} \begin{bmatrix} z_1 \\ z_2 \end{bmatrix} = \begin{bmatrix} 0 \\ 1 \end{bmatrix} f(t) \qquad (1)$$

with the diagonal mass matrix, the stiffness matrix structured as shown and the bearing force f(t). The example, simple as it is, contains the important features of an actual flexible mechanical system. The coordinate z_2 acts as a "window" giving access to the mechanical system (Fig. 2) :

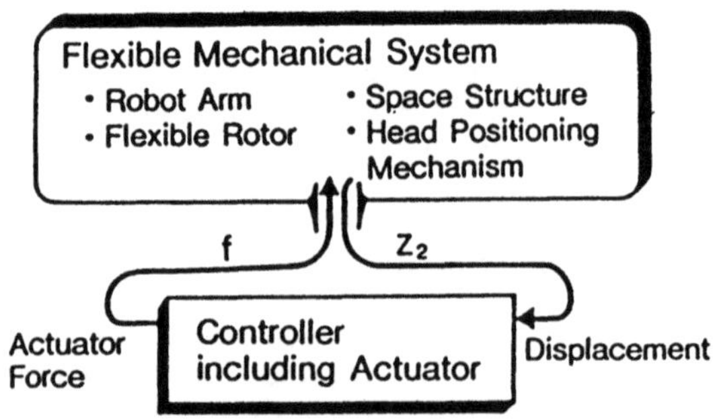

Fig. 2: The measured mechanical variables z_2 and the actuator forces f(t) act like an access "window" to a mechanical system

The FEM program used for the reference model is the HIROT-package /1/ based on the stiffness matrix method. For an actual machine rotor, the number of degrees of freedom of the FEM-model often becomes so large, that it is not suitable even for advanced control layout methods. Therefore, a reasonably reduced model is sought, which still accurately describes the first few vibration modes.

2.2 Modal Reduction

The well known modal reduction method is based on mode shapes. The mode shapes depend on the reaction force f(t). This reaction force on the right hand side of equation (1) is replaced by a pure stiffness K_b in the bearing, as produced e.g. from displacement feedback:

$$f(t) = -K_b z_2(t) \tag{2}$$

This yields a homogeneous conservative equation. The eigenfrequencies (Fig.3) are computed in function of bearing stiffness K_b. For modal decomposition, a fixed and realistic value of K_b is selected, e.g. 30 DN/mm. The number of corresponding mode shapes V_i is truncated arbitrarily. The transformation for the first three modes is:

$$\begin{bmatrix} z_1 \\ z_2 \end{bmatrix} = \sum_{i=1}^{3} \begin{bmatrix} V_{i1} \\ V_{i2} \end{bmatrix} S_i \tag{3}$$

with the modal coordinates S_i and the mode shape vectors V_i structured as the z-vectors in equation (1) . The modal equations of motion are:

$$M^* \ddot{S} + K^* S = R_b f(t) \quad \text{where } M^* = \mathrm{diag}(m_i^*),\ K^* = \mathrm{diag}(k_i^*) \text{ and } R_b = \begin{bmatrix} V_{12} \\ V_{22} \\ V_{32} \end{bmatrix}$$

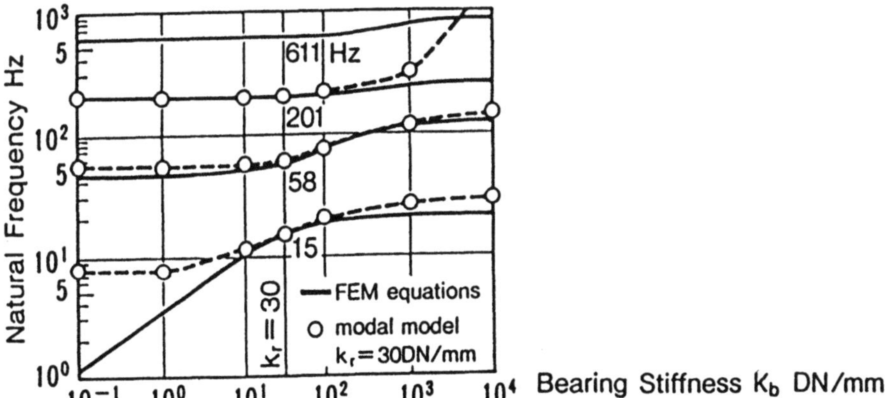

Fig. 3 Eigenfrequencies of the FEM model and modal reduced system of the rotor example in function of bearing stiffness K_b with the limit cases for free and fixed boundary condition for small resp. large K_b .

Numerical values for the example of Fig. 1 are:

$$m_i^* = \begin{bmatrix} 13.5 \\ 7.6 \\ 0.3 \end{bmatrix}, \quad \omega_i = \begin{bmatrix} 15 \\ 58 \\ 201 \end{bmatrix} \text{Hz where } k_i^* = m_i^* \omega_i^2 \quad \text{and } R_b = \begin{bmatrix} .47 \\ .8 \\ .23 \end{bmatrix}$$

In fig. 3, the eigenfrequencies of this reduced model are plotted in function of the bearing stiffness, as was done for the FEM reference model. There is exact agreement only for the selected value of the bearing stiffness. The large deviations (specially of the first mode) are an incentive to look for a more suitable reduction method than modal truncation.

2.3 Quasi-modal Reduction

In analogy to mode synthesis /2/ and in accordance with our earlier publications /3/, a different reduction matrix is proposed here. The aim is a reduced model, that can be used for various bearing characteristics. Therefore, the method should be independent of boundary conditions (stiffness or damping) at the actuators, i.e. from controller parameters. This is achieved by separating the rotor (inner system) from the bearing to be designed (outer system.).

The **inner system** is characterized by the mode shapes with fixed boundary conditions, as shown in Fig. 4a. These modes, independent of the bearing parameters, are called Φ_i .

For the connection with the outside, i.e. to provide access to the relevant bearing variables (displacement and force), mode shapes obtained from unit force displacement at the bearing have to be used. The input- and output matrices then contain only ones (at the bearings) and zeros (elsewhere). These modes are called δ_i. For our example with only one variable z_2, there is only one δ-mode, the rigid-body mode of Fig. 4b, equal to the free boundary condition first mode.

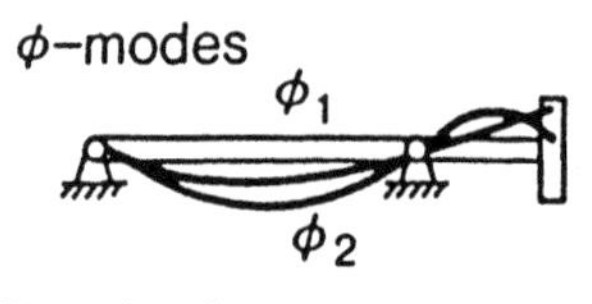

Fig. 4a Inner system mode shapes (only two shown), eigenfrequ. ω_{pi}

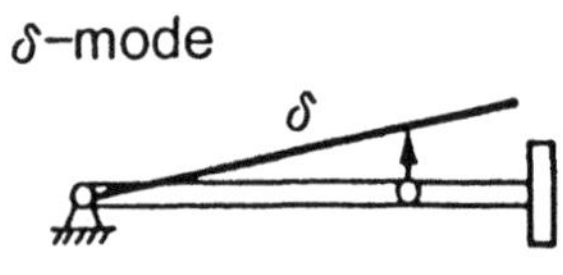

Fig. 4b Deflection mode shape

Fig. 4 Mode shapes for the quasimodal transformation.

The transformation matrix is analogous to the modal method, but now using the Φ- and δ-modes. A reduced system of 3rd order is obtained by using the δ-mode and two Φ-modes in this way:

$$\begin{bmatrix} z_1 \\ z_2 \end{bmatrix} = \begin{bmatrix} \phi_1 & \phi_2 & \delta \\ 0 & 0 & 1 \end{bmatrix} \begin{bmatrix} S_1 \\ S_2 \\ z_2 \end{bmatrix} \tag{4}$$

In opposite to the modal reduction, physical bearing displacements z_2 remain in the variable vector of the reduced system. Thus, the controller is easily connected to the mechanical system. The quasi-modal reduced system has the following general structure:

$$M^* \begin{bmatrix} s \\ z_2 \end{bmatrix}^{\bullet\bullet} + K^* \begin{bmatrix} s \\ z_2 \end{bmatrix} = R_b f(t)$$

The system reduced to 3rd order is:

$$\begin{bmatrix} m_1^* & 0 & m_{c1} \\ 0 & m_2^* & m_{c2} \\ m_{c1} & m_{c2} & m_\delta \end{bmatrix} \begin{bmatrix} s_1 \\ s_2 \\ z_2 \end{bmatrix}^{\bullet\bullet} + \begin{bmatrix} k_1^* & & \\ & k_2^* & \\ & & 0 \end{bmatrix} \begin{bmatrix} s_1 \\ s_2 \\ z_2 \end{bmatrix} = \begin{bmatrix} 0 \\ 0 \\ 1 \end{bmatrix} f(t)$$

where $M^* = \begin{bmatrix} 16.7 & 0 & 16.2 \\ . & 8.08 & -4.8 \\ sym & . & 19.6 \end{bmatrix}$ $\qquad k_i^* = m_i^* \omega_{pi}^2$

$\omega_{p1}^2 = 22\,\text{Hz} \quad \omega_{p2}^2 = 126\,\text{Hz}$

The obtained reduced matrices are not diagonal. Due to the orthogonality of the Φ-modes, however, they are partially diagonal. Due to the simple structure of the input- (and output-) matrix, the bearing stiffness can easily be included or excluded in matrix K*. This is an advantage over the modal transformation.

The most important advantage of the quasi-modal reduction is the good agreement of the eigenvalues of the reduced system and FEM-model for the full range of the stiffness K_b (fig.5) . The agreement is best for the lowest modes, as seen clearly when comparing fig. 3 and fig. 5.

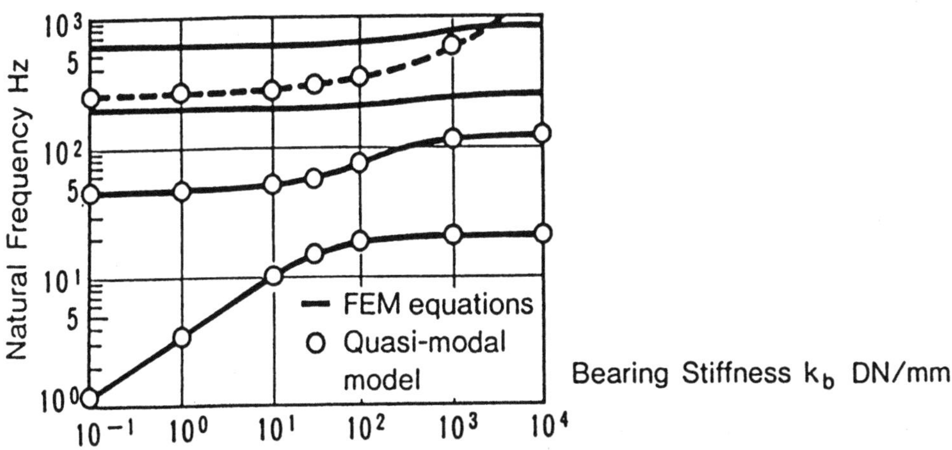

Fig. 5 Eigenfrequencies of the FEM model and quasi-modal reduced system of the rotor example in function of bearing stiffness K_b with the limit cases for free and fixed boundary condition for small resp. large K_b .

The same advantage of better accuracy of the low modes is also present for the modal damping ratios in function of bearing damping (i.e. velocity feedback). This is shown in fig. 6:

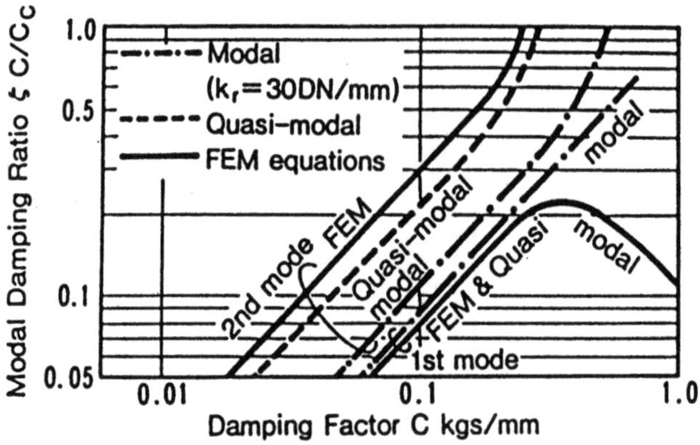

Fig. 6 Modal damping of the FEM model and the two reduced systems in function of bearing damping (and constant stiffness K_b = 30 DN/mm). **The damping ratio saturation, important for bearing controller design, is present in the FEM model and the quasi-modal reduced system, but not in the modal reduced one.**

The superior accuracy of the quasi-modal reduced system in function of widely varying bearing parameters is due to the use of the combination of Φ-modes and δ-modes.

3. Equivalent Quasi-modal System

3.1 Why the equivalent transformation?

The quasi-modal reduced system as presented in the previous chapter is characterized by a non-diagonal mass matrix, i.e. mass coupling, and a diagonal stiffness matrix. In some cases, a physically better understandable structure with stiffness coupling and diagonal mass-matrix is preferred.

The displacement vector of the quasi-modal system contains the time-variable weighting values S_i for the bending modes. These variables must be interpreted as relative displacements. Absolute displacements z^*_i are introduced for each bending mode:

$$S_i = a_i \, (z^*_i - z_2) \tag{5}$$

A diagonal mass matrix is obtained, by applying this transformation to the quasi-modal system and defining the constants a_i as : $a_i = m_{ci} / m^*_i$.

The structure of the transformed system (shown only for third order) is:

$$\begin{bmatrix} m_{eq1} & 0 & 0 \\ 0 & m_{eq2} & 0 \\ 0 & 0 & m_{eq\delta} \end{bmatrix} \ddot{\begin{bmatrix} z_1^* \\ z_2^* \\ z_2 \end{bmatrix}} + \begin{bmatrix} k_{eq1} & & -k_{eq1} \\ & k_{eq2} & -k_{eq2} \\ sym & & k_{eq\delta} \end{bmatrix} \begin{bmatrix} z_1^* \\ z_2^* \\ z_2 \end{bmatrix} = \begin{bmatrix} 0 \\ 0 \\ 1 \end{bmatrix} f(t) \tag{6}$$

where $M_{eq} = \begin{bmatrix} 15.7 & . & . \\ . & 2.85 & . \\ . & . & 25.8 \end{bmatrix}$, $k_{eqi} = m_{eqi}\, \omega_{pi}^2$, $k_{eq\delta} = \sum_i k_{eqi}$

This system can be interpreted directly in the physical mass-spring system shown in fig. 7, along with a summary of the calculation procedure.

Coupling factor $a_i = \frac{m_{ci}}{m_i^*} = \frac{\phi_i^t M \delta}{\phi_i^t M \phi_i}$

Equivalent mass $m_{eqi} = a_i^2 m_i^*$

Equivalent residual mass $m_{eq\delta} = m_\delta - \Sigma m_{eqi}$

★ Equivalent Reduced Model

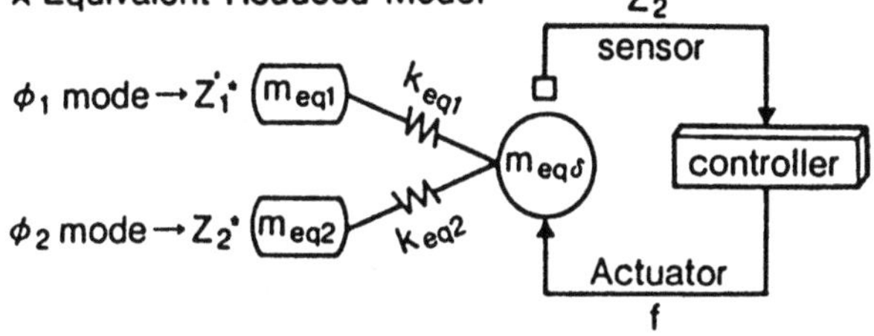

Fig. 7 Equivalent quasi-modal reduced system and its physical interpretation.(only two bending modes shown)

3.2 Example of a model with two bearings

In this discussion, a most simple example has been used. It is however a quite general procedure, not restricted to collocation or only one sensor-actuator. A slightly more general example, with two bearings and three bending modes in the reduced model (5 degrees of freedom per radial direction), is shown in fig. 8 and fig. 9 . A gyroscopic matrix can also easily be included.

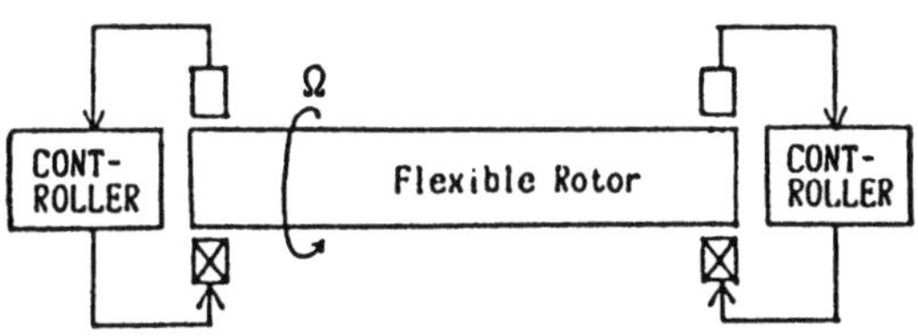

Fig. 8 Example of a rotor with two radial bearings and sensors.

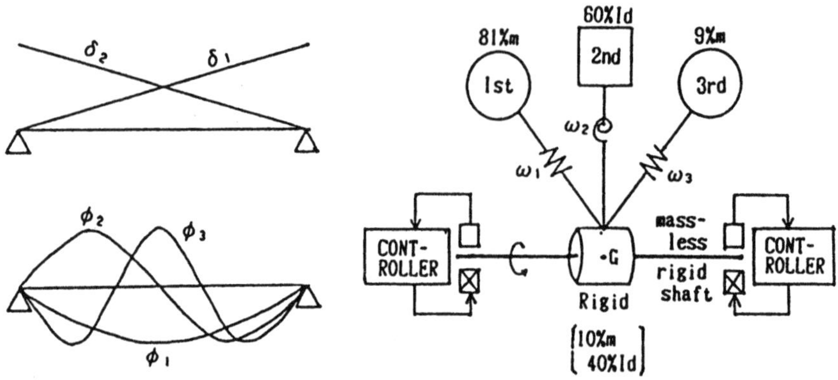

Fig.9 Equivalent reduced model for the rotor with two bearings of fig.8. and five degrees of freedom per radial direction.

4. Transfer Function

The quasi-modal reduced system can be represented in form of a transfer function relating bearing-force as input variable with bearing displacement as output variable. This is shown in fig. 10 for the simple rotor example of fig. 1 and the corresponding reduced model.

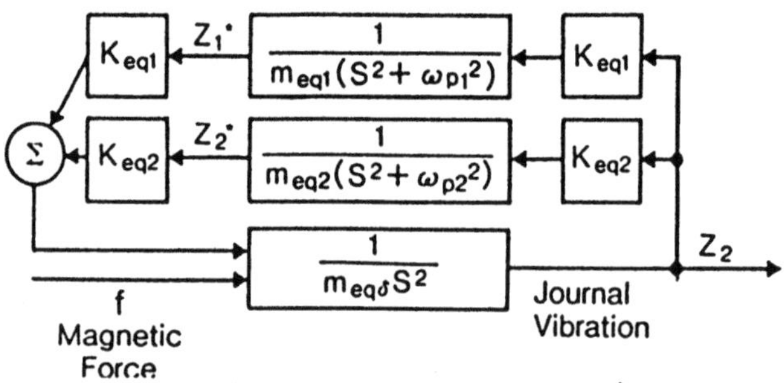

Fig.10 Block diagram of the transfer function of the equivalent quasi-modal reduced system for the example of fig. 1 and 7 . It is based on equations (6) ; the variable s is the Laplace domain variable.

The block diagram of fig. 10 can be reduced to the following transfer function, using the parameter definitions given with equ. (6) and fig. 7:

$$G(s) = \frac{z_2}{f} = \frac{(s^2 + \omega_{p1}^2)(s^2 + \omega_{p2}^2)}{m_{eq\delta}\, s^2 (s^2 + \omega_{f1}^2)(s^2 + \omega_{f2}^2)} \tag{7}$$

where ω_{f1} and ω_{f2} are natural frequencies of the system with free boundaries (no actuators)

$m_{eq\delta}$ is the equivalent mass for the main rigid body

ω_{p1} and ω_{p2} are natural frequencies of the inner system restricted by the pin boundaries,

Note that this transfer function reflects the correct behavior of the eigenfrequencies when the bearing stiffness moves from free to pinned boundary condition.

5. Luenberger Observer Based Bearing control

This control scheme basically consists of state-feedback combined with a model of the mechanical system used to estimate ("observer") the unmeasured state variables in real-time. The description given in control theory textbooks is usually in state-base description. Here, the equations of motion notation is used in order to provide some additional physical insight into the observer dynamics. A complete system is shown in fig. 11:

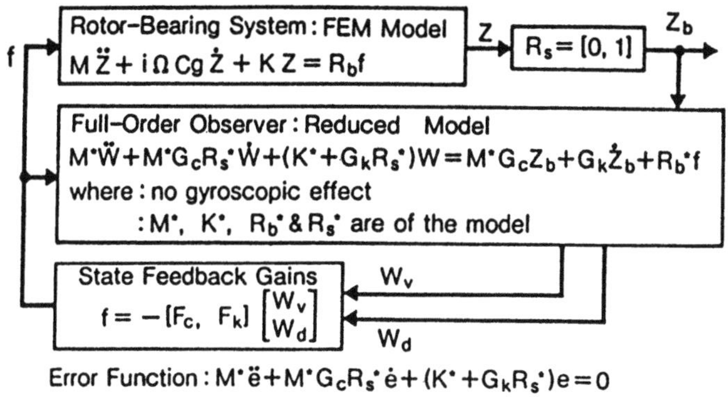

Fig. 11 Closed-loop mechanical system with observer-based state feedback. The observer dynamics is determined by matrices Gc and Gk which can be interpreted as stiffness and damping respectively

6. Conclusion

The proposed quasi-modal reduction method overcomes some clear disadvantages of the modal truncation reduction method, namely its dependence on boundary conditions.

The higher effectiveness, accuracy and easier physical interpretation of the quasi-modal method are demonstrated and are now being tested for the control layout of vibrational systems like active magnetic bearing support of elastic structures.

7. References

/1/ O. Matsushita et al.: Survey of Rotordynamics Analysis Program HIROT, ICVPE, Xian, July 1986, p. 817

/2/ R.R. Craig et al.: Coupling of Substructures for Dynamic Analyses, AIAA J. 6-7, July 1968, p. 1313

D.A. Glasgow et al.: Stability Analysis of Rotor-Bearing Systems Using Component Mode Synthesis, J. Mech. Design ASME, Vol.102, p.352

G.M.L. Gladwell: Branch Mode Analysis of Vibrating Systems, J. Sound & Vibration, 1964 , p. 41

A. Ookami et al.: Jap. Measurement and Control Soc.(in Japanese), Vol. 26-10, Oct. 1987, p. 19-28

H. Kanoh: Jap. Measurement and Control Soc. (in Japanese), Vol. 26-11 Nov. 1987. pp 968 - 976

/3/ O. Matsushita et al.: Vibration Analysis of Rotor Bearing Systems by Quasi-modal Transformation, Bul.JSME, 26-213, March 1983, p.414

O. Matsushita et al.: Application of Quasi-modal Concept to Rotational Response Analysis & New Balancing, IMechE, C319/84, 1984,p.427

/4/ O. Matsushita et al.: An Equivalent Reduced System for Coupled Vibration Analysis, J. JSME (in Japanese), 54-499-C, March 1988, p.587

Optimal Design of Structure Predefined Discrete Control for Rotors in Magnetic Bearings (SPOC-D)

R. Larsonneur
R. Herzog
Institute of Mechanics
Swiss Federal Institute of Technology
Zürich - Switzerland

ABSTRACT

Many layout techniques of time-discrete control algorithms for magnetic bearings are based on quasi-continuous approximations of traditional time-continuous controllers, although this approach does not provide the full dynamic range and may even lead to instability.

State-feedback methods require an observer since the number of measured signals is usually less than the number of states in the mathematical rotor model. In case of flexible rotor structures this observer approach leads to a high-order fully coupled controller and often shows untolerable parameter sensitivity.

The goal of this paper is to present a layout method for optimal discrete dynamic compensators with structural constraints typical in magnetic bearing applications, i.e. a predefined controller order or a decentralized feedback structure, in order to fill the gap between well known PD-algorithms and state-LQ-schemes. Similar to the latter method the optimal feedback coefficients are obtained by minimization of a quadratic performance index. Both the performance index and the corresponding vector gradient can be computed easily for every set of feedback parameters. Quick convergence can be achieved by a powerfull numerical optimization routine.

Results of a **SPOC-D** (Structure-Predefined Optimal Control for Discrete systems) layed out simple magnetic bearing system are presented and compared with the system properties obtained by standard controller design methods.

1. INTRODUCTION, GOAL OF THIS PAPER

Active electromagnetic bearings show a number of interesting characteristics and are more and more applied for new solutions to classical machine dynamic problems. Due to the rapid development in microprocessor technology, digital control makes it nowadays possible to take advantage of the wide range of achievable magnetic bearing parameters (i.e. damping and stiffness). Even changes in the structure of the controller can easily be made by varying the controller software instead of loosing much time altering the controller hardware.

The choice of the controller structure depends on several aspects of the magnetically borne rotor system and has to be carefully evaluated to achieve the desired performance properties. These aspects are:

- rotor and corresponding rotor model,
- number of controlled bending modes and overall dynamic properties,
- number of available output signals,
- performance of the microprocessor.

A fundamental property of magnetic bearing control is the fact that the number of measurement signals is usually less than the number of states in the corresponding model description. Therefore optimal pure state-feedback is not possible.

Basically, there are two ways to handle the problem of too few outputs and of getting satisfactory controller performance: one possibility is the use of a complete state-feedback with a full or reduced-order observer to estimate the missing velocities and, in case of higher system orders, all the residual states. This approach leads to a high-order fully coupled controller with the consequence of lower sample rates due to the given computational power of the controller hardware. In addition to this disadvantage untolerable parameter sensitivities might occur implementing high-order observers (/1/).

A second way to cope with the problem of too few output signals is the predefinition of the dynamic compensator structure, i.e. the predefined discrete controller consists of a given number of time lag elements with given interconnections. This output feedback approach leads, for example, to a decentralized control where every output path is directly connected to a specific input path without any further interconnections. Decentralized continuous controllers are a classical feedback structure. Optimization of these controller types has been investigated by LEVINE/ATHANS (/2/) and SENNING (/3/) and specifically for their application in magnetic bearing systems by BLEULER (/4/).

The practical application of decentralized time-discrete control schemes is an often followed design approach. The necessary control parameters are usually obtained by a quasi-continuous approximation of the corresponding continuous parameters. In case of magnetic bearings, however, it can be shown that much better sets of control parameters can be found if a parameter optimization is done directly in the time-discrete state space and not based on quasi-continuous approximations. Furthermore, even better optimization results can be achieved if the controller structure is slightly changed, so that a quasi-continuous approach to determine the feedback coefficients is not possible any more.

The **SPOC-D** method (Structure-Predefined **O**ptimal **C**ontrol for **D**iscrete systems) presented in this paper provides a way of designing controllers by parameter optimization. Similar to state-LQ-methods the optimal feedback coefficients for the structurally constrained system are obtained by minimization of a quadratic performance index involving both the system states and the control inputs. The numerical minimum search can be done quite easily due to the fact that both the performance index and the corresponding vector gradient are given in an analytic form for every set of feedback parameters. An additional feature of the SPOC-D method is the possibility of optimization under several constraints as prescribed static bearing stiffness or consideration of symmetries. Furthermore, SPOC-D can minimize a multi-model performance index to ensure stability over a wider range of changing plant parameters.

A simple example of a magnetically borne elastic rotor system is presented (fig. 3.). A quasi-continuous controller design method is compared with the SPOC-D method for a layout of an optimal set of control parameters. Results are illustrated by simulations of the system behaviour.

2. Optimizing Structure-Predefined Discrete Control

2.1. Plant, Controller Structure and Performance Index

A general time-discrete state space description of a linear dynamic system is given by

$$x_{k+1} = A\,x_k + \sum_i b_i\,u_{ki} \tag{1}$$

$$y_{ki} = C_i\,x_k \qquad (i = 1...) \tag{2}$$

In case of structure predefined control the state vector of the plant is augmented by the states of the dynamic compensator to a global state vector x_k. The global system matrix A describes the dynamic behaviour of the plant and includes the predefined interconnections between plant and compensator. The scalar input signals u_i are called *control stations*. For each control station a single input column vector $\mathbf{b}_i$ describes its influence on the dynamic system. Furthermore a special *observation station* y_{ki} is assigned.to each control station

The structure predefined control loop will be closed by the unknown output feedback parameters d_{ij} included in the single row vectors $\mathbf{d}_i$.

$$u_{ki} = \mathbf{d}_i\,y_{ki} = \mathbf{d}_i\,C_i\,x_k \qquad (i = 1...) \tag{3}$$

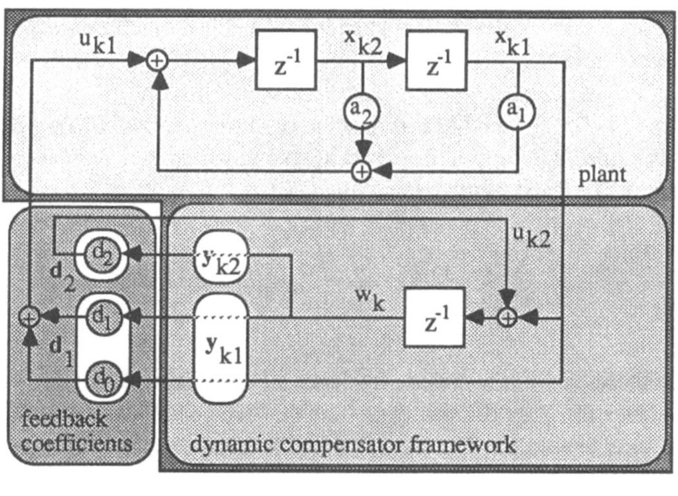

fig. 1. plant augmented by a structure-predefined dynamic compensator for direct output feedback

Fig 1. shows a simple example of a dynamic system as shown in (1, 2, 3): the plant description of order two (x_1, x_2) is augmented by a predefined dynamic compensator of order one (w). Two output signals (y_1, y_2) are available for the controller specified by the three unknown parameters (d_{10}, d_{11}, d_{22}). For simplicity, the station index *i* for the coefficients d_{ij}

will be omitted in all further examples (as already shown in fig. 1.). The special form of (1, 2) for the given example is shown in (1a, 2a).

$$\begin{bmatrix} x_1 \\ x_2 \\ w \end{bmatrix}_{k+1} = \begin{bmatrix} 0 & 1 & 0 \\ a_1 & a_2 & 0 \\ 1 & 0 & 0 \end{bmatrix} \begin{bmatrix} x_1 \\ x_2 \\ w \end{bmatrix}_k + \begin{bmatrix} 0 \\ 1 \\ 0 \end{bmatrix} u_{k1} + \begin{bmatrix} 0 \\ 0 \\ 1 \end{bmatrix} u_{k2} \qquad (1a)$$

$$y_{k1} = \begin{bmatrix} 1 & 0 & 0 \\ 0 & 0 & 1 \end{bmatrix} x_k \; ; \; y_{k2} = \begin{bmatrix} 0 & 0 & 1 \end{bmatrix} x_k \qquad (2a)$$

The closed-loop structure for the given example is an outer feedback u_{k1} for the plant and an inner feedback u_{k2} for the dynamic compensator. Thus we get

$$u_{k1} = \begin{bmatrix} d_0 & d_1 \end{bmatrix} y_{k1} \; ; \; u_{k2} = \begin{bmatrix} d_2 \end{bmatrix} y_{k2} \qquad (3a)$$

with the unknown controller coefficients d_0, d_1 and d_2 that have to be determined by optimization of a performance index in order to guarantee overall stability and satisfactory dynamic behaviour of the closed-loop system (1, 2, 3). For this purpose it is assumed that the dynamic system (1, 2) is completely observable and controllable.

For the subsequent optimization of the described output feedback it will be very important that the parameters d_{ij} of each feedback vector $\mathbf{d}_i$ must be independent of each other: each coefficient d_{ij} can appear only once at one given place in one feedback vector $\mathbf{d}_i$. Furthermore, the dynamic compensator structure should be formulated in a suitable normal form and should not be over-parametrized.

As described in (/2/, /3/) the performance index PI for the closed-loop system (1, 2, 3) is formulated as a quadratic form in the states $\mathbf{x}_k$ and in the control stations u_{ki} involving the weighting matrix $Q \geq 0$ and the scalar weighting factors $r_i > 0$:

$$PI = \sum_{k=0}^{\infty} (\mathbf{x}_k^T Q \mathbf{x}_k + \sum_i u_{ki}^T r_i u_{ki}) \quad \text{--->} \; \min \qquad (4)$$

Fig 2. now shows the four necessary fundamental equations ((5)-(8)) for optimality of the performance index (4). They differ slightly from the time-continuous case. The derivations are briefly explained below, but not shown in detail in this paper. The exact steps will be presented in (/5/).

At first the performance index is brought into an equivalent form with the use of the initial state $\mathbf{x}_0$ and the closed-loop matrix A_{clsd}. The infinite sum can then be replaced by a Lyapunov equation whose solution P determines PI together with the initial state $\mathbf{x}_0$. Applying small perturbations $\mathbf{dd}_i$ to the feedback vectors $\mathbf{d}_i$ a perturbed performance index PI + dPI is found which finally leads to an analytic form for the vector gradient of the performance index together with a dual Lyapunov equation.

$$A_{clsd} = A + \sum_i b_i d_i C_i \tag{5}$$

$$A_{clsd}^T P A_{clsd} - P + Q + \sum_i C_i^T d_i^T r_i d_i C_i = 0$$
$$PI = trace(PX_0) \tag{6}$$
$$X_0 = x_0^T x_0$$

$$A_{clsd} X A_{clsd}^T - X + X_0 = 0 \tag{7}$$

$$\frac{\partial PI}{\partial d_i} = 2(b_i^T P A_{clsd} + r_i d_i C_i) X C_i^T = 0 \quad (i = 1...) \tag{8}$$

fig. 2. necessary optimality equations to minimize the performance index PI

2.2. Minimization of the Performance Index Under Constraints

The prescription of the static bearing stiffness is typical for magnetic bearing applications, so that the output feedback coefficients d_{ij} *are no longer independent* of each other.

Any parameter interdependences of this or a different kind can be considered in SPOC-D by including additional control stations in (1), the feedback coefficients of which are linear or nonlinear functions of the residual set of independent parameters d_{ij}. The effect on the optimality equations ((5)-(8)) are additional terms for the closed-loop matrix A_{clsd}, the Lyapunov equation for P and for the vector gradient $\partial PI/\partial d_i$.

Note that all constraints are considered in the modified necessary equations for optimality and not by a special algorithm for minimization under constraints.

Although these modified optimality equations are of much more interest to the field of technical applications, they are not shown in this paper, but will be well described in (/5/). All results for the example below, however, have been found with the modified equations for optimization under constraints.

3. Numerical Minimization Procedure

The necessary matrix equations for optimality of the performance index ((5)-(8)) have to be solved simultaneously. They are coupled and highly nonlinear. Since it is not possible to find a closed form for the minimum of the performance index PI the solution has to be found by means of a powerful numerical method including the ability to solve Lyapunov equations, which is found in many available software packages for matrix handling (e.g. CTRL-C). The fact of having the vector gradient in an analytic form is of great advantage for the minimization.

The numerical procedure used to find the results given below is a special *Quasi-Newton-Method* called *Davidon-Fletcher-Powell-Method* including self-scaling and restarting. It is described in (/6/). The method requires an additional one-dimensional minimum search algorithm which, in our case, is implemented as a cubic fit.

4. RESULTS

The figure below shows a simple magnetically borne rotor system. It is assumed that the elastic rotor can only move in one plane. Thus the total number of degrees of freedom is three if the mass of the elastic shaft is neglected. Gyroscopic effects are not considered. The system is geometrically symmetric, though the shape of the rotor movements may not be symmetric.

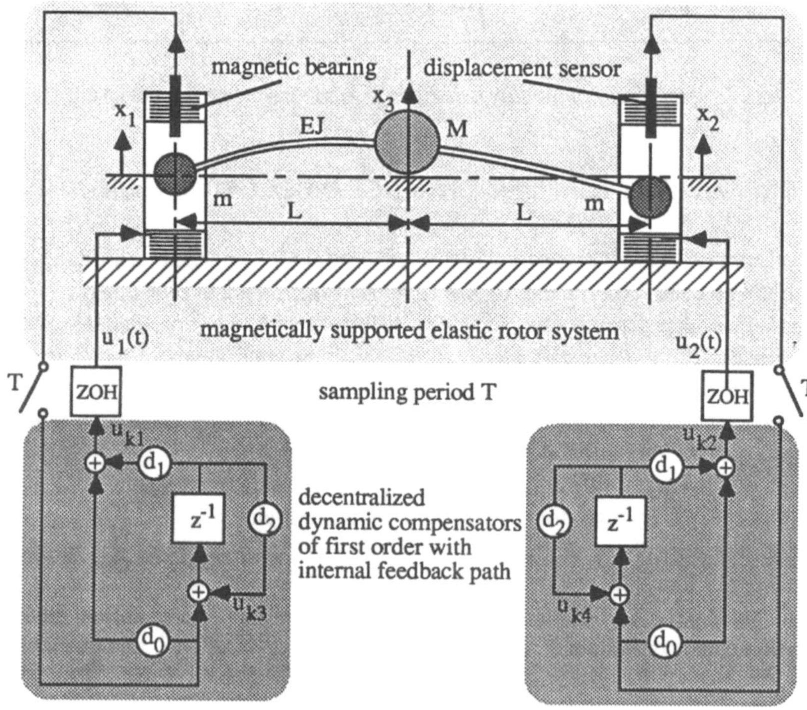

fig. 3. elastic magnetically borne rotor system augmented by two time-discrete decentralized dynamic compensators of predefined structure

The rotor system has two input signals: the force or electric current in each magnetic bearing. Only two sensors measure the rotor position in each bearing, which is most frequently the case in technical applications.

The state space dimension of the plant description is six; thus four system states are not measured. The control loop is closed augmenting the plant description by two decentralized dynamic compensators to stabilize the global system and to achieve a "good" dynamic performance. Each decentralized controller can be described by a general first-order transfer function involving the coefficients d_0, d_1 and d_2. This controller structure predefinition does not correspond with a reduced observer which would be of order four and fully coupled.

The equations of motion are not derived in this paper. However, the physical data are given as follows: masses (m, M), rotor stiffness (k_R) and negative magnetic bearing stiffness (k_S) due to the pre-magnetization current:

$$m = 1\text{kg} \qquad k_R = \frac{EJ}{L^3} = 206\ \text{kN/m}$$

$$M = 5\ \text{kg} \qquad k_S = -150\ \text{kN/m}$$

To give an idea of the physical behaviour of this system the time-continuous eigenvalues are given below (the transmission zeroes are not listed here). Note that the open loop system is unstable due to the negative bearing stiffness k_S.

$$\lambda_{1,2} = \pm 35 + 0\,i\ \text{Hz} \qquad \lambda_{3,4} = \pm 62 + 0\,i\ \text{Hz} \qquad \lambda_{5,6} = 0 \pm 139\,i\ \text{Hz}$$

The time-discrete system equations (1, 2) for this example are obtained via the transition matrix and its integral. The sample rate T is chosen to correspond with a sample frequency about 5 to 10 times higher than the highest system frequency. In this example T equals 1 millisecond.

For magnetic bearing applications it is important that a certain closed-loop static bearing stiffness k is achieved. Typically, k lies in the range of the absolute value of k_S. Thus, the three coefficients d_0, d_1, d_2 of each decentralized controller are not independent (see paragraph 2.2.). For the given example (fig. 3.) the static bearing stiffness parameter interdependence can be formulated as follows:

$$d_0 + \frac{d_1}{1 - d_2} + k - k_S = 0 \tag{9}$$

4.1. Quasi-Continuous Determination of the Feedback Parameters

The traditional approach to determine the values of the feedback coefficients is by a digital approximation of well-known continuous control laws (/7/, /8/). Very often found in magnetic bearing applications is the PD-algorithm which corresponds with a spring-damper element in classical rotor dynamics. For this case the digital first-order causal approximation of the corresponding transfer function G(s) leads to the following discrete controller transfer function G(z):

$$G(s) = P + D\,s \quad \Longleftrightarrow \quad G(z) = d_0 + \frac{d_1 z^{-1}}{1 - d_2 z^{-1}} = P + \frac{D}{T}(1 - z^{-1}) \tag{10}$$

Note that d_2 is set to zero by this quasi-continuous approach, and that therefore only one feedback coefficient is freely choosable after selection of the static bearing stiffness k (9).

In the following figure k is chosen to equal the absolute value of k_S. The closed-loop step response to a unit force applied from outside in only one bearing (asymmetric load case) is simulated for the first 150 milliseconds. The rotor displacements in each bearing (x_1, x_2) and the corresponding bearing forces (u_1, u_2) are plotted for three different controller layouts respecting (9, 10). Note that one bearing force has to be twice the unit force due to the negative bearing stiffness k_S.

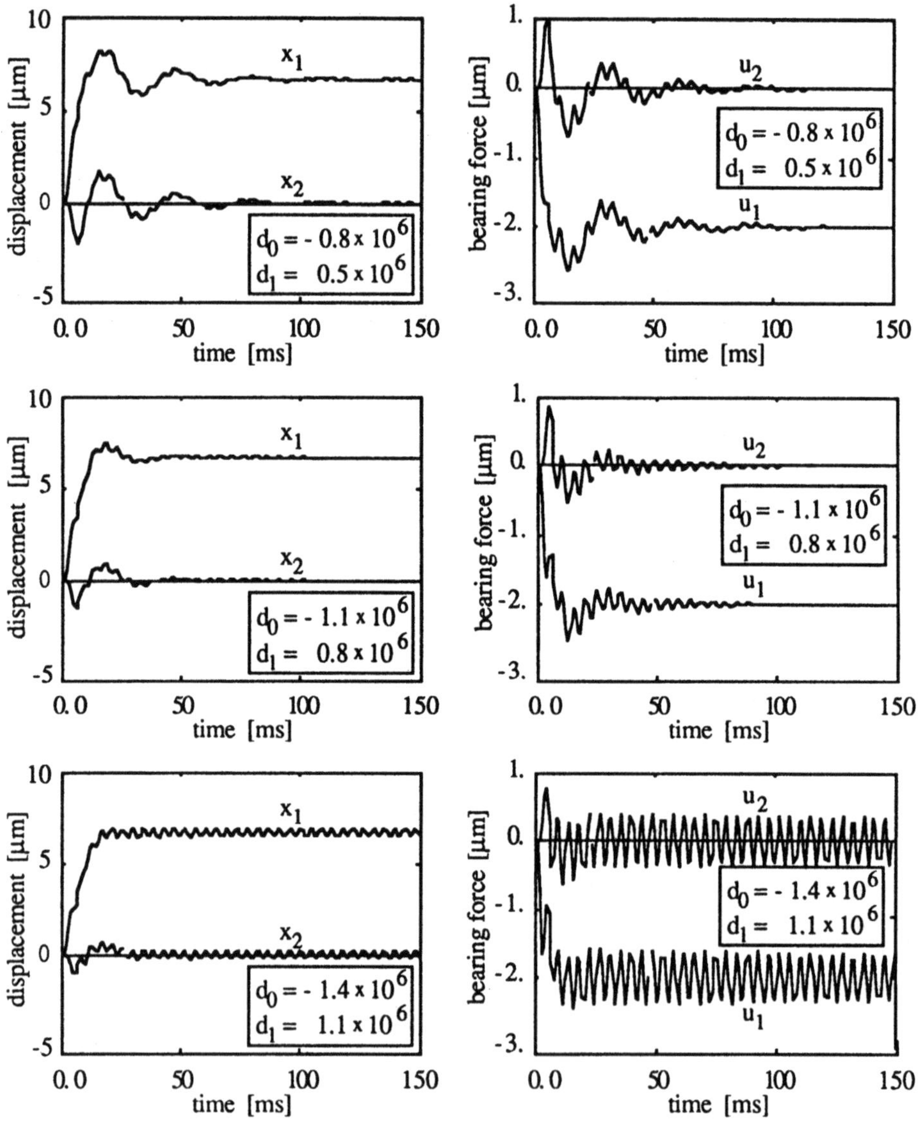

fig. 4. simulations of the step response for a controller layout based on a quasi-continuous approach

Fig. 4. shows the very interesting effect that for "low damping" (d_1 small) neither the low frequency rigid body modes nor the high frequency elastic mode are influenced in a satisfactory way. Applying "more damping", i.e. changing the one only free parameter, the rigid body mode will tend towards a good behaviour whereas the elastic mode, however, tends to get unstable.

Thus it is not possible to achieve a satisfactory closed-loop system performance using a first-order decentralized controller with a given static bearing stiffness layed out by a quasi-continuous approximation of a PD-controller.

4.2. SPOC-D OPTIMIZED FEEDBACK PARAMETERS

Much better results are obtained by application of SPOC-D. The results of the parameter optimization underlying the constraint of the static bearing stiffness differ in a very crucial way from those given above: the feedback coefficient d_2, set to zero by the traditional approach, turns out to play an important role in improving the system performance. Fig. 5. shows the step response for the same conditions described above but for the SPOC-D optimized controller.

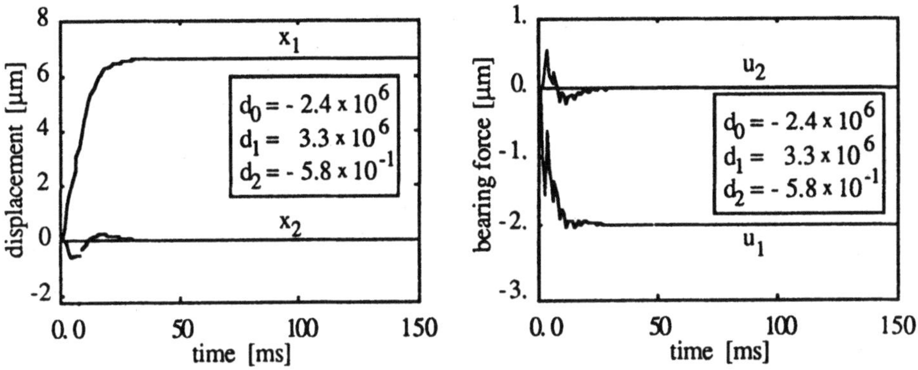

fig. 5. simulations of the step response for a controller layout based on SPOC-D

The robustness of the optimized closed-loop system with respect to changing plant parameters is not investigated in detail here. However, a comparison of the simulation results between the quasi-continuous layout and the optimized controller is shown in fig. 6. for some altered values of the mass M *but for unchanged controller coefficients*. Only the displacement x_1 is plotted.

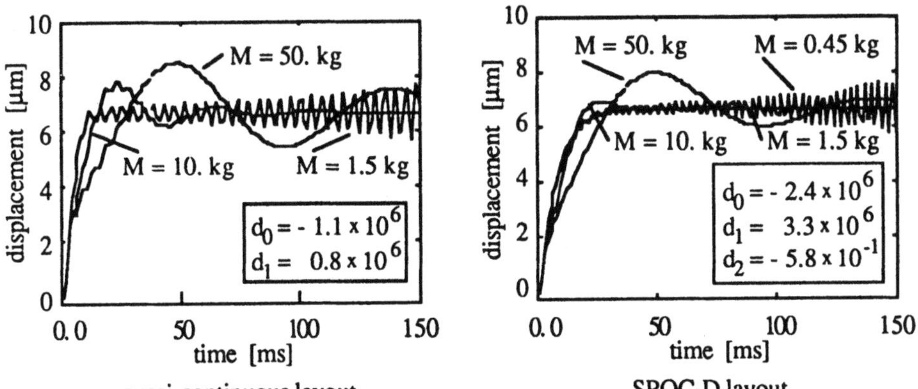

fig. 6. comparison of the step responses for different values of M

5. CONCLUSION

The layout of digital controllers with predefined structure is often based on a quasi-continuous approach. It can be shown, however, that the time-discrete approximation of a "good" time-continuous control algorithm does not always lead to comparably "good" results.

An optimization of the controller coefficients directly in the time-discrete state space leads to much better results. A reason therefore is the fact that the control parameters are optimized based on a performance index including the plant characteristics; thus the optimized controller is better adapted to the specific plant and is less sensitive to changing plant parameters.

The proposed way of optimization is very flexible. A large number of well posed problems with any kind of control structure predefinition can be handled. Any constraints of the controller coefficients, linear or nonlinear, can be considered in the necessary equations for optimality. The choice of the weighting matrices and factors used in the performance index is not problematic if the system equations are brought into a suitable normal form. A certain amount of experience, however, is necessary to find good solutions in a short time.

REFERENCES

/1/ N. M. LEHTOMAKI, N. R. SANDELL, M. ATHANS, "Robustness Results in Linear-Quadratic Gaussian Based Multivariable Control Designs", IEEE Transactions on Automatic Control, AC-26,1981, p. 75-93

/2/ W. S. LEVINE, M. ATHANS, "On the Determination of the Optimal Constant Output Feedback Gain for Linear Multivariable Systems", IEEE Transactions on Automatic Control, 1970

/3/ M. F. SENNING, "Feasibly Decentralized Control", Thesis ETH Nr. 6321, Swiss Federal Institute of Technology, Zürich

/4/ H. BLEULER, "Decentralized Control of Magnetic Rotor Bearing Systems", Thesis ETH Nr. 7573, Swiss Federal Institute of Technology, Zürich

/5/ R. LARSONNEUR, "Very High Speed Rotor in Magnetic Bearings", Thesis ETH, Swiss Federal Institute of Technology, to appear in 1989

/6/ D. G. LUENBERGER, "Linear and Nonlinar Programming", Addison-Wesley Publishing Company, Massachusetts, second edition 1984

/7/ G. HOFFMANN, "Digitale Approximation kontinuierlicher Übertragungssysteme (QTA-Verfahren)", Automatisierungstechnik, 35. Jahrgang, Heft 4/1987, p. 148-155

/8/ O. FÖLLINGER, "Lineare Abtastsysteme", R. Oldenbourg Verlag, München, 2. Auflage 1982

Digital Control of Magnetic Bearing with Rotationally Synchronized Interruption

Yohji OKADA, Bunshu NAGAI and Takeshi SHIMANE

Department of Mechanical Engineering
Ibaraki University
Nakanarusawa 4-12-1
Hitachi, JAPAN 316

Summary

For the purpose of clean and high speed rotation, an actively controlled magnetic bearing has been developed. Recently, a digital control system has been used to get high performance by using the state variable feedback. However, the gyroscopic or inductive mutual effect of a rotating disc sometimes makes the high speed rotor unstable.

This paper introduces a new digital control scheme which utilizes the rotational interruption. The standard PID control is carried out with the constant time interval interruption, while the rotational interruption subroutine carries the static error cancellation and the cross coupling feedback to compensate for the undesirable mutual effect. This scheme is applied to a single rotor bearing system, and its capability is tested.

THE EXPERIMENTAL ROTOR AND ITS CONTROL SYSTEM

The Equations of Motion of the Rotor The equations of motion of the rotating shaft supported by the magnetic bearing are written in the following matrix form:

$$m\ddot{x} - \Omega J_p \dot{y} + k x = f_x - f_{bx} \tag{1}$$

$$m\ddot{y} + \Omega J_p \dot{x} + k y = f_y - f_{by} \tag{2}$$

where m and k are the mass and stiffness matrices for the x and y directions, and J_p is the gyroscopic moment matrix. The f_x and f_y are the external force vectors and f_{bx} and f_{by} are the supporting bearing force vectors /1-5/.

The Electro-Magnetic Force Positional feedback is used to maintain the neutral position of the shaft, while derivative and integral operations are widely used to improve the dynamic and static properties.

However the rotor is usually made of electro-conductive

material which produces the eddy-current causing the undesirable Fleming force which is called as the inductive force. This effect produces the brake force to the rotation and the undesirable cross coupling effect. Hence, the disc is recommended to be made of laminated plates or ferrite material. For experimental convenience, we use only the solid iron to make the rotor, so that the inductive force cannot be neglected. Let us approximate the inductive force to be proportional to the relative velocity multiplied by the magnetic field. Then the electro-magnetic force can be written by

$$f_{bx} = K_p x + K_I \int x \, dt + K_D \dot{x} - K_1 \Omega y + K_2 \dot{x} \qquad (3)$$

$$f_{by} = K_p y + K_I \int y \, dt + K_D \dot{y} + K_1 \Omega x + K_2 \dot{y} \qquad (4)$$

where the first to third terms are the PID algorithm, while the fourth and fifth terms indicate the inductive force. The fifth term is the damping and can be neglected compared to the third term. However the fourth term indicates the cross coupling effect and affects adversely to the system stability /2,5/.

The Stability Analysis Equations (1) and (2) are applied to the experimental one degree-of-freedom rotor system. Hence, the displacement vector x and y can be simplified by the scalar. Applying the PD control of eqs. (3) and (4), we have

$$m \ddot{x} - \Omega K_J J_p \dot{y} + K_D \dot{x} + (k + K_p) x - K_1 \Omega y = 0 \qquad (5)$$

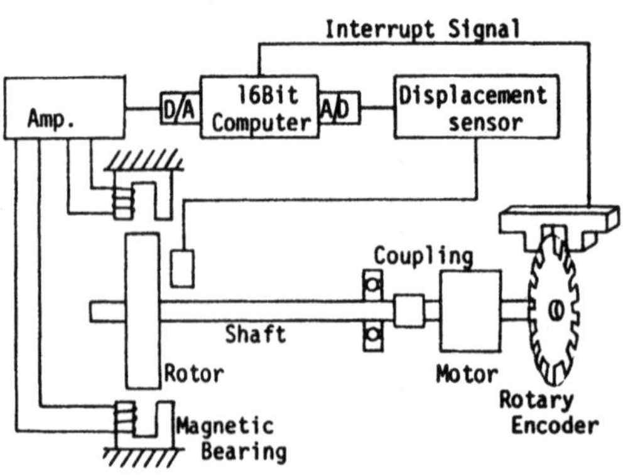

Fig. 1. The scheme of experimental setup

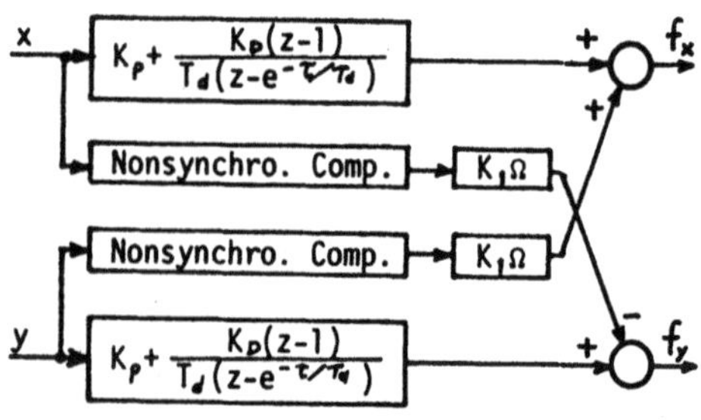

Fig. 2. The block diagram of the proposed digital control

$$m\ddot{y} + \Omega K_J J_p \dot{x} + K_D \dot{y} + (k + K_p) y + K_I \Omega x = 0 \qquad (6)$$

where K_J is the gyroscopic factor determined by the vibrating mode. The velocity feedback K_D improves the damping of the system. Hence the fundamental pole of $\sqrt{(k+K_p)/m}$ will be damped by increasing the gain K_D.

However, the cross coupling effects are apt to make the system unstable. The gyroscopic effect will increase the frequency of one of the duplicated poles and decrease the other, and are recognized as the forward and backward precessional motion. When the backward precessional frequency approaches zero, the system is apt to be unstable. A cross feedback technique has been introduced to correct this difficulty /5/. A more harmful mutual effect is the inductive force which will move the duplicated poles away one to the right and the other to the left-half plane. This means that increasing $K_I\Omega$ makes the system unstable. Cross feedback is highly desirable to cancel out the undesirable cross coupling effects.

DIGITAL CONTROL SYSTEM

The PID Control The displacement of the rotor can be measured without phisical contacts, and fed back to the magnetic bearing to maintain the neutral shaft position. The derivative operation improves the dynamic stability and integral feedback decreases the static error. Hence the following analog transfer function is widely used:

$$G_c(s) = \frac{K_I}{s + 1/T_I} + K_P + \frac{K_D s}{1 + T_D s} \qquad (7)$$

where T_I and T_D are the time constants of the derivative and integral circuits.

The equivalent discrete transfer function is given by

$$G_c(z) = \frac{K_I}{z - e^{-\tau/T_I}} + K_P + \frac{K_D(z-1)}{T_D(z - e^{-\tau/T_D})} \qquad (8)$$

where τ is the sampling interval. This digital PID control is carried out by the constant time interrupt subroutine which is shown in Fig. 3.

The Rotationally Synchronized Interruption As mentioned before, the cross coupling effects will affect adversely to the high speed stability of the rotating shaft. The most preferable technique is the use of cross feedback

$$G_m(s) = (K_J J_p s + K_1)\Omega \tag{9}$$

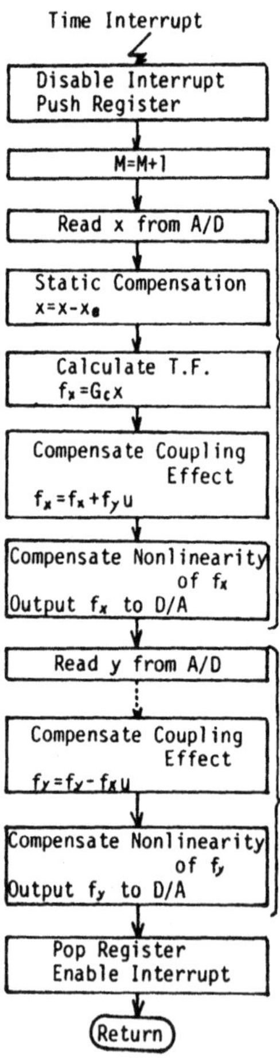

Fig. 3. The time interrupt subroutine

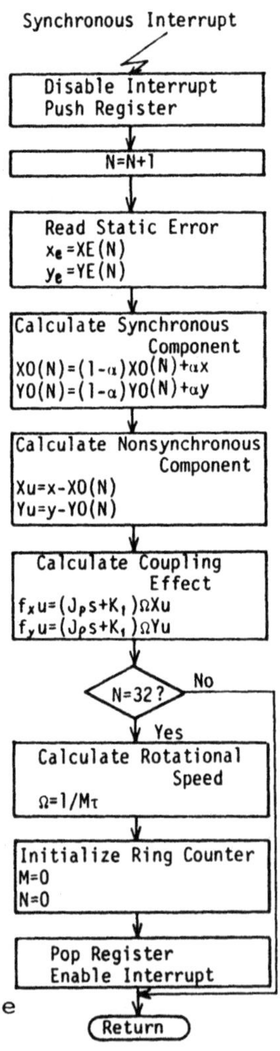

Fig. 4. The rotational interrupt subroutine

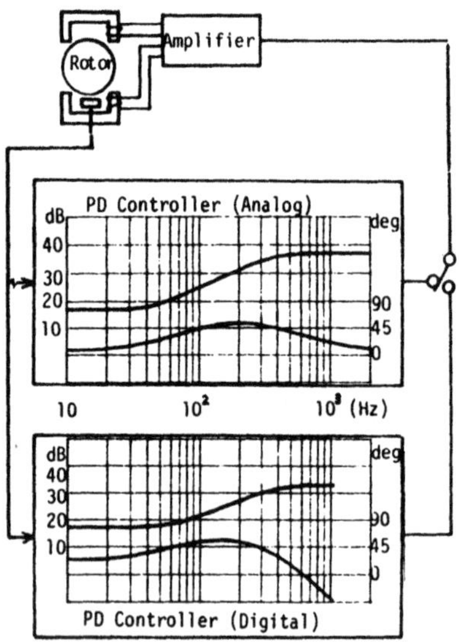

Fig. 5. The comparison between the analog and digital control

to cancel out the undesirable coupling effects. This means, however, that the feedback gain should change proportionally to the rotating speed Ω. The digital controller can easily measure the rotating speed by means of rotationally synchronized interruption: that is, the number of time interruptions M is counted for one revolution of the shaft. Then the rotating speed can be calculated by $\Omega = 1/M\tau$.

Another important technique is to separate the synchronous and asynchronous components from the measured signal. The synchronous component can be used to compensate for the static error of the measured signal. The eddy-current displacement sensor sometimes produces the synchronized error caused by the nonhomogeneity of the measured surface material. This error is premeasured and stored in the ring memory XE(N). For each interruption N (in this case N=1, 2, $\cdots$,32), the displacement x is subtracted by this error XE(N) to produce the true displacement.

The cross coupling effect increases when the shaft runs at very high speed. Hence, the unstable frequency is usually several times lower than the rotating speed. The cross feedback signal should preferably be made from the asynchronous low frequency component. In the interrupt subroutine, the synchronous component X0(N) can be made by exponentially averaging the measured signal.

$$X0(N) = (1 - \alpha)\, X0(N) + \alpha\, x \qquad (10)$$

where α $(0<\alpha<1)$ is the averaging constant. Then the asynchronous component Xu can be made by subtracting the synchronous component X0(N) from the measured signal x.

$$Xu = x - X0(N) \qquad (11)$$

This signal is used to make the cross feedback signal. The rotational interrupt subroutine is shown in Fig. 4. The block diagram of proposed digital controller is shown in Fig. 2.

EXPERIMENTAL RESULTS AND CONSIDERATIONS

The Experimental Apparatus The scheme of the experimental magnetic bearing system is shown in Fig. 1. The shaft is made of

4 Φ × 200 mm stainless steel with an iron rotor of 78.5 g. This solid rotor is supported by a magnetic bearing which is converted from a stator of a 4 phase stepping motor. Hence, the inductive force cannot be neglected. The rotor has a 0.7 mm air gap and can run up to 1047 rad/s (10000 rpm) by a DC motor. A 32 step interrupt encorder is mounted on the other side of DC motor.

PD Control The critical speed of the rotating shaft was analyzed by means of the transfer matrix method and the results are shown in Fig. 6. The bearing stiffness is 2600 N/m, which causes the first critical of 180 rad/s (1720 rpm) and the second of 2530 rad/s (24100 rpm). This paper uses PD control of x and y directions separately. The frequency responses of the analog and digital controller(τ=1 ms) are shown in Fig. 5. The unbalance response is shown in Fig. 7 and the orbital trajectory where the shaft is running at 1043 rad/s with PD controller is shown in Fig. 8 (a). These experiments show that the shaft can be stabilized and can rotate up to 1043 rad/s.

Simular results are obtained by the digital PD controller.

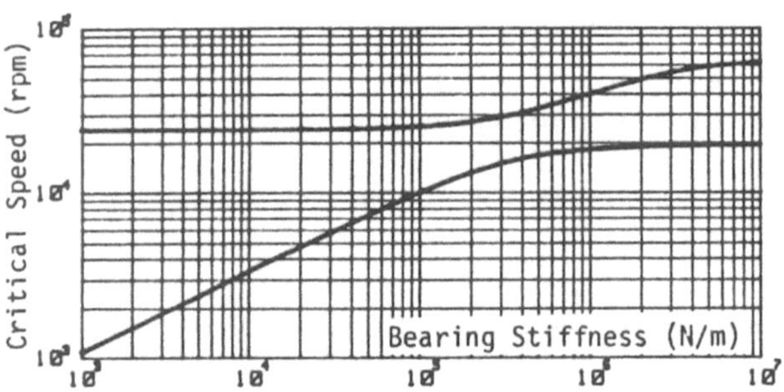

Fig. 6. The critical speed versus bearing stiffness

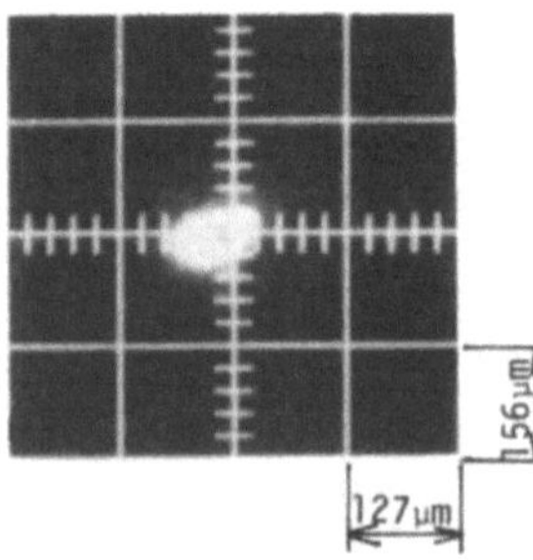

(a) Analog control

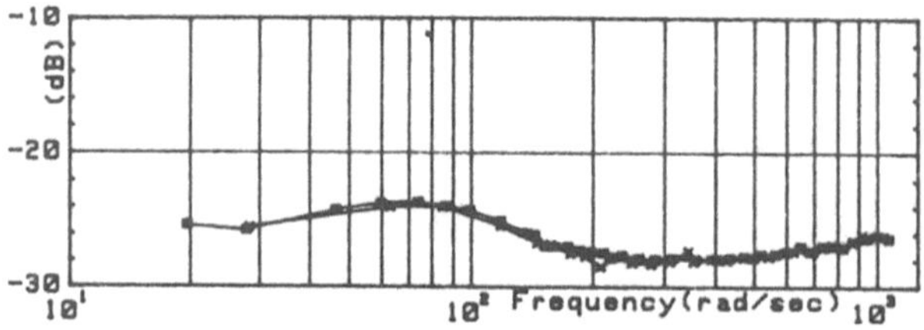

Fig. 7. The unbalance response with analog PD control

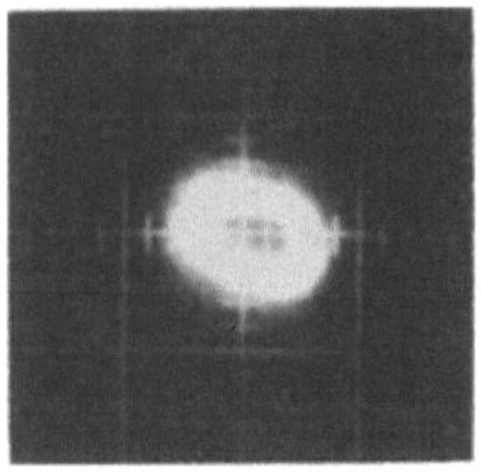

(b) Digital Control

Fig. 8. The orbital trajectory with the PD control

The orbital trajectory at 1043 rad/s is shown in Fig. 8 (b), which is a little larger than the analog controlled case. The frequency response is shown in Fig. 9. In this case, the rotor is running at 280 rad/s. The fundamental rotating peak at 280 rad/s and the second one at 560 rad/s can be recognized. However, the response indicates high stability of this system.

The Rotational Interrupt Control In addition to the standard PD control, rotationally synchronized interruption and cross feed-back control is applied to the magnetic bearing. The system is apt to be unstabilized by the electro-inductive force. The example of orbital trajectory which indicates the subharmonic vibration is shown in Fig. 10. Decreasing the velocity feedback gain, a lower frequency unstable vibration occurs. Case (a) indicates a 100 rad/s unstable vibration with analog control, while (b) shows the 175 rad/s unstability with digital control.

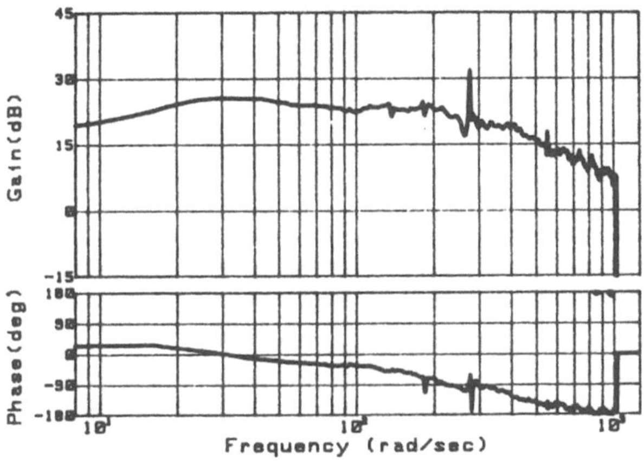

Fig. 9. The frequency response with digital PD control

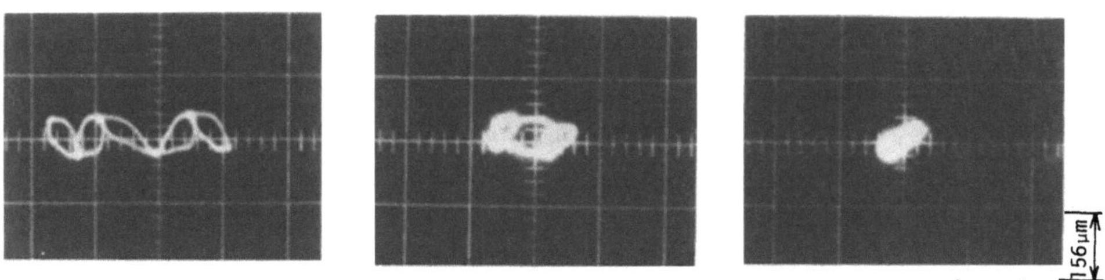

(a)Analog control (b)Digital control (c)Rotationally interrupt control

Fig. 10. The comparison of the orbit trajectories

A rotational interrupt control is carried out with a standard 8086 microprocessor which accepts 32 interrupt pulses per revolution as shown in Fig. 1. For each interruption, a subroutine calculates the asynchronous component of the measured signal which is crossly fed back to the rotor to cancel out the undesirable inductive force. The stabilized orbital trajectory is shown in Fig. 10 (c).

CONCLUSIONS

A simple one degree-of-freedom magnetic bearing rotor system is controlled by a digital control algorithm installed in a 16 bit microprocessor. The following conclusions are obtained:

(1) A digital PD controller can improve the dynamic stability of the rotating shaft as well as the analog one. Soft support and high damping control is preferable.

(2) A new digital control algorithm is applied which can compensate for the undesirable cross-coupling effect of the radial directions. A rotationally synchronized interrupt and its subroutine can measure the rotating speed and asynchronous component which can be utilized to compute a cross feedback signal.

Further work is continuing to apply this cross feedback technique to the multi degree-of-freedom rotor system and to clarify its ability to compensate gyroscopic effects.

REFERENCES

(1) Brunet, M.; Contribution of Active Magnetic Bearings Spindles to Very High Speed Machinery, S2M Habermann, H. and Brunet, M., ASME, 84-GT-117, 1

(2) Nagai, B., et al., Int. Cof. on Rotordynamics, JSME & IFToMM, Tokyo, Sept., (1986), 415

(3) Allaire, P. E., et al., 4.Workshop on Rotordynamics Instability Problem in High-Performance Turbomachinery, Texas A&M Univ., June-1986

(4) Nonami, K., Fleming, D. P., Int. Conf. on Rotordynamics, JSME & IFToMM, Tokyo, Sept. (1986),429

(5) Matsushita, O., et al., Int. Conf. on Rotordynamics, JSME & IFToMM, Tokyo, Sept. (1986), 421

Signal Processors and Applications

Active Control of a Flexible Rotor by an Active Bearing

Yoichi KANEMITSU, Masaru OHSAWA, Katsuhide WATANABE

EBARA RESEARCH Co.,LTD.
4-2-1 Honfujisawa,Fujisawa-shi,251,JAPAN

SUMMARY

This paper details the effects of analog and digital compensation circuits used when passing through critical speeds of a flexible rotor suspended by an active magnetic bearing. Several kinds of filters are used for rotation tests and vibration characteristics of the rotor are measured. A digital IIR filter implemented by a DSP could attenuate the vibration of the rotor at critical speeds.

INTRODUCTION

A magnetic bearing which uses DC magnets consists of proximity sensors measuring relative diplacement between rotor and stator, electric circuits regulating current to keep input signal from proximity sensor constant and DC electlic magnets as actuator.

We have made a five axis active magnetic bearing as an experiment and controlled it by using an analog circuit, a digital FIR filter circuit and a digital IIR filter circuit as a compensator in order to support a flexible rotor. A digital compensator is suitable for adaptive or optimal control because the properties of the compensator can be easily changed by modifying it's software. The validity of these circuits is confirmed by rotation tests of the rotor operating beyond the bending natural frequencies of rotor.

MAGNETIC PULL OF RADIAL POLE

A radial magnetic bearing(RMB) has 4 electric magnets as shown in FIg.1. The magnetic pull between the rotor and stator is calculated from the magnetic flux density in the clearance between the rotor and stator and also the magnetic flux density excited by the stator pole coil is calculated by the permeance method based on the ralation between magnet-motive force IN and magnetic intensity H in each magnetic pole given by Ampere's law.

Fig.2 shows a comparison of the relations between magnetmotive force, gap and magnetic pull calculated by the permeance method and by a 2-dimensional magnetic field analysis using the finite element method. In this figure, magnetic pull by the permeance method becomes large as the magnetmotive force increases and saturates near 600N. The value from FEM is slightly smaller than that calculated by the permeance method because leakage flux must take place and be neglected in the permeance method.

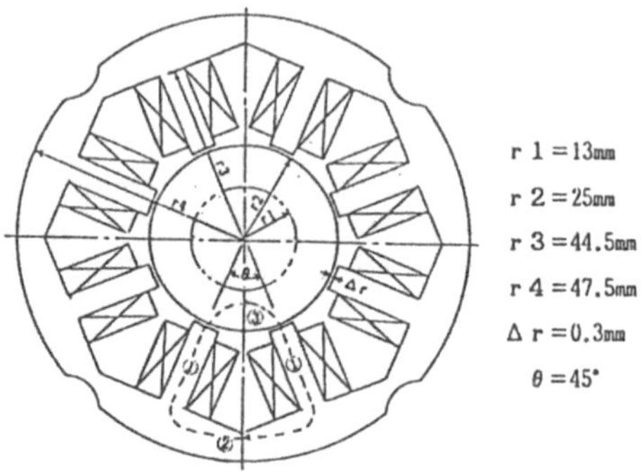

Fig.1 Shape of Radial Magnetic Bearing

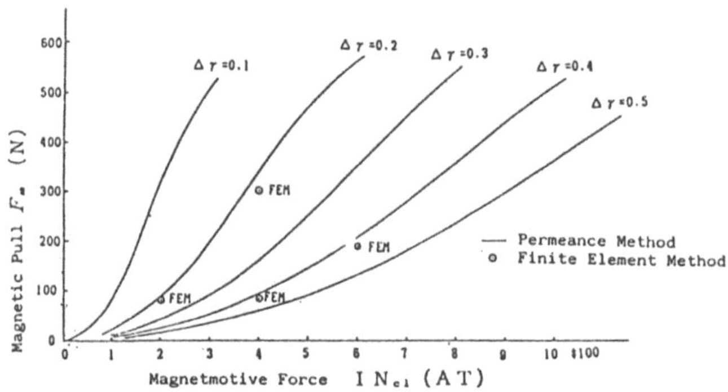

Fig.2 Magnetic Pull of Radial Magnetic Bearing

CONFIGRATION OF CONTROLLER

Each axis of the active magnetic bearing(AMB) is regulated individually and it's controller is made up of a shift circuit of a displacement signal, a compensation circuit, two driver and two power amplifiers. The block diagram of the system is shown in Fig.3. The shift circuit of the displacement signal corrects an error of signal and retains the rotor at the desired position. The compensation circuit is made up of a integration circuit to increase static rigidity, a phase lead circuit to increase damping force of the rotor in the frequency range above rigid mode natural frequency and a low pass filter to attenuate the gain in high frequency.

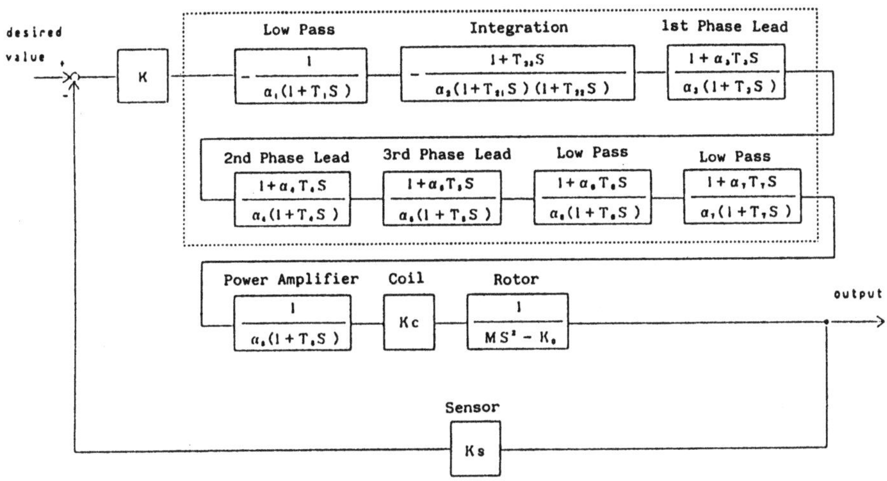

Fig.3 Block Diagram of System

TEST APPARATUS

Fig.4 shows a test apparatus for attesting the ability of the analog compensator and digital compensator. The shaft is 531mm in length, and 26mm in maximum diameter. A motor rotor is located in the middle of the shaft and two RMB rotors are situated on both sides of the motor rotor. A thrust magnetic bearing(TMB) is located at left end of shaft. Eddy current type proximity sensors with two mounting holes are prepared for the RMB sensors at both sides of the bearing.

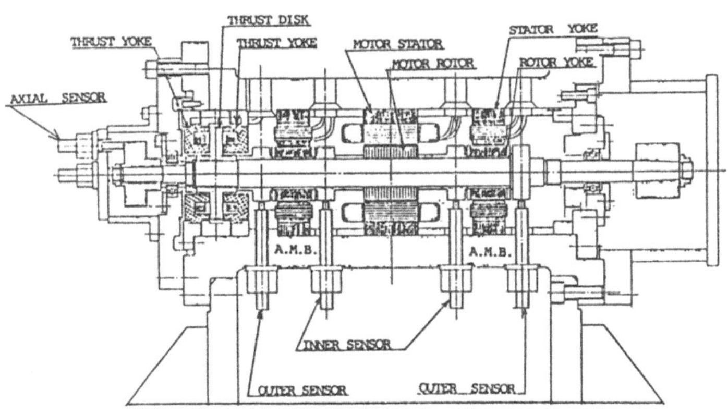

Fig.4 Test Apparatus

The natural frequencies and modes of the test rotor measured in a free-free state are presented in Fig.5. According to this figure the outer sensor of the left bearing coincided with a nodal point of 1st mode so that the observability of this mode cannot be achieved by this sensor.

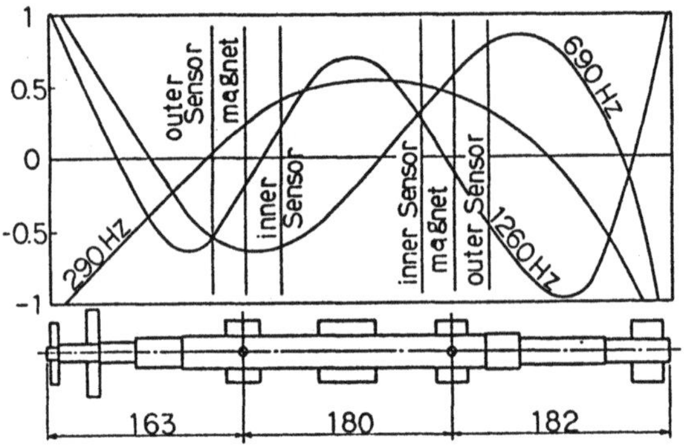

Fig.5 Natural Vibration Modes of Rotor

VIBRATION DURING ROTATION TEST WITH ANALOG COMPENSATOR

For an analog compensator, a compensation circuit(A) containing a low-pass filter at high-frequencies and ones(B) without it were made and their measured transfer functions are as shown in Fig.6. Design of the compensation circuit without a low pass filter was based on the idea that the gain at high frequencies would be reduced by the mass effect of the rotor to be controlled and that it would be better in the interest of stability to advance the phase up to the highest frequency possible.

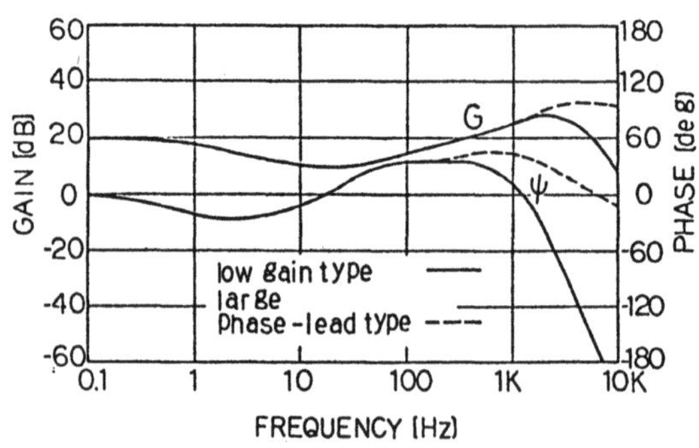

Fig.6 Transfer Fanction of Analog Compensator

Fig.7 and Fig.8 show waterfall diagrams of displacement vibration using the inner sensors at the inner proximity sensor. When the outer sensors were used, unstable vibration occured near the first critical speed and the rotating speed could not be increased.

Fig.7 is the test result using the inner sensor and the analog compensator(A). Here the rotor reached 32,000rpm, but 1st natural vibration occured at this speed.

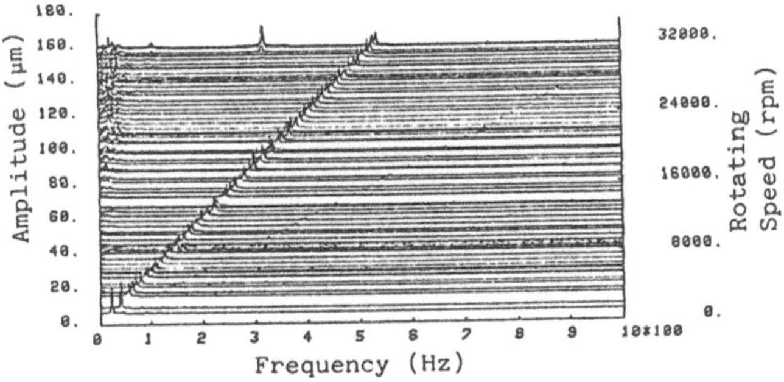

Fig.7 Radial Vibration of Rotor with Compensator(A)

Next, Fig.8 shows results of a rotating test using the compensator(B) which has a larger phase-lead than compensator(A) as shown in Fig.6. The rotating speed reached 40,000rpm by compensator(B) without unstable vibration because of the increase in phase-lead. Maximum rotating speed in a series of rotating tests using the inner sensor is 40,000rpm which is just below the second critical speed of 42,000rpm. As each test has no rigid mode resonance, and small resonance of 1st bending mode, we proved that the analog compensator has sufficient damping action in these frequency ranges. But this compensator could not give the rotor sufficient

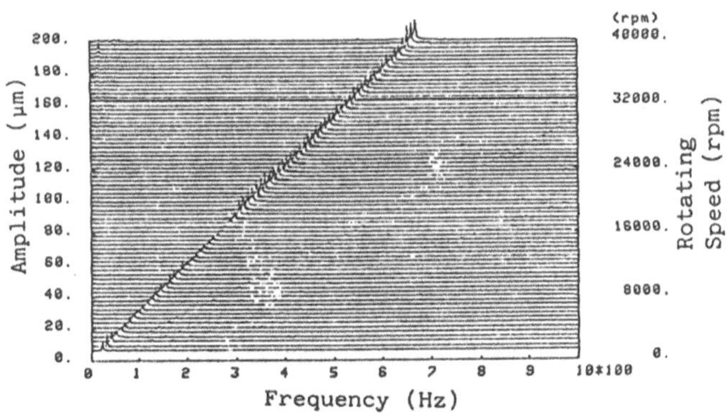

Fig.8 Radial Vibration of Rotor with Compensator(B)

damping force to suppress the 2nd natural frequency, so that the rotor vibration became larger near this natural frequency. The DN value of radial AMB is about 2×10^6 and it's circumferential velocity is 104m/s, which is faster than that of a conventional bearing.

CONTROL BY DIGITAL FIR COMPENSATOR

Using the transfer function of compensator H(s), we get the impulse response of compensator h(t) as follows.

$$h(t) = \frac{1}{2\pi}\int_{-\infty}^{\infty} H(j\omega)\, e^{j\omega t}\, d\omega \qquad \text{—— (1)}$$

The output y(t) is obtained by following convolution integral because the system responds to the past input alone.

$$y(t) = \int_{0}^{\infty} h(\tau)\, x(t-\tau)\, d\tau \qquad \text{—— (2)}$$

Where x(t) is input. In discrete system of which sampling period is Δt, y(t), h(t), and x(t) are expressed as y(n), h(n), x(n) at $n\cdot\Delta t$ respectively. Eq(2) can be described as follows.

$$y(n) = \sum_{p=0}^{\infty} h(p)\, x(n-p) \qquad \text{—— (3)}$$

The impulse response of a real filter converges within finite time, so that we substitute as follows.

$$h(p)=0 \quad (p>m) \qquad \text{—— (4)}$$

A FIR compensator of 128 taps (m=127) requres 128 times multiplication, addition and memory access. In order to calculate this FIR compensator by a general purpose MPU(micro processing unit) such as Z80, A 2-byte addition needs 135 cycles and A 2-byte multiplication needs 327 cycles. When frequency of clock is 6MHz, the sampling period is 1/101 { $=128\times(135+327)/6\times10^6$ }. Because sampling frequency of system with a general MPU would be almost 100Hz even if time of access to memory was ignored, a MPU is unsuitable for controlling a high speed AMB. For that reason, we used a 16×16 bits multiplier-accumulator NPC/SM5810-BD(pin-compatible with TRW/TDC1010J) which is able to calculate multiplication and addition within 65ns. Time for calculation of the FIR compensator with SM5810BD is as follows.

(1) A/D conversion (DATEL ADC-HS12) 15μs

(2) reading filter coefficients(SRAM HM6116) 200ns×128=25.6μs

(3) multipliscation and accumulation 65ns×128=8.3μs

(4) imput signal shift 200ns×128=25.6μs

(5) D/A conversion (DATEL DAC – Hz12) 3μs

Total time of (1)–(5) is 77.5μs, so that sampling frequency can be raised to 12kHz and this filter may be applied to an AMB. We put the desired transfer function of the compensator into a personal computer and obtained the impulse response of the compensator by the inverse fast fourier transform. Time series datum of impulse response is equivalent to the coefficients of the FIR compensator. After these coefficients are sent to memory in the FIR compensator board from the host computer, the FIR compensator begins to regulate rotor motion.

Fig.9 shows the block diagram of hardware for the FIR compensator.

For the purpose of testing the FIR compensator hardwere, we replaced coefficients by

$$h(i)=1 \quad (i=0)$$
$$=0 \quad (i=1\text{-}127) \qquad \text{—— (5)}$$

and measured the FIR compensator property. When the system clock frequency is chosen at 2MHz, the sampling frequency results in 7812Hz.

Fig.10 shows the measured transfer function of the FIR compensator with 7812Hz in the sampling frequency. The characteristics of gain G(f) are flat up to 4kHZ, but the phase becomes -2π at 3906Hz which is Nyquist frequency. Therefore we may conclude that this hardware has a sufficient function.

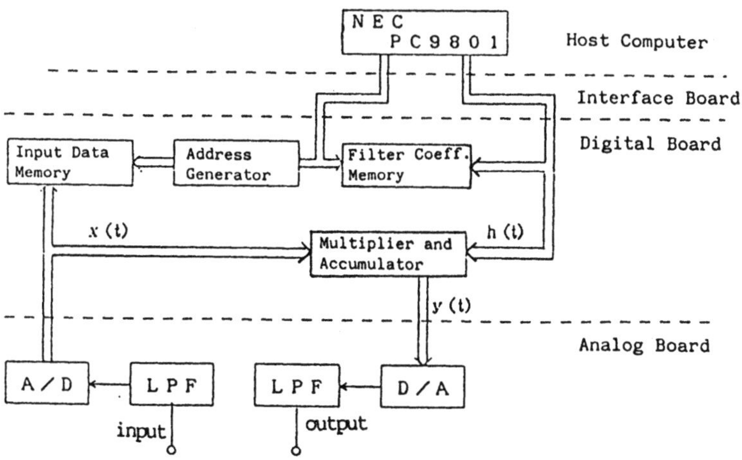

Fig.9 Block Diagram of FIR Filter Compensator

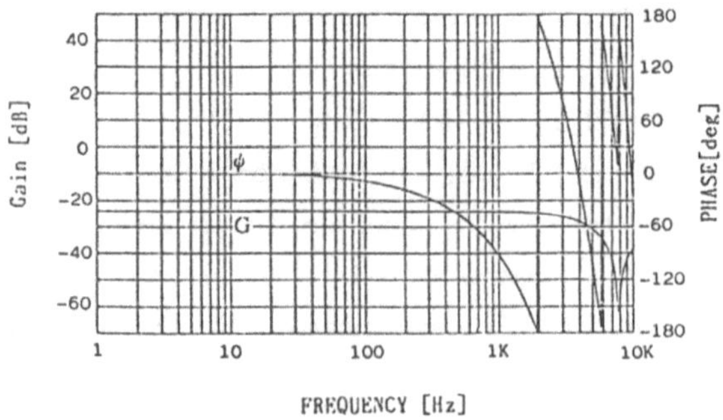

Fig.10 Measured Transfer Function of FIR Filter (Flat)

As the PID filter is considered as a type of FIR filter, PID control for AMB is feasible with the hardware shown in Fig.9. Firstly this compensation circuit is applied to control of axial direction. The filter coefficient was set so as to attain a phase lead of 45° at 130Hz where a resonance frequency of the axial direction is present. The transfer function of PID type and PD type filters are shown in Fig.11.

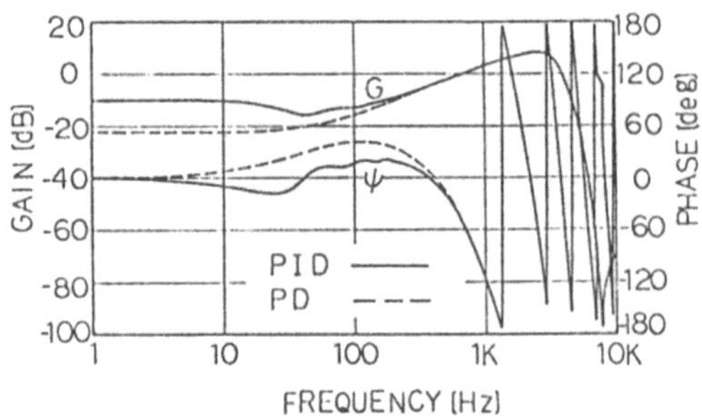

Fig.11 FIR Filter Characteristics

When a PID filter was used as a compensation circuit, low frequency vibration occured. The phase-lag of 5-40Hz of the filter was thought to be the cause of these vibration and the coefficients were changed to a PD type filter to advance the phase in this frequency region. The results of rotation tests using this filter for axial direction control are shown in Fig.12.

In this test the critical speed in the axial direction was 5,000rpm and vibration of 20 μm was observed there. After the critical speed the vibration was reduced to 10 μm or so and the speed was increased to 30,000rpm.

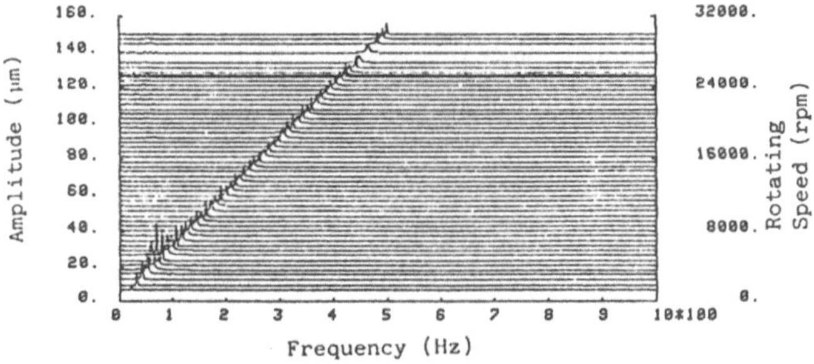

Fig.12 Axial Direction Control by FIR Compensator(PD)

Next this filter was installed in place of an analog compensation in the horizontal direction of a RMB on the free end and a rotation test was carried out. The results are shown in Fig.13. Vibration of 30μm or so was only observed near the critical speed of 2,000rpm in the rigid mode and the speed could be increased in small amplitudes up to near the 1st bending critical speed of 18,000rpm. As the speed approached this critical speed, synchronous vibration increased abruptly so that it was impossible to increase the speed beyond this critical speed, because almost no phase lead took place in the frequency region near 300Hz in Fig.11.

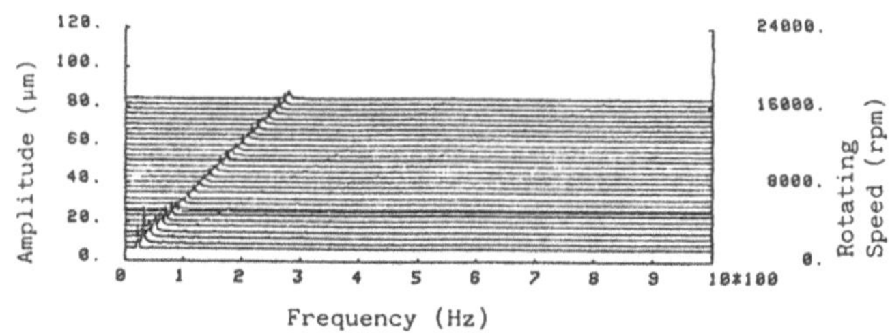

Fig.13 Radial Control with Fir Compensator(PD)

CONTROL USING DIGITAL COMPENSATOR WITH DSP

Based on an analog compensator where the speed was raised beyond the 1st bending natural frequency of the rotor, an IIR filter compensator was prepared. The analog compensator must have parts changed when it is necessary to change their filter characteristics. With digital compensators, their filter characteristics are altered by changing the contents of the filter coefficient memory. Therefore, if it is possible to change the memory contents during operation, an optimal or adaptive control may be realized. As the first step toward this objective, an IIR filter based on a known analog compensator was prepared. When the transfer fanction G(s) on s-plane is given, the transfer function D(z) on z-plane is found by bilinear transformation expressed below

$$D(z)=G(s)\Big|_{s=\frac{2}{T}\frac{1-Z^{-1}}{1+Z^{-1}}} \qquad (6)$$

Where T is the sampling period. As the relation between input X(z) and output Y(z) of the n stages second-order IIR filter can be expressed as follows,

$$\frac{Y(z)}{X(z)} = D'(z)=\prod^{n}\frac{U(z)}{E(z)}=\prod^{n}\frac{b_0+b_1Z^{-1}+b_2Z^{-2}}{1+a_1Z^{-1}+a_2Z^{-2}} \qquad (7)$$

the filter output Y(z) can be found by calculating the output of each stage sequentially. The output u(n) of each stage at t=n·T is found as follows.

$$\begin{aligned}u(n) &= b_0\ e(n)+b_1\ e(n-1)+b_2\ e(n-2)\\ &\quad -a_1\ u(n-1)-a_2\ u(n-2)\end{aligned} \qquad (8)$$

With the FIR filters a high-speed multiplier and accumulator SM5810 was used for high speed repetition of simple multiplication and addition. But with the IIR filter, the number of multiplication and additions performed is fewer than with the FIR filter, and the handling of numerical values is more complicated. As a result, a digital signal processor TI/TMS320C25 is used. With this DSP, a high-speed A/D converter(DATEL ADC817) and 3 D/A converters were used to make 3 compensators consisting of 7 stages 2nd-order IIR filters. At the sampling frequency of 4000Hz for each axis (12kHz in total), it operated at an instruction time of 125ns. Up to 665 instructions can be executed during this sampling interval. Actually 366 instructions were necessary to operate this filter, whose block diagram is shown in Fig.14. The AMB system with this filter shown in Fig.15 converts analog signals from the sensor to digital signals and stores them in the IIR filter. It also changes the compensated digital signals to analog signal for output.

A comparison between the transfer function of the analog and digital filters is shown in Fig.16. The IIR filter(1) is designed to have the same properties as the analog filter and the IIR filter(2) has an additional

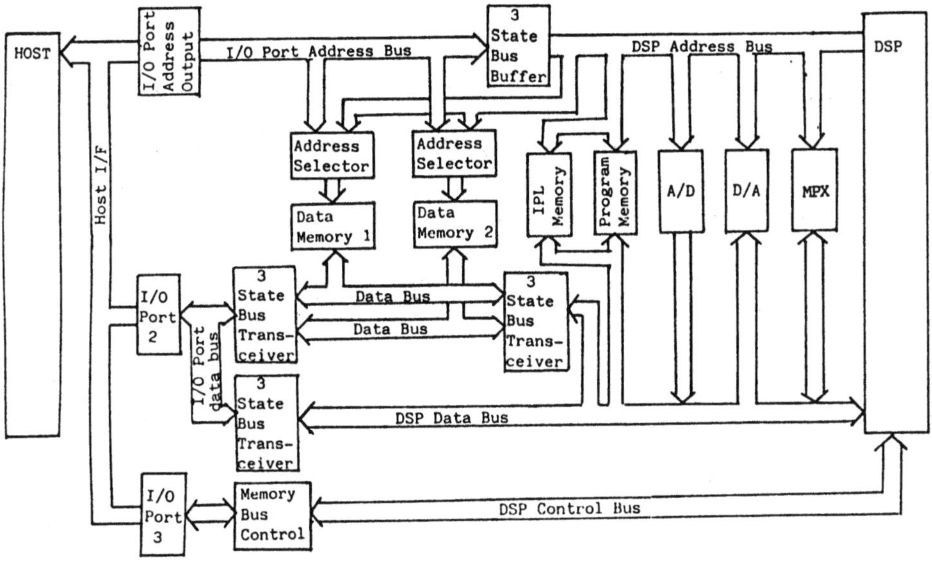

Fig.14 IIR Filter Block Diagram

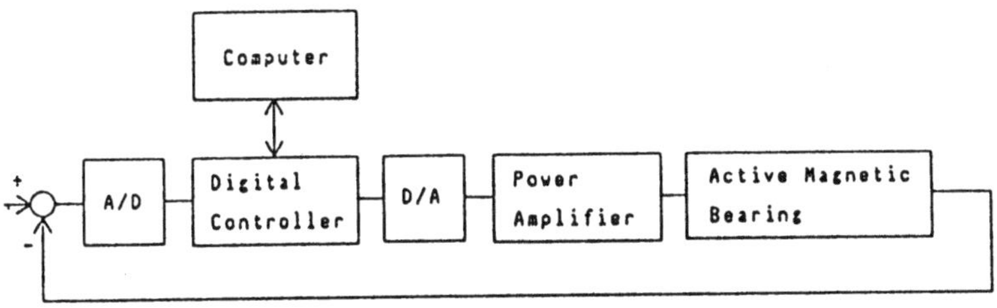

Fig.15 Digital Control Systm

second-order phase lag-lead stage. The gain G of the IIR filter(1) agrees with the gain of the analog filter up to 700Hz. Beyond this frequency, however, the gain of the IIR filter(1) increases and falls abruptly beyond a certain frequency. A comparison of the phase reveals an agreement below 70Hz. But the phase of the IIR filter(1) begins to delay gradually beyond 70Hz. At 1kHz it delays 50° from the analog filter. With the IIR filter(1) the phase is negative above 800Hz. Especially the phase lead is 25° or so at the 1st bending natural frequency of 300Hz. Thus it is smaller than the control design target of more than 30°.

In order to attain an adequate phase lead at the natural frequencies of 300Hz and 700Hz, another IIR filter(2) was made by adding a 2nd-order phase lag-lead stage to IIR filter(1). This filter attained phase leads of about 50° and 35° at the 1st and 2nd natural frequencies of 300Hz and 700Hz. However, this modification enlarged the phase lag in the low frequency region. A waterfall diagram of radial vibration in the rotation test using IIR filters is shown in Fig.17. With the IIR filter(1), the radial vibration shown in Fig.17(a) increased near the 1st critical speed of 18,000rpm. Therefore, the speed was not increased beyond this range. With the IIR filter(2) whose phase for the frequency range including the natural frequency of the rotor is advanced sufficiently, the response characteristics were attained as showned in Fig.17(b). As a result the response is fully damped even at the critical speed. This means that the vibration of the rotor is fully controllable by means of the phase compensation. With this filter the speed was raised to 28,000rpm.

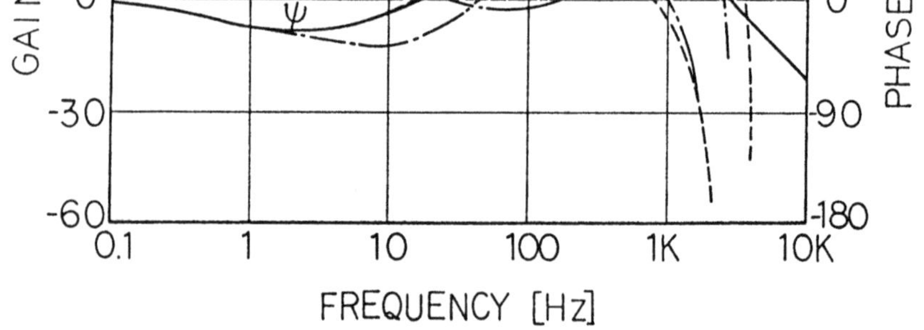

Fig.16 Comparison of Transfer Function between Analog Compensator and Digital(IIR) Compensators

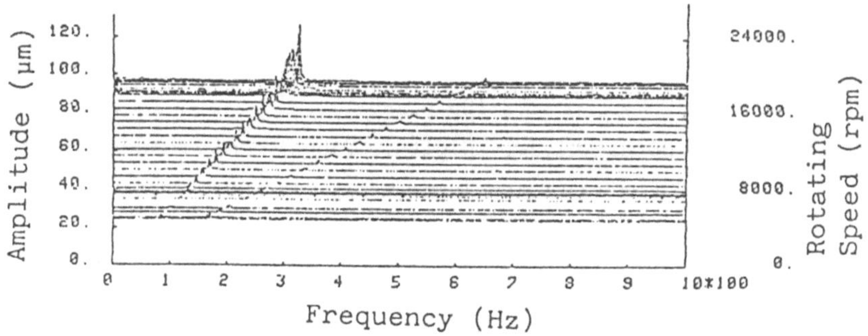

(a) In Case of IIR Filter(1)

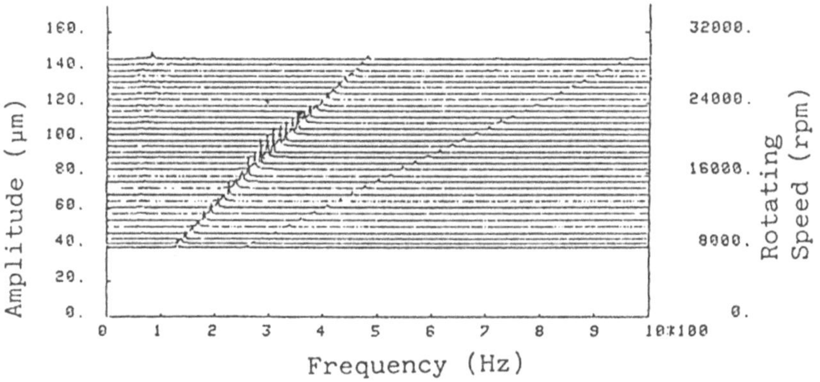

(b) In Case of IIR Filter(2)

Fig.17 Vibration in Rotor

CONCLUSION

Rotation tests were conducted by suspending the flexible rotor with the 5-axis magnetic bearing that uses DC electromagnets. With these tests analog and digital compensators were used for feedback control and vibration characteristics of the rotor were measured. Since sufficient attenuation was attained by the analog compensator, the rotor could easily pass through the 1st critical speed. The digital FIR filter was used for PD control to increase the rotor speed immediately before the first critical speed. With the digital IIR filter whose phase for the frequency range including the bending natural frequencies of rotor is advanced sufficiently, the vibration response is fully damped even at the 1st critical speed.

REFERENCES

(1) Kido,Kenichi, Introduction, to Digital Signal Processing, Maruzen (in Japanese), Tokyo, pp.183-189 (1985)

(2) Slivinsky,C., Borninski,J., Digital Signal Processing Applications with the TMS320 Family (Texas Instruments) pp.689-724 (1986)

Controlling Magnetic Bearing Systems with a Digital Signal Processor

Hannes Bleuler

Institute of Mechanics,
ETH Zürich (Swiss Federal Institute of Technology)

Outline

First, the succession of magnetic bearing systems designed at the Institute of Mechanics in Zürich is retraced briefly with special emphasis on the control aspects. The most recent examples make use of the digital signal processor.

The main body of the paper contains a detailed description of the magnetic bearing control of our first DSP controlled system, a highly elastic rotor. Experimental results permit to assess the advantages of the new control device. Future possibilities of magnetic bearing control layout are shortly indicated.

1. Short Survey of Magnetic Bearing Projects at the Institute of Mechanics

The succession of magnetic bearing projects at the Institute of Mechanics in the last few years can serve as an example for some typical trends. The list is not complete.

Our first project was a rigid rotor built for purely experimental purposes. The system includes analog power amplifiers bought as standard laboratory equipment, analog control and premagnetizing coils. All these features contribute to give a relatively simple, easy to handle system.

Different control strategies were tried for this project. They include state feedback with differentiators for the velocity signals, Luenberger observer for the velocities and decentralization. The results proved that such a drastically simplified control can give very satisfactory system performance. Furthermore, the robustness of the observer for velocity was clearly demonstrated for this rigid body application. A similar type of observer is also used for magnetic levitation vehicles.

The next system is the centrifuge for liquid-cristal epitaxy experiments. It was presented by Bauser et al. /1/ in this symposium. The one attribute changed here is mainly the elongated rotor shape, resulting in non-neglectable structural elasticity.

The bearing-sensor arrangement and the stiffness requirements of the bearing were such, that the simple analog control of the first rotor could be applied here as well. However it became apparent that for other applications the control would have to be more powerful and more sophisticated

The first digitally controlled magnetic bearing system was then designed. This was again a rigid rotor. The 8-bit microprocessor handles two decoupled control channels at a sampling rate of about 1 kHz. New switched-type power amplifiers were employed, we also experimented with digital gap sensors based on CCD arrays. The bearing characteristics was linearized on the software side instead of the hardware side. This means that special linearizing coils and the constant current power supplies could be eliminated.

For later projects, the controller power was increased by introducing 16-bit microprocessors. This development made it possible to consider more demanding control algorithms. Therefore, a digital controller was used for the experimental setup of the project on elastic rotors. It uses two 16-bit processors. This project is described below in detail. The same rotor-bearing system was later used for the first DSP application for magnetic bearing control. To complete this survey, some of the other recent projects are mentioned.

An elastic rotor-bearing system was built for a university in Hamburg. Improvements mainly on the software side made it possible to achieve the desired results in a short time. The specifications required operation at several critical speeds. This rotor with a shaft thickness of only 8 mm can operate at five bending critical speeds up to its operational speed of 7200 rpm . The control is with two general-purpose microprocessors.

Fig. 1: Elastic rotor: Shaft thickness 8mm, 7200 rpm, 5 critical speeds.

For other projects now under way, attention turned to the sensors and the amplifiers. Switched power amplifiers are used for the milling spindle project, where a high bearing load is required. The rotational speed reaches 40000 rpm. The controller was first implemented with single-chip processors and is now being changed to a DSP. The final results and evaluations for this project are not yet available.

Due to the good results of the first DSP control, the same controller hardware will be used for the very high speed rotor project.

New sensing principles have been or are successfully being tested, one of them being the flux-density measurement in the bearing. This also opens up new possibilities for the mechanical rotor-system design as well as for the control strategy. We also aim at integrating the amplifier properties in the system modelling and control design.

2. The Elastic Rotor Project

The advantages of magnetic bearings for vacuum technology or high rotational speed are obvious. Less obvious, but just as interesting are the possibilities for the damping of structural vibration modes. To demonstrate this capability, a highly elastic test rotor was equipped with contact-free electromagnetic bearings.

The critical speed vibrations can be directly influenced through the relative damping of the supporting bearings. A thorough investigation of the mathematical, mechanical as well as the control theoretical aspects of this problem is available in the thesis of Salm /2/.

The controller was first implemented on two 16-bit microprocessors with a sampling rate of 400 microseconds and a time-lag between input and output signal of 250 microseconds. The effect of time-lag inherent to any digital controller is now shortly discussed.

The Effect of Time-Lag:

The primary goal is to produce a damping force. The measured signal, from which this force must be produced, is a position signal. This means that the controller transfer function will have to have a positive phase between input and output signal, no matter what control strategy is applied. Naturally, time delay in the controller always makes a negative contribution to the phase angle which increases with the frequency. The shorter this time-lag, the higher the frequency up to which a damping force can be exerted.

For the high-frequency range, excitation is avoided by providing a low gain of the controller. If the gain is sufficiently low, the natural damping present in all mechanical system will overcome the residual exciting force component.

Therefore the microprocessor system used first with the time lag of 250 microseconds, as mentioned, made it necessary to provide a carefully designed analog low-pass filter with a cutoff frequency at about 400 Hz. A speed of about 5000 rpm could be reached.

The time lag of the DSP controller of about 40 microseconds (including AD-conversion time) made it possible to do without this filter. It then became possible to reach a rotational speed of 10000 rpm with the unbalanced elastic rotor and to operate it at four critical speeds. The greater part of this 40 microsecond time-lag is not due to the DSP, but is used for the AD-converters.

3. The Digital Signal Processor (DSP)

As indicated by the name, the digital signal processor is a special purpose microprocessor. It can perform the specific tasks it is designed for faster by a factor of about 10 to 20 compared with a typical general purpose microprocessor. The power of a single processor is also superior to a single transputer, whose advantages become effective mainly in multi-processor systems. There are now integer and floating-point DSP's available.

The relative importance of DSP's versus general purpose microprocessors is growing rapidly in a wide range of applications. DSP's are also used in many low-cost mass products like CD-players or toys. The applications cover not only the obvious fields telecommunications and image processing, but more recently also the area of controllers for industrial machinery.

There are already some compilers for high-level language programming of DSP's available, but the resulting code is not yet optimal Therefore, at this time we have to rely on assembler programming for the fast parts of the control algorithm.

Summing up what has been said up to here, it seems to be a logical step to apply the DSP for magnetic bearing control as we have done in the project now being described is some more detail.

4.1 The Control: Theory

It is sensible to start a discussion on magnetic bearing control with the basic concepts of mechanical **stiffness** and **damping**. These two force components will somehow show up in the input variable of the control plant, i.e. the rotor to be supported.

The most obvious way to obtain these forces is to feed back a position signal with a classical **proportional-differential** controller. For the selection of stiffness and damping coefficients, one can use engineering judgement or refer to some of many publications on magnetic bearings, e.g. /2/, /3/ or /4/.

The main difficulty is, as in the continuous case, the **differentiating** path. As minimal time-lag is a fundamental requirement, algorithms with small computation-needs are favoured. The simple **difference sequence** can often be applied successfully. The main drawback of this method is the relatively large amount of noise present in such a differentiating path, especially for short sampling periods. There is a very simple and obvious method to overcome this particular problem: The time step for the difference sequence can be increased to an integral **multiple of the sampling period.** It can be noticed that this most simple method can give good results for many of applications.

Of course, in some cases **higher order** algorithms can be necessary. Which method is best will depend on a given practical application. Polynomial approximations of second order have so far been applied successfully in some of our projects, in the present one as well.

But even in the case of a first order algorithm, i.e. an algorithm using only one time-lag element, there are additional possibilities. The algorithms mentioned up to now are all so-called finite-impulse-response (**FIR**) algorithms. The parameters are all in the feedforward path of the controller (fig. 2), which corresponds to the numerator of the z-transfer function.

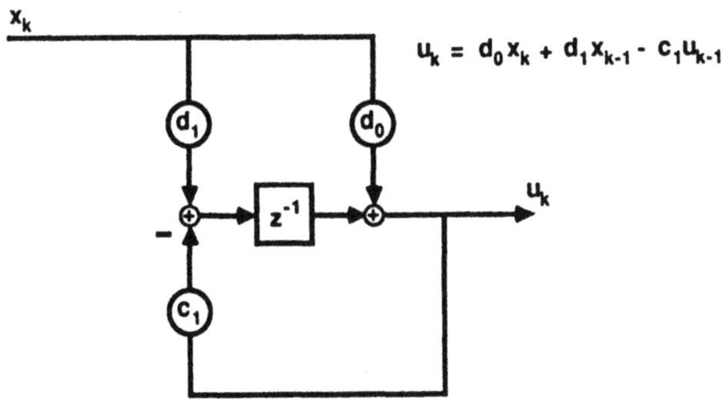

Fig.2: General first-order compensator

The parameters in the internal feedback path have not yet been used. For a general case implementation, the parameter called c_1 in Fig. 2 is not zero. This results in a system which the filter specialists call "infinite impulse response" (**IIR**). The name is self explaining. For such algorithms, the parameters cannot be interpreted as easily as before in terms of stiffness and damping, except in some special cases.

One such special case is the digitally implemented **PID** algorithm. The integrating feedback path will have the effect of a position independent of the load, i.e. a statically infinite stiffness up to the maximum bearing force.

Another way to obtain an IIR algorithm by applying the stiffness-damping concept is by using a **local Luenberger-observer** to obtain an approximation for the velocity signal. This can even be done with a first-order algorithm.

A third method is to **transform a continuous** PD or PID controller of any order from the Laplace-domain to the **z-domain** and to obtain the corresponding difference equations. Many textbooks on sampled control can be consulted on this method. Frequency domain methods are also presented by Kanemitsu et al. /6/.

A most interesting and much more generalized method has been presented by **Larsonneur and Herzog** /5/ earlier in this symposium.

Up to now, we have only discussed **single-channel** control. The control systems engineer of course first sees the **multi-input multi-output** system represented by any magnetic bearing problem with more than one degree of freedom. We also investigate the corresponding control strategies. For the case of continuous control, the question is treated in depth in /2/. Most results can be generalized to discrete-time controlled systems. I will therefore not go into details here and only mention some possibilities.

State-feedback gives of course very nice results. One simplification which can usually be done at neglectable cost to the controller performance is the **decentralization** of the feedback.

The main problem is the fact, that not all state variables are measured. The derivation of velocities has been treated with the single-channel control. A fully coupled **Luenberger observer** can be used to derive the unknown state variables from a mathmatical model of the plant. Such an observer could also be used to obtain the amplitudes of the **bending modes** from a reasonably reduced model of an elastic rotor.

In this way, it would be possible to actively and selectively exert damping for the system modes. We have made first simulations with this kind of controller. The results so far indicate that a relatively parameter-sensitive system is obtained. Furthermore, the number of on-line calculations for such an observer increases by an order of magnitude and seems to be at the limit of what our current fixed-point arithmetic DSP can do. However, research in this area is being pursued, especially in conjunction with interesting new model reduction techniques /7/.

A very promising solution seems again the approach of a generalized **optimal dynamic feedback** compensator with predefined structure, as presented in /5/. It is a considerable extension and generalization to sampled systems of continuous-time control methods described in /3/ and in /8/. Application is expected in the near future.

For the current project, the goal was not sophisticated control, but on testing the new hardware and software with a simple, well known algorithm. It seems that some decisive advantages of the new controller can be made out clearly even when from the control design point of view a very simple design is used.

4.2 The Control: The Algorithm

The DSP is designed to compute summation expressions as they appear in the following general digital control formula of n-th order.

$$u_k = \sum_{i=0}^{n} d_i * x_{k-i} - \sum_{i=1}^{n} c_i * u_{k-i} \qquad (1)$$

where k is the numbering of the sampling steps, u_k is the output variable at the time T_s*k , x_k is the input variable and d_i are constant coefficients.

Address calculation and data multiplications operate simultaneously. This brings the time for one multiplication down to one processor cycle (or 200 nanoseconds in our case). The operands are 16 bits each, the accumulator is 32 bits, the data from and to the converters is 12 bits only. Therefore, no scaling problems have occured so far in our magnetic bearing applications.

The high sampling rates achievable are above 50 kHz (20 microseconds). For this application, a single DSP is used for four or five channels. The converters need about 25 microseconds conversion time. Faster converters are easily available when needed. Therefore the minimal sampling time for five channels is roughly **80 microseconds** (12 kHz). This is high enough to actively control the first few mechanical vibration modes of such rotor systems.

As mentioned before, short sampling times in conjunction with simple difference-sequence algorithms make it necessary to consider multiples of the sampling time for the step-length. Therefore, the formula (1) is generalized to

$$u_k = \sum_{i=0}^{n} d_i * x_{k-i*m} - \sum_{i=1}^{n} c_i * u_{k-i*m} \qquad (2)$$

with the integer m.

This control algorithm has to be programmed in assembler. In order to implement this quite general formula for a varied set of parameters, **ringbuffers** for the values of input and output variables (x and u respectively) are used. These ringbuffers can also serve for analysis purposes on-line or off-line. They contain a record of the system and control behavior, also after termination of the control program. The length of the ringbuffers is limited only by the available memory which in our case goes up to 64 kbytes .

The use of such ringbuffers has proved to be very valuable both for the program flexibility and for data analysis, which can be done e.g. by graphics.

4.3 The Control: Hardware & Software

Our colleagues at the Institute of Electronics have developed a **DSP card** for IBM compatible PC's with the Texas Instruments DSP, a dual port RAM, interfaces and a **software development package**. All of this is now commercially available, along with custom-made analog interface cards. Separate converters are used for each channel. We did not have to develop any electronic circuits ourselves.

The **service program** loading all the parameters and the programs to the DSP card can be written in Pascal on the PC. The two programs, the service program for the PC and the assembler routine for the DSP are now shortly described.

From the **screen display** of the service program the user can choose various functions like starting or stopping the DSP, loading another program, changing or computing new control parameters including the sampling time, calibrating the sensors by directly addressing the converters, examining the ringbuffers etc. From this list, it should become understandable that much more programming effort is put into the Pascal program than into the assembler routine.

The **assembler routine** consists mainly of a loop computing a new output for every channel and every time-step. One important feature is the fact, that the time lag is minimized by making all the calculations not using the newest input value **before or during the AD conversion**. This means that after the reading of the AD-converter value, only one more multiplication and addition has to be done to get the newest value of the output variable.

The mathematical output variable u_k can be used for the ringbuffer. For the actual output to the DA-converter, some additional processing is necessary. This consists mainly of adding the **offset currents** for linearizing of the force-current characteristics, checking for saturation of the actuators and selecting the proper output channels according to the sign of the output variable.

All these tasks for five channels and the converter operation are interlocked in such a way as to **minimize the time-lag**. Therefore the channels are not sampled simultaneously, as is usually assumed for the theoretical system description. This must be taken into account when simulating the system behaviour. On the other hand it would certainly not be the sensible thing to do to degrade system performance by increasing time-lags just in order to get an easier theoretical system description.

5. Results

The control has been realized as described above. Hardware and software have fully met the expectations. A controller with two general purpose microprocessors and a time-lag of 250 microseconds has been replaced by a system with a single DSP and a time-lag of 40 microseconds. This represents an increase in power by a factor of about 12.

The elastic rotor system could be operated without analog filter, which made it possible to reach 10000 rpm compared with about 5000 rpm in the earlier implementation. These results will be presented in detail at the coming conference on rotordynamics /4/.

Our institute subsequently presented magnetic bearing applications with this type of controller at two industrial fairs. The DSP controller has been chosen for several current projects.

6. Conclusions

Digital signal processors are **well suited for magnetic bearing** applications. They are built specifically for such a control task. In the case of magnetic bearing control, the requirement of minimal time-lag seems more important than minimal sampling time. Therefore a single DSP can be used for many control channels.

The assembler programming has not been a major drawback, as the control function is relatively simple. For future applications with more sophisticated control algorithms it will be highly desirable to use high level programming languages, which produce efficient code. Such languages are now being developed. New projects will use floating point processors, which are now becoming widely available. **Industrial applications** can become feasible as the cost for stand-alone applications are falling.

It seems too early, to decide whether transputers will replace DSP's in the field of magnetic bearing control. There are indications that transputers show advantages for other mechatronic applications like robotics, where multi-processor systems and high-level programming are crucial.

7. References

/1/ E. Bauser, G. Schweitzer, H.P. Strunk, A. Traxler: "Centrifuge for Epitaxial Growth of Semiconductor Multilayers", First Int. Symp. on Magnetic Bearings, ETH Zürich, June 1988

/2/ J. Salm: "Eine aktive magnetische Lagerung eines elastischen Rotors als Beispiel ordnungsreduzierter Regelung grosser mechanischer Systeme", Prom ETH 8465, Zürich 1987 / VDI-Fortschrittberichte, Reihe 1, Nr. 162,1988

/3/ H. Bleuler: "Decentralized Control of Magnetic Bearings", Prom ETH 7573, 1984

/4/ H. Bleuler, J. Salm: "Active Electromagnetic Suspension and Vibration Control of an Elastic Rotor with a Signal Processor", 4th Int. Conf. on Vibrations in Rotating Machinery, Heriot-Watt University, Edinburgh, Sept. 1988

/5/ R. Larsonneur, R. Herzog: "Optimal Design of Structure-Predefined Discrete Control for Rotors in Magnetic Bearings (SPOC-D), First Int. Symp. on Magnetic Bearings, ETH Zürich, June 1988

/6/ Y. Kanemitsu, M. Ohsawa, K. Watanabe: "Active Control of Flexible Rotor Suspended in Magnetic Bearings", First Int. Symp. on Magnetic Bearings, ETH Zürich, June 1988

/7/ O. Matsushita, H. Bleuler: "Modelling for Flexible Mechanical System", First Int. Symp. on Magnetic Bearings, ETH Zürich, June 1988

/8/ W. S. Levine, M Athans: "On the Determination of the Optimal Constant Output Feedback Gain for Linear Multivariable Systems", IEEE Trans. on Automatic Control, Vol. AC-15, febr. 1970, pp 44-48

T. Someya (Ed.)

Journal-Bearing Databook

With contributions by T. Someya, J. Mitsui, J. Esaki, S. Saito, Y. Kanemitsu, T. Iwatsubo, M. Tanaka, S. Hisa, T. Fujikawa, H. Kanki

1988. 506 figures, 76 tables. XIII, 323 pages.
ISBN 978-3-642-51726-6

This **Databook** presents the data on static and dynamic characteristics of journal bearings and their applications to analysis and design of rotating machinery. Journal bearings with their oil film not only support the weight of the rotating shaft, but they also represent almost the only machine element that is able to suppress various exciting forces acting on the shaft. As these exciting forces have increased in recent years, knowledge of the exact characteristics of the bearing force on the oil film has become necessary. However, the amount of reliable data now applicable to practical design is limited.

Through the activity of the "research subcommittee on dynamical characteristics of journal bearings und their application" established in the Japan Society of Mechanical Engineers, these data together with the static characteristics have been calculated and also measured using a number of new test rings. These data, covering almost all practically used types of journal bearings, are the principal contents of the present databook.
Moreover, it contains detailed explanations of the characteristics of oil film force and also examples for their practical application so that it may serve as a useful handbook for either experts or beginners, as well as for researchers or designers.

Springer-Verlag Berlin
Heidelberg New York London
Paris Tokyo Hong Kong

MIX
Papier aus verantwortungsvollen Quellen
Paper from responsible sources
FSC® C105338

If you have any concerns about our products,
you can contact us on
ProductSafety@springernature.com

In case Publisher is established outside the EU,
the EU authorized representative is:
Springer Nature Customer Service Center GmbH
Europaplatz 3, 69115 Heidelberg, Germany

Printed by Libri Plureos GmbH
in Hamburg, Germany